AF685733

GOLD MINING IN THE 1990's

THE COMPLETE BOOK OF MODERN GOLD MINING PROCEDURE

By

Dave McCracken

OTHER BOOKS AND VIDEOS BY DAVE McCRACKEN

BOOKS

OTHER BOOKS AND VIDEOS

ADVANCED DREDGING TECHNIQUES
Professional Gold Dredger's Handbook. $7.95

ADVANCED DREDGING TECHNIQUES, VOLUME 2, PART I
Finding and Recovering Paystreaks. $7.95

ADVANCED DREDGING TECHNIQUES, VOLUME 2, PART II
Succeeding at a Gold Dredging Venture. $7.95

GOLD MINING IN THE 1990's
The Complete Book of Modern Gold Mining Procedure. $19.95

SAFETY IN GOLD DREDGING $3.95

VIDEOS

MODERN GOLD MINING TECHNIQUES $49.95 (VHS)

SUCCESSFUL GOLD DREDGING MADE EASY $49.95 (VHS)

COMMERCIAL GOLD DREGDING TECHNIQUES $49.95 (VHS)

Available from: Keene Industries
9330 Corbin Avenue
Northridge, California 91324
(818) 993-0411

Autographed copies available from:
New Era Publications
P.O. Box 47
Happy Camp, California 96039
(916) 493-2062

GOLD MINING IN THE 1990's

Chapter heading artwork done by Floyd York of Graphictech West
Most photos of Equipment courtesy of Keene Industries
All other photos by author unless otherwise marked
Original technical artwork done by Leanne Lockhart McKay
Final artwork completed and arranged by Karen Davis, New Era Publications
Book design and layout by Maria McCracken, Modern Gold Miner's Association
Final Editing by Doyen Salsig and Dean Davis
Technical editing by Marcia Stumpf and Gordon Zahara
Copyright © 1993 David C. McCracken

ALL RIGHTS RESERVED. No portion of this book--written text, photographs, line drawings, or hand sketches--may be reproduced by any means of reproduction, in full or in part (except in the case of brief passages of literary review), without the express permission in writing from the author.

PUBLISHED BY

KEENE INDUSTRIES
9330 Corbin Avenue
Northridge, California 91324

(818) 993-0411

PRINTED AND BOUND IN THE UNITED STATES OF AMERICA.

TABLE OF CONTENTS

ACKNOWLEDGEMENTS

This book is, by far, the most comprehensive work available on today's methods being used to locate and recover gold deposits on a small scale. The book, while simple to read and understand, represents an enormous amount of work--and not just on my part for writing it.

My success during the early days came from a large volume of hard work on my own part. Success today is largely the result of hard work on the part of others who also give themselves completely to my projects.

Special acknowledgement is due to Doyan Salsig who dropped everything on short notice each time another (and another) chapter was sent for final editing.

A special thank you is also due to Electronic Prospecting Specialist, Gordon Zahara, in his advise and encouragement as a technical editor.

Marcia Stumpf, Vice President and General Manager of Pro-Mack Mining Supplies, who already *''has too much to do,''* was always willing to drop everything else to help me finish up this project--which took months and months to finish.

And, finally, to my wife, Maria, who is always there for me, through thick and thin, good and bad--and worse, work and more work. I just wrote the book; she put it in final form.

This book was written for those small-scale miners, past, present and future, who put their hearts into it.

Dave Mac

INTRODUCTION

ECONOMICS

GOLD is a medium of exchange which throughout recorded history has maintained--or increased its own trading value.

In a large civilization it is necessary to have some sort of exchange medium; because while it's nice to be able to trade three loaves of bread for a sack of potatoes, bartering becomes inefficient as a total means of exchange in a modern world. Without some kind of medium, the advancements of civilization would have been tremendously slowed, or might never have occurred at all. It takes a lot of time to haul your products all over the place in order to trade a small portion of them to each person who has something you need. In today's world you could spend more time running around trading your products than in actually producing them. And what about the guy who makes electronic thingamageeks? Perhaps there are only two or three other parties on the entire planet who need such things. So it's necessary for them to be able to give him something of value, which he can then re-exchange for the things that he needs to get on with life.

Since the earliest days of civilization, gold has been considered valuable and so has been used as such a medium. Farmer Jones had a crop of corn. Some of the local people would come and trade gold for some of his corn. He needed a cow and so traded some of his gold to a cattle rancher for one. The cattle rancher needed wire to keep his stock on the range, so he took some of his gold over to the nearest trading post and traded for wire, and so on.

It doesn't really matter why gold is considered valuable. It's enough to know that it has been considered valuable to mankind for as long as written history tells, and it is likely to be considered valuable for a long time to come. The reason for this falls under the heading of economics.

Thorndike Barnhart's advanced dictionary defines economics as the *''science of the production, distribution and consumption of goods and services.''* A basic fundamental of economics is that in order to survive, any group or individual must produce some kind of goods or services which can be exchanged with others for more than they cost to produce. The key word is exchange. In today's free world, and in tomorrow's, exchange is necessary.

Money by itself has no true value of its own. It's paper. All the value a paper currency has is that which is attached to it by the agreement of the people who use it, and their trust in the government which prints it.

Originally, the purpose of paper currency was a substitute for the use of gold and silver. Gold is heavy to carry around in large amounts. It also takes time to weigh it out each time it passes hands, and you can never be certain of its purity without an assay by a competent technician. Some ruling body probably came up with the idea of paper currency being used in place of gold. This was a good idea--as long as paper was directly backed by the real thing. A dollar was worth one pinch of gold. The fellow traded in some of his gold for dollars because they were easier to deal with. This was fine as long as he could trade the dollars back for gold if he wanted to. There was never any worry about the gold's exchange value because it had always been safe. So there was no need to worry about the exchange value of paper currency as long as it was backed by gold (or silver) and could be traded back for the same amount of gold or silver at any time.

Over the years, for one reason or another, the system of "having so many dollars represent so many units of gold, which was held safe by the government" dropped out; and a new system was set up in which paper currency was printed with nothing at all to back it up and hold it stable--other than the authority of the government.

With the new system, a loaf of bread could be traded for a dollar, but why? The dollar has no value of its own. The answer is that the dollar can be traded for a loaf of bread because the authority which makes the currency says that it can. So it takes a lot more trust in the government in this new system because you are using a medium of exchange that has no intrinsic value, and has no actual valuable commodity backing it.

A paper currency which is not backed by anything of value, has nothing really to hold it stable as a currency. Such currencies can inflate (become worth less in their trading value) because of a number of reasons, a main one being that the world becomes less trustful of the government which is standing behind the currency. Therefore as an exchange medium, it becomes less trustworthy, and thereafter will become worth less as a bargaining commodity.

If a person has all of his assets in paper money form, as in a bank or in bonds, and the government which puts out that currency folds up or is overrun, or goes suddenly unstable for some reason, that person's assets can become a lot smaller--or disappear altogether.

Another way for a currency to inflate, meaning to become worth less in its trading value, is for the government to print up a bunch of it and just give it away without demanding in return a fair exchange of what it is supposed to be worth. In this case, you have on the one hand, the government printing up money and saying that it is worthy of being used as an exchange medium, and on the other hand, the government is just giving it away and demanding nothing in return. This activity lessens the currency's exchange value, and is a reason why the currency of any welfare state is guaranteed to inflate.

One way to watch the rate of inflation of currency over an extended period of time is to watch the exchange market value of gold for that currency. Once, in the USA, gold sold for $17 per troy ounce. Today, it sells in excess of 20 times as many more dollars ($335). The appearance of this is that gold has increased its value 20 times more than it was worth during the mid--1800's. Actually, this is not the case. While today gold may be a touch more valuable than it once was, because it's in such great demand as a stable investment during such unstable economic times, the truth of the matter is that the dollar has depreciated in its exchange value to the point where it takes 20 times (or more) as many of them to equal the same amount of exchange value it once had. To further show this point, there was once a time in the U.S. when you could buy a loaf of bread for a nickel. Today, a good loaf of bread on the average store shelf is selling for $1.30. Is bread worth 20+ times more than it used to be? No! Bread is bread, and if anything, it should take less to buy a loaf--because there is no scarcity whatsoever.

A gold miner today can cash in an ounce of gold and go out and buy about the same amount of goods and supplies as an old-timer could by cashing in an ounce and shopping in a large town during the mid--1800's. You have to remember that a guy could get a haircut for a nickel and a steak dinner with all the trimmings for a quarter in those days.

If that same old-timer were using his dollars on today's market, he would need in excess of 20 times as many of them to buy the same amount of goods, and that was only about 120 years ago. That's a lot of inflation! Yet, if that miner brought his own gold to the present, he'd be able to buy the same amount of goods--or more--as he could back in the 1800's.

So why the lesson in economics? Just to show that gold is very stable as an exchange commodity, is likely to be considered as valuable for a long time to come, and to show how the inflation of paper currencies has adversely affected the gold mining industry, until recently. . .

GOLD MINING

There seems to be a prevalent idea in modern society that during the gold rush days the old-timers mined out most or all of the gold. This is simply not the case. While the earlybirds probably did get most of the exceptionally easy to find and recover bonanzas, they did not come anywhere near to finding all of--or even most of the gold.

This idea that most of the gold has already been found and recovered stems partially from the almost total lack of gold mining activity since the early 1940's--that is, until recently. The reason is as follows:

During the early rush days, it was made mandatory, by law, in the USA, to sell all gold that was mined in the USA to the government at a set price, which was arbitrarily fixed by law. This fixed price was probably O.K. during the early days. But over the years, inflation caused the dollar to rapidly decline in its exchange value--yet the fixed exchange rate for gold required it to be bought by the same amount of dollars. As inflation caused the costs of operating a gold mine to become more in dollars, and the fixed rate of exchange for gold remained the same in dollars, it became necessary to find and recover more and more gold in order to keep an operation running without a loss. Many mines which were once run at good profit eventually had to be shut down due to the effects of inflation, along with the fixed exchange rate of gold. In 1934, the fixed rate was raised from $20.67 to $35 per troy ounce of pure gold. This started a minor gold rush. Many old mines were re-opened and worked at a profit using newer mining techniques and modern equipment. But the fixed rate remained the same, inflation continued, and eventually all but the richest of producing gold mines had to be shut down.

In 1941, most of the mines that were still in operation were shut down due to the war effort. After the war was over, very few gold mines were re-opened because of the increase in minimum wages and the many jobs that were available elsewhere. Plus, inflation had continued and the fixed rate remained at $35/ounce.

Since that time, and until recently, very little actual gold mining has occurred. For the reasons laid out above, there was little interest in it. After all, who would want to mine gold, which had depreciated in its exchange value at a rate of 20 times in the last 120 years? And that was the case as long as the gold exchange value remained arbitrarily fixed and inflation rates continued to bleed the buying value out of the dollar. This is the primary reason for the idea that *''all the gold has already been mined up,''* because nobody was mining!

In 1974, a bill was passed through Congress which made it legal for individuals to own gold again, and to buy and sell it on the open market. With the arbitrarily fixed exchange rate finally being taken off gold, the currency exchange value for gold went through the roof. Consequently, a great many mines that were shut down years and years ago are presently being re-opened at a good profit. Many areas which could not be worked viably at an earlier time can now be mined by the individual and give excellent returns.

A tremendous amount of progress has been made in earth-moving equipment and gold recovery methods over the past hundred years--especially within the past 40 years, while the field has been dormant. Today, it's possible for the individual to process ten, or even a hundred cubic yards of gold-bearing materials, whereas in the same or similar areas, he may have been able to move just a few yards during the early days.

Today, supplies and equipment are readily available to the modern-day prospector at relatively low cost, compared to what the old-timer had to pay for tools and supplies on the early frontier--if he could get them at all.

One thing about the old days is that it was often necessary for the early miner to drag, haul and hoist his equipment over some very rough terrain. He also had to hunt much of his food and fight off the Indians--all the while trying to find and recover acceptable paying quantities of gold. Today, there are highways, roads and trails that have been made all over gold country which pass through or at least nearby the same spots that it took the earlybirds weeks or even months to get into, along with their gear.

A great amount of natural (and man caused) erosion has taken place within the last hundred years or so, which has placed volume amounts of new gold into the present rivers and streams. This is evidenced by the many valuable deposits that are being regularly turned up out of the same streams and rivers which were supposed to have been thoroughly mined by the 49'ers or those who followed them.

Yes, the 49'ers got many of the easiest rich deposits, and these were some rich deposits indeed! How could they help it when people were tripping over nuggets lying on the streambanks! However, most of the deposits that were taken by the earlybirds were those that were easily exposed and mined during their time. But by no means did they find all of the easy deposits, because new ones are being found at this writing.

Present-day geologists say that 99 percent of all the gold on or inside the earth's crust has yet to be mined. Actually this varies; some experts say that not twenty percent has yet to be mined. However, they all seem to agree that most of it still remains to be found and recovered. True, much of this gold is down beneath tons and tons of sediment or down into the earth's hardrock interior and is all but inaccessible to us without the most sophisticated (and expensive) mining equipment available today. Yet, there is still an amazing amount of gold that is accessible to the small-scale operator, and a good many bonanzas are yet to be had.

With the steadily increasing market exchange value for gold, and with experts in the economic field agreeing that it will see the $1,000/ounce mark--and higher within the next few years, and with runaway U.S. deficit spending by Congress occurring at this very moment, and with the working man's wages being taxed more and more all the time, a successful gold mining venture can be a very comforting activity to be involved with today. More and more people are turning to gold mining as a professional or hobby activity.

An ounce a week, or maybe a little more, would be acceptable wages today. With the modern equipment which is readily available at relatively low cost, namely the sluicing device and the suction dredge (both covered later in detail), and some hard work and experience, an ounce a week average (on top of expenses and taxes) is not too difficult to attain. As a matter of fact, if you are willing to put in the time, study, and effort to gain some experience and a greater understanding, much higher averages can be, and are being, attained by the small-scale operator.

There is a gold rush occurring today, in the 1990's, which is of a size that is comparable to the rush which occurred during the 1800's. Today's gold rush is not nearly as dramatic; because in those early days if one was going to go gold mining, it had to be a total commitment. Those were rough times on the early frontier and survival was at stake.

Today, the modern prospector can go out on the weekend and sample various locations, using his modern-day methods, and cover ground that might have taken weeks to prospect by the methods used during the early days--or might have been impossible to work at all.

Some tremendous discoveries were made during the earlier rushes. Some miners became wealthy. Yet, most of those who went to mine gold were unable to find enough to even make wages. The reason for this was that the equipment and methods used in the early days enabled the small-scale operator (the one or two-man operation) to process only a small amount of earth, which made it necessary to find the richest deposits of gold, and easiest to mine, because of their limitations. These deposits were few, and finding them meant a considerable amount of prospecting and hard work.

When a hot spot was found, it was very exciting and dramatic.

Conditions are different today. Routes have opened up. Recovery methods have improved. And processing equipment is available to the individual which will allow him to do far more than the earlier prospectors could. As a result, the success rate among miners in finding paying quantities of gold today is actually higher than it was during the earlier rushes.

Perhaps there are not as many prospectors running around in the hills as there once were. But it's a fact that a much greater percentage of today's prospectors do far better than the early guys did--and with much less investment too.

Take it from me, I've been out in them thar hills and streams and have taken a fair amount of gold myself. I plan to get lots more! There's plenty to be had.

I'm not saying it's all roses and whipped cream. It takes hard work, study, and persistence to learn the hows and the wheres of it, and to be continuously successful--whether as a hobby or as a profession. But it's being done, and on a broader scale every day.

Oh, don't worry, there's still plenty left for you, and there will be for some time to come.

A lot of mining has gone on over the years, and with the almost total lack of it during the past half century or so, much of the data has been lost. A lot of it was never written down. The material I cover in this volume is the best that I personally have run across--and IT WORKS! It works for me; my various partners have done well with it; and it has worked for many others whom I have been associated with in the gold mining field who have been successful. My purpose here is to give you something to go on that WORKS, so that you too can get started and do well at it. After you have been into it for a while, you'll get your own ideas of the best methods and equipment for the job--which is good. If you run across something in the field that seems to work better than the way I have outlined it in this volume, by all means write me a letter and tell me about it. I'll be doing a revision on this manual every once in a while, and if you point out something to me which is better, I'll add it in so that everyone can benefit.

And don't forget, the real treasure of gold mining is being able to enjoy wonderful adventures in the great outdoors. Any and all of the gold you are able to recover is just a bonus!

This book has been designed to give you a no-nonsense account of gold mining basics. I have attempted to cover everything possible that you will need and want to know about the successful fundamental procedures being used to find and recover gold in the 90's.

If you are interested in mining gold and want to succeed at it, if you are willing to stick with it, and keep at it, no matter how tough the going gets, I'd like to see you succeed too--and this book was written for you.

So here's the scoop on gold mining. Have at it. I'll see ya out there.

Good Luck

Dave Mac

GOLD--
WHAT IT IS

CHAPTER 1

Author demonstrates 15 1/2 ounce nugget found in his gold dredging operation on the Klamath River in northern California.

All things of the physical universe which can be sensed by man are made up of one or more chemical elements. An *''element''* is a basic simple substance which cannot be further broken down by chemical means. Elements are the basic building blocks of the material universe in which we live. Scientists have discovered over 100 basic elements, and have laid them out in order on the *''periodic table of elements''* according to their atomic structures.

Gold is one of the 92 naturally occurring elements found on earth. There is no known natural substance that can destroy gold. It can be dissolved by chemical means, but even then it remains as gold--only in a more widely dispersed state.

For eons, man has dwelled on where gold might originally have come from--that is, its native source. Scientists have recently discovered that gold can be artificially produced by the atomic bombardment of lead, which is another basic element. However, the process is very expensive, more so than the monetary value of the resulting gold. So it is not artificially produced on a commercial scale. However, this discovery has brought on the theory that the gold being found on our earth might originally have been manufactured in the nuclear furnaces of stars which have long since vanished, our planet being part of the remaining debris.

The scientific symbol for gold is Au. It is number 79 on the periodic table of elements. Gold is not magnetic, but it is an excellent conductor of electricity. Its melting point is 1945° F. Gold is not corroded or tarnished by moisture, or oxidized (rust) by the effects of oxygen and water, or affected by ordinary acids, as most other metals are. Deposits of gold that have lain inside a mountain or under a streambed or even on the ocean's bottom will remain there and be rather unaffected until moved by the natural forces of the earth--or taken by man.

Gold is a very soft metal, being 2. 3 on a hardness scale of 10, which is one of the factors giving gold its tremendous malleability--meaning that it can be pounded, twisted, rolled and/or squeezed into all kinds of different shapes without breaking apart. In fact, the yellow metal can be pounded so thin that it is translucent, and yet still remain intact as a solid sheet of gold. It has been said that such sheets of gold can be produced so thin that it would take a quarter of a million of them stacked one on top of the other to make a pile one inch tall! Thin sheets of gold such as these have the distinctive quality of allowing sunlight to pass through, yet they will reflect off a large portion of the sun's infrared rays (heat). For this reason, thin layers of gold are being used in the window glass used in many of today's modern skyscrapers to help save on the tremendous costs of energy necessary to keep the interior of such buildings cool during the hot summer months. Similar films of gold have also been used in the face shields of astronauts' helmets to reflect off much of the increased

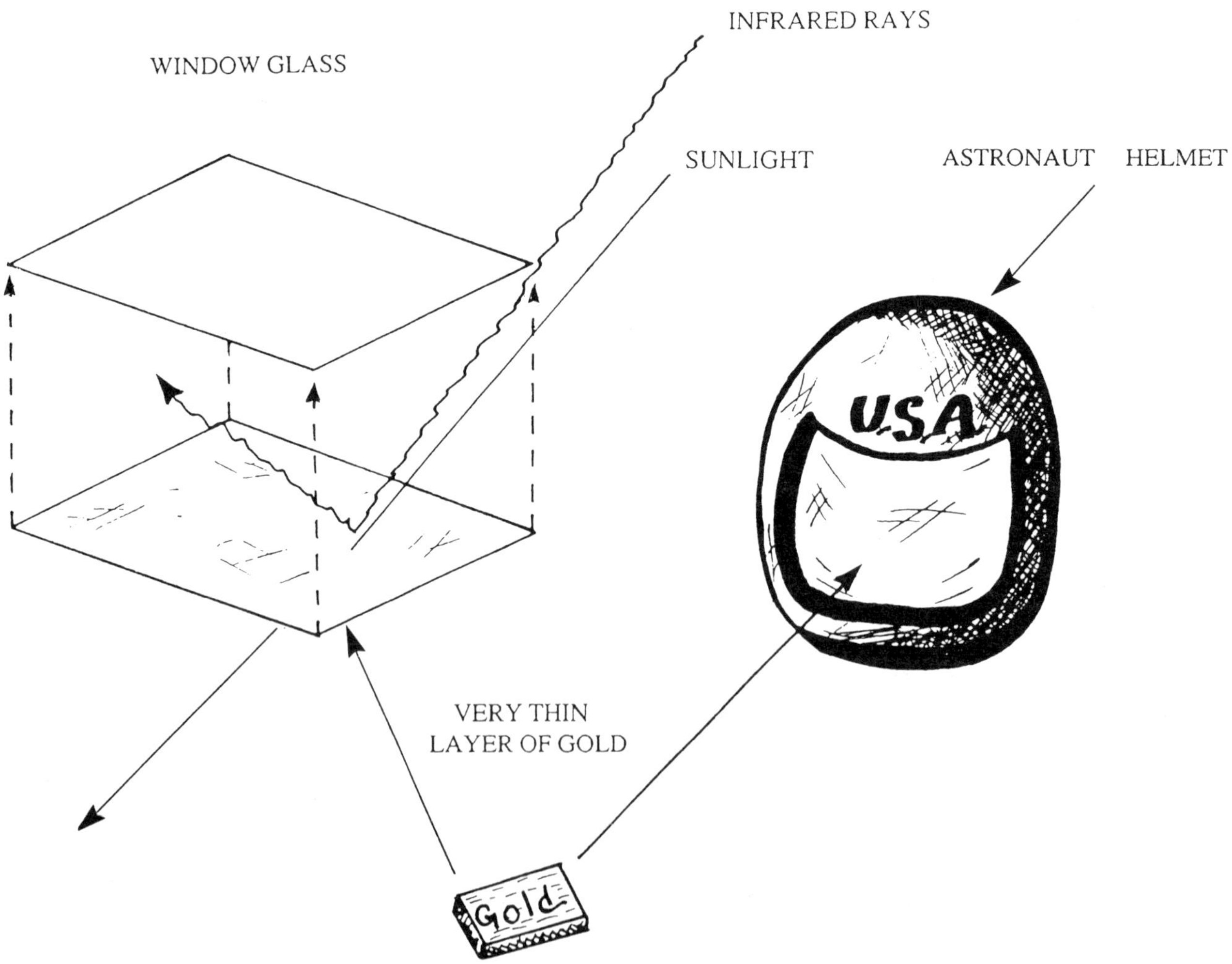

Fig. 1-1. Gold has many uses--one of which is that in very thin sheets, it will reflect off a large portion of the sun's infrared rays, but still allow sunlight to pass through.

bombardment of infrared rays which occurs out in space.

Gold is extremely ductile--meaning that it can be drawn out into wire or threadlike forms without becoming brittle and breaking. It is said that gold can be drawn out so thin that a single ounce could be made to stretch a continuous length of 35 miles. That would make the thread very thin indeed, but that's exactly what is needed in today's electronics industry, where entire circuits are being placed in chips the size of a pinhead. Because of its high electrical conductivity, its extraordinary resistance to deterioration and its ductile qualities, gold is in great demand by the electronics industry. So it's not uncommon to find gold being used in the circuits of many of today's common everyday electronic devices--like TV's and calculators, not to mention in some of the more sophisticated electronics of today's world.

Gold is also commonly used by dentists and is widely used to make jewelry.

Gold's use in the making of coins has greatly increased within the past few years as an effort by some nations to stabilize their inflating economies.

Gold has another distinctive quality, which is perhaps most important to the prospector (other than its value) and that is its weight. Gold is extremely dense, one of the heaviest of all metals. The specific gravity of gold is 19.3, meaning that it weighs 19.3 times more than an equal volume or mass of pure water. Iridium (one of the platinum group metals) is one of few metals that has a higher specific gravity than gold, being 22.6. Gold is 2-1/2 times heavier than silver, and around 8 times heavier than the quartz rock which it is commonly associated with when gold is found in hardrock form. A single cubic foot of gold will weigh approximately 1187 pounds. It is this quality of gold, having a superior weight factor over the other materials usually found along with it, that is used in gold recovery methods. This will be covered in detail within the following chapters.

Gold is not the most valuable metal, but it is extremely valuable and probably the most sought after of all the valuable metals found on earth. At today's market exchange value, that same cubic foot of pure gold would be worth in the neighborhood of 6-1/2 million dollars. A cubic inch of pure gold would be worth about four thousand dollars. So gold is valuable, very valuable indeed, and it does not take very much of it to accumulate a considerable amount of wealth.

There is one other distinctive quality of gold worthy of mention. In its natural form, gold is a very rich and beautiful substance to look at. In fact, this is so true that there is a saying among experienced miners, and those individuals who handle a great deal of the yellow metal, that it is not a good idea to look at any substantial amount of raw gold for very long at any given period of time. This is because it has a tendency to bring on a condition referred to as *''goldfever.''* It's true!

''Gold Fever'' affects different people in different ways. While it might make one person want to buy the gold at almost any price, it could just as easily make another want to steal it--at any price. However, the ''fever'' tends to always make a person want to have the gold for himself, and more of it if possible, with the means of getting it depending on the character of the individual. This condition, (gold fever) is something to take note of for anyone who is planning to get involved with

mining or dealing with gold. This is nothing to laugh off, for it has been the cause of a great many deaths, failures, wars, enslavements, and loss of friendships. It has altered the course of a significant amount of history-- much of it being worse for the others involved. It's true that for many, gold is the thing that dreams are made of. Therefore gold has a tendency to strike below the social behavior in a person and bring out some of the stronger passions which lie underneath. It is well to keep this strongly in mind during the stage in which you are considering whom to take on as a partner (or as employees) in a gold mining venture of any size.

One common characteristic of a person who has been touched, just mildly, with a case of gold fever is that he tends to throw all good business sense to the wind and dive in head over heels, much as a young child might do if he found a tub full of his favorite candy or ice cream. It is this very same factor that the conman stirs up and plays off of. And if you think that there aren't a few good ones in the gold mining field--think again. The vast majority of failures in gold mining ventures are the result of this same loss of good judgment which sometimes occurs when dealing with the valuable metal.

Perhaps the most successful precaution against being struck to any harmful degree with the "fever" is in honestly taking on the viewpoint that *"anything worthwhile is worth lots of time and effort."* It's the guys who intend to get rich quickly without much energy output on their part who most often fail in the business of gold mining. If, upon examining your own intentions, you find you are interested in getting rich quickly, without having to work for it, it's almost a certainty that you've caught at least a touch of the fever. On the other hand, if you're interested in going out into God's country to see if you can find some of the yellow metal as an adventure, and/or perhaps to see if it can be done as a viable business venture, you're probably on the right track and you are more likely to succeed. And who knows, maybe you will strike a big one; it happens often enough! Just remember that the finding and recovering of gold is similar to any other business venture. It takes a fair amount of time and work to get consistently good at it. Take it on as such and you will have less trouble--and fewer losses.

SOURCE OF GOLD

When considering the source of gold on this planet alone, it is necessary to study the earth and take a look at some of its more recent geological history.

Scientists believe that the earth is an extremely solid mass, which grows more and more dense towards its center.

It is believed that the gold which is found on the earth's surface and in its outer crust was once deep down inside the earth's molten mass and was carried up to the surface by the effects of volcanic activity.

In hard-rock form, gold is generally found associated with quartz in the form of *veins* which protrude through the general mountainous rock--referred to as *"country rock"* by the geologists. In

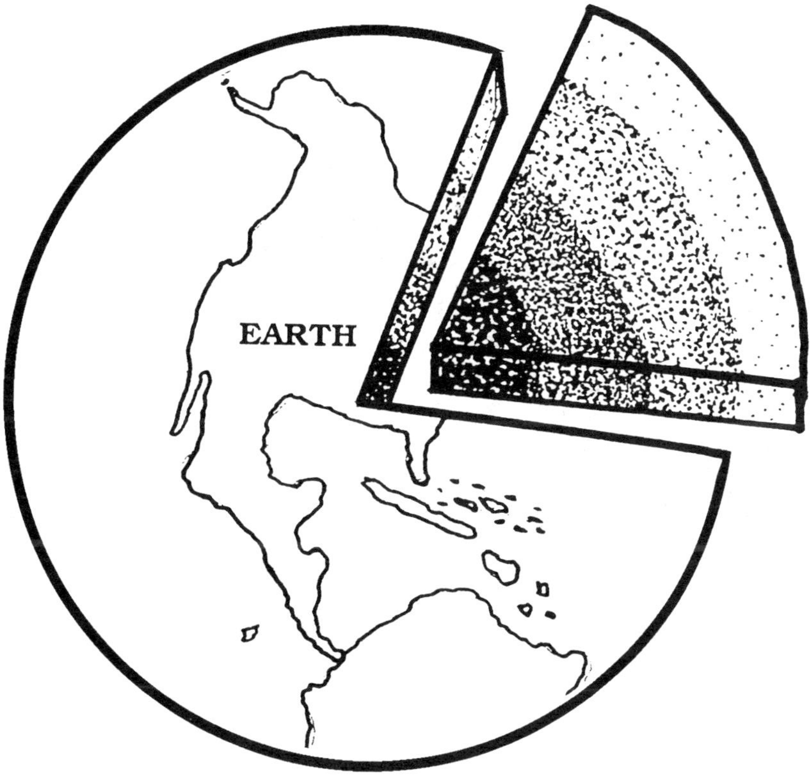

Fig. 1-2. Scientists believe that the earth is solid, and becomes more and more dense towards its center.

the early days of gold mining, it was generally believed that these quartz/gold veins were once pushed up out of the lower bowels of the earth as volcanic magma. Upon this principal it was thought that a vein should become richer as one followed it deeper into the earth. However, this was not always the case.

More recently it has been discovered that quartz veins were not formed during the same time period as the country rock that surrounds them, but later. The cooling of the earth's outer crust (country rock) apparently caused many cracks and fissures from which the gasses and superheated steam could escape out of the earth's molten interior. These water vapors also carried minerals with them through these avenues of escape, with one of the predominant minerals being silica--which forms quartz. The water vapors carried other minerals upward too, of which gold, silver, iron and platinum are just a few. Silica has distinctive characteristics of its own, one of which is that it tends to trap heavier elements when they are passed over and through it in a dissolved form. So while the water vapors pushed the heavier minerals upward towards the earth's surface through cracks and fissures, they often combined with silica and formed vein-like structures (see Figure 1-3).

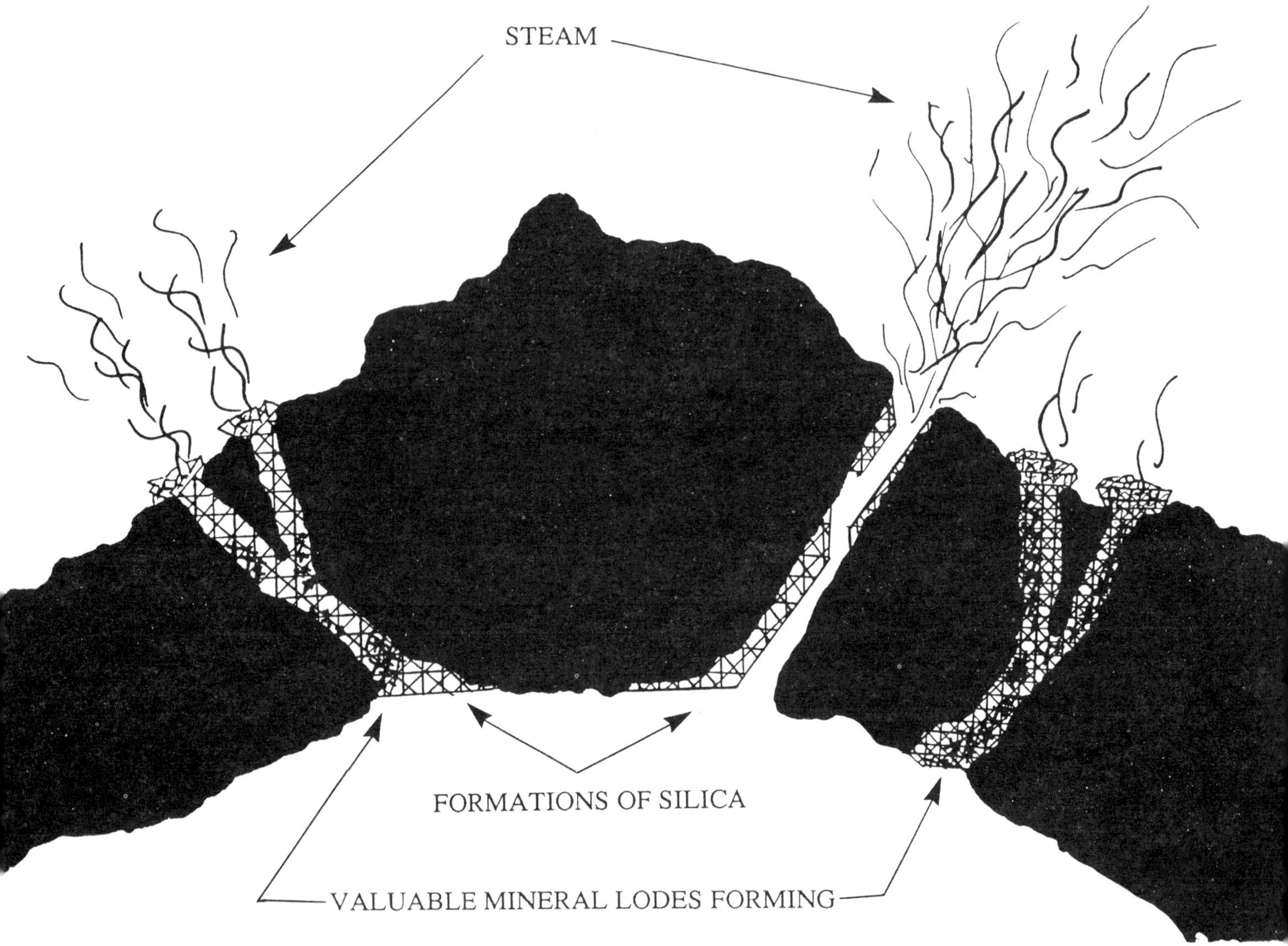

Fig. 1-3. Formation of mineral lodes.

Generally, gold is associated with quartz. However, quartz is not always associated with gold, because there are countless quartz veins that have no gold--or at least not enough gold in evidence to make the mining of the vein profitable. Veins which contain valuable minerals are referred to as *"lodes."*

"Ore" has been defined as any deposit of rock from which a valuable metal or mineral can be profitably extracted.

When a valuable lode is found, in order to mine it, the ore is usually blasted out of the vein and is crushed down into a very fine powder from which the gold, silver and other valuable minerals can be extracted by any number of chemical or mechanical procedures. This entire process is called *"lode mining"* or *"hardrock mining"* and is covered in more detail in Chapter 9 of this volume.

Millions of years passed after the rich mineral lodes were formed, during which time a large amount of weathering was caused by heat and cold, animals and vegetation, rain and wind, snow and ice, glaciers--and their resulting runoffs, earthquakes, and ocean tidal changes as great as 800 feet in elevation for each tide-- due to the moon revolving closer to the earth during an earlier period. So, after the rich veins were formed, a tremendous amount of disturbance and erosion took place, which washed many of the exposed rich mineral lodes out of the mountainous rock and into the stream and river systems which flowed during that time period.

The steady heavy flow of water over a streambed creates a continuous movement of the streambed materials, causing a natural sorting of the various minerals by their different sizes, shapes and weights. Gold, being extremely heavy in relation to most of the other materials that end up in a streambed, tends to be deposited in the various common locations where heavier materials can become trapped because of their greater weight. Deposits of gold and other valuable minerals which have been washed away from their original lodes and redeposited by water in streambeds are called *"placer deposits"* (pronounced "plaster"--without the t).

Finding and recovering placer gold deposits requires an understanding of where heavy sediments will collect while being transported by the forces of water, which much of the remainder of this manual will outline for you.

Gold directly from a lode is crystalline in structure (see Figure 1-4), and is usually referred to as *"rough gold"* because of the coarseness of its surface. Once washed from its original lode and swept away by the forces of nature, gold tends to become pounded flat and rubbed smooth.

Often an experienced prospector can get a pretty good idea of how far a piece of gold has traveled from its lode by the degree of roughness on its outer surface.

Gold directly out of a lode is crystalline in structure.

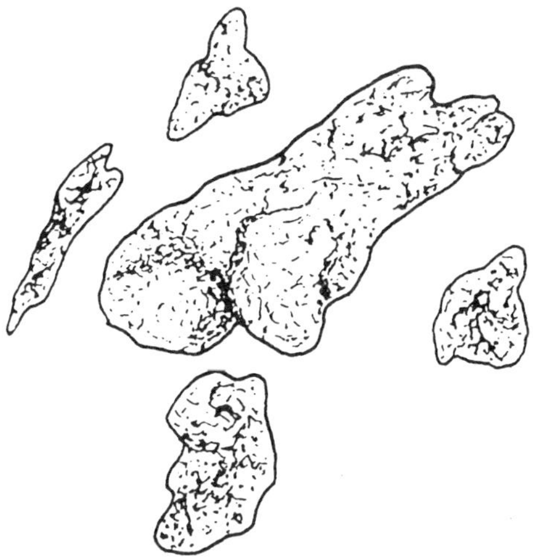

Gold tends to get pounded smooth as it is being washed further from its original lode.

Fig. 1-4.

PURENESS OF NATIVE GOLD

In its native state out of a lode, gold is almost never 100 percent pure, but has a percentage of other metals along with it. These other metals contained with gold, whether of value or not, are called *''impurities.''* The impurities with gold most often consist of silver, copper and a little bit of iron, platinum and cadmium in differing amounts. The proportions of these other minerals change from lode to lode, which gives the gold coming from one location different colors, qualities, and value from the gold extracted from another location.

When someone finds gold--either in placer or lode form--it is not uncommon to have the gold tested *(assayed)* to find out what percentage of impurities are present, and exactly what they are. The actual gold content in native gold just out of a lode or placer deposit changes from location to location. But a reasonably safe average (at least in California) would be to say that 80% is gold and 15% is silver and/or copper.

Gold which contains 20% or more of silver is called *''electrum.''*

Pieces of placer gold, and those pieces which have eroded from a lode, come in a wide variety of sizes and shapes, ranging from large pieces *(nuggets)* as great as 200 pounds in weight (very rare) to *''flakes,''* and smaller *''grains,''* and even smaller *''dust,''* down to pieces so microscopic in size that it would take perhaps 8 million particles combined in order to accumulate enough gold to value one dollar.

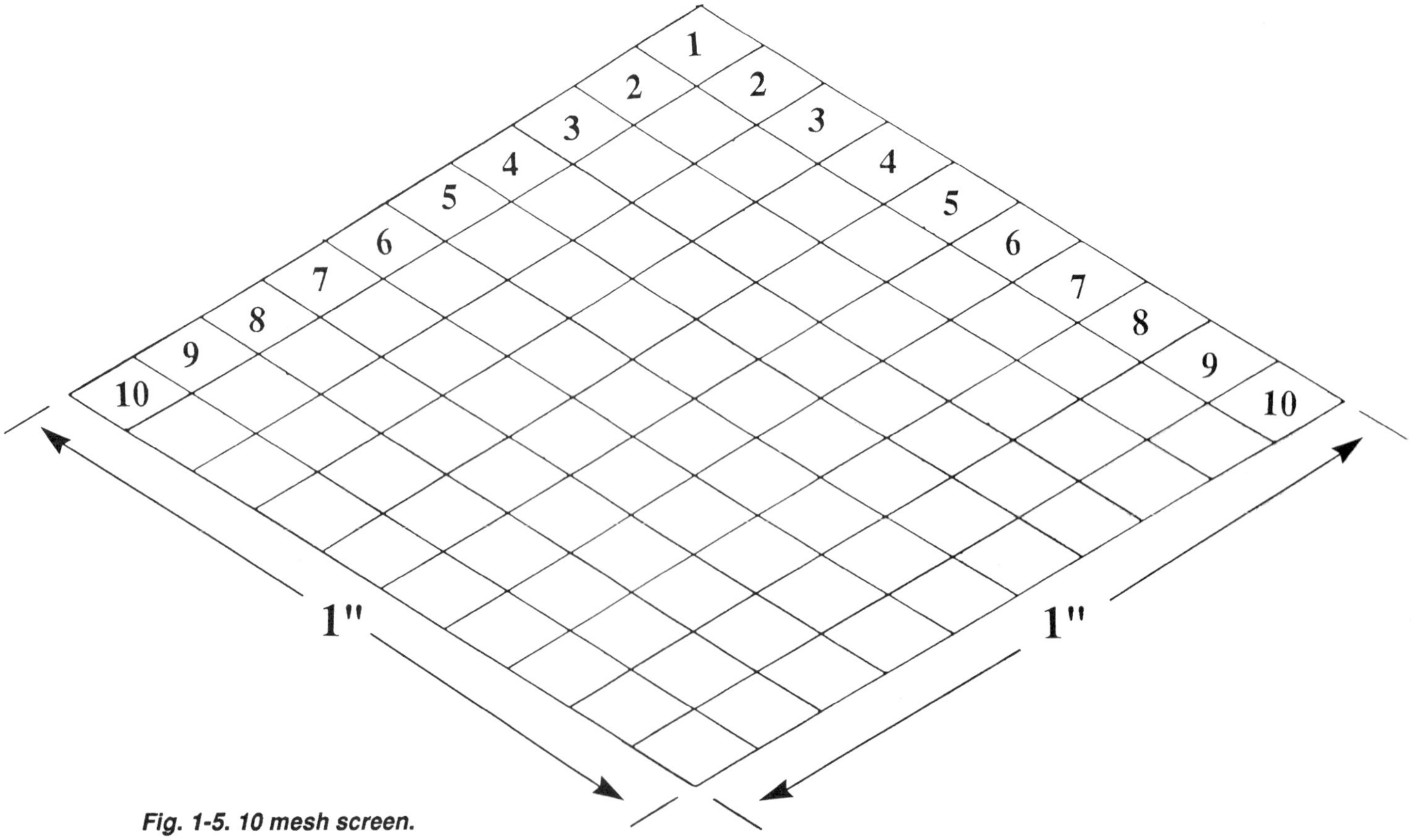

Fig. 1-5. 10 mesh screen.

The general breakdown commonly used to classify the size differences of gold is done with the use of mesh screen. "*Mesh*" signifies the number of openings contained in a lineal inch of screen or wire cloth. For example, a screen labeled "10 mesh" would contain 10 openings per lineal inch, or 100 openings per square inch. "Twenty mesh" would have 20 openings per lineal inch, or 400 openings per square inch, and so on (see Figure 1-5).

Those pieces of gold which will pass through 10 mesh (1/16" sized openings), yet will not pass through 20 mesh (1/32" sized openings), are classified as "10-20 mesh." Gold passing through 20 mesh which will not pass through 40 mesh (1/64" sized openings) is classified as "20-40 mesh," and so on (see Figure 1-6).

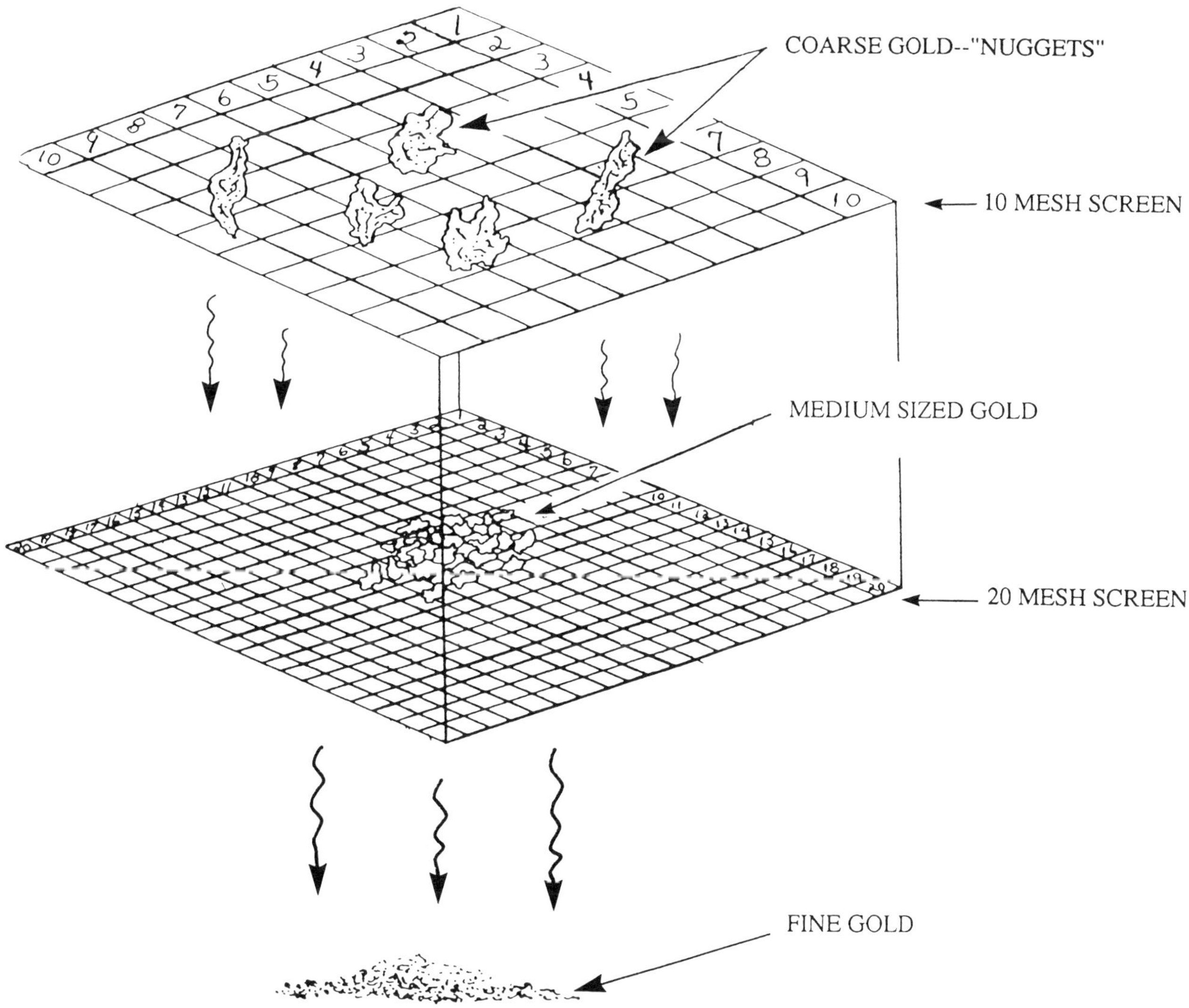

Fig. 1-6. Gold is classified and labeled as to its size.

Among some miners, the various sized pieces of gold are labeled as to their different mesh sizes. "*Coarse gold*" or "*nuggets*" are agreed to be any pieces of gold which will not pass through a 10 mesch screen . Medium sized pieces of gold, flakes and so forth, are of the 10-20 mesh range, of which it would take an average of 2,200 separate pieces (colors) to make up a troy ounce. *''Fine gold''* is of the 20-40 mesh size, of which it would take an average of 12,000 separate colors to make a troy ounce. *''Flour gold''* or *''dust''* includes all pieces which are smaller than 40 mesh, including the microscopic-sized particles.

Because gold is so malleable, as it is pounded, rubbed, and pushed along by the forces of nature, the gold will tend to hold together, and some of its impurities will be pounded and washed free. So generally, pieces of gold become more pure as they are pounded and worked by the different forces of nature--especially in a streambed where it can be heavily pounded by rocks, boulders and such, along with the steady flow of water to help wash away the impurities. This pounding also has the effect of breaking the gold down into smaller pieces, from which more impurities can be washed free. And so it is generally found that smaller pieces of gold are of richer gold content than the larger pieces from the same source. For example, the fine and flour gold recovered out of many gold-bearing streams in the Western U.S. will be found to have a gold content greater than 90%.

Nuggets recovered out of the very same deposits are often found to have a gold content of less than 80%. At first glance, this might seem to indicate that fine gold has a greater value than the coarse gold, and this is true as far as the actual gold value is concerned. However, the larger pieces have *''jewelry specimen value,''* due to their own unique and natural characteristics, and so can bring in a greater monetary return than their actual gold value. Fine gold is usually sold to a refiner, who melts it down and refines it into pure bullion form, to be eventually sold on the world gold market.

INDICATION OF GOLD CONTENT

There are two ways of labeling the gold content of raw and unprocessed gold which are commonly being used in the field today. The *''fineness,''* or *''percentage system,''* is used most often among miners and refiners. This system breaks down the purity into thousandths in order to label the ''fineness'' (purity) of the gold sample. For example, a specimen that contains 90% gold and 10% impurities would be labeled .900 fine; 88% gold content would be labeled .880 fine; 75% gold content would be indicated as .750 fine, and so on. In this system, .999 fine is used to indicate pure gold. The fineness system is often used among assayers to indicate the gold content in the samples they test.

The other system sometimes used to indicate the gold content in your samples or specimens is the *''carat system.''* This method is most commonly used the jewelry business, and is based on a 24-point system with 100% gold content being 24 carats, 50% gold being 12 carats, 75% gold being 18 carats, and so on.

MEASURING GOLD BY WEIGHT

There are also two separate measuring systems used to weigh gold. The most commonly used is the *''troy system''* which is as follows:

24 grains = 1 pennyweight
20 pennyweight = 1 troy ounce
12 troy ounces= 1 troy pound

Scales having these measuring increments are usually available wherever gold mining equipment is sold. The troy system, as laid out above, is most commonly used by miners in the field and when gold is sold on the open market.

A troy pound is equal to .3732 kilograms, whereas a standard pound (avoirdupois) is equal to .4536 kilograms. So you see that they are not the same, a standard pound weighing considerably more than a troy pound. Also take note that there are only 12 troy ounces to a troy pound, as opposed to the 16 ounces it takes to make up a standard pound. The result is that a troy ounce weighs slightly more than a standard ounce.

The other system of measuring gold is by grams. Most triple beam balance scales measure in terms of grams instead of troy increments. So it is not uncommon to find a gram scale being used to measure gold. In this case the conversion scale is as follows:

TROY--GRAM CONVERSION TABLE

1 troy pound = 373.248 grams
1 troy ounce = 31.104 grams
1 pennyweight = 1.555 grams
1 grain = 65 milligrams

FOOL'S GOLD

It is not unusual for a beginner to wonder about the difference between gold and the other materials found in a streambed or lode deposit. Sometimes a beginner will puzzle over shiny rocks, and quite often iron pyrites *(fool's gold)* are mistaken for the real thing. In fact, this is so much the case that there is a story of an entire shipload of iron pyrites having been shipped over to England during the 1500's--the yellow stuff having been mistaken for gold. So you can understand where it gets the term *''fools gold.''*

Gold is a brassy yellow metal. Once you have seen a bit of it in its natural form a few times, you will no longer have much difficulty in distinguishing it from the other materials that are commonly associated with it. Gold does not look anything like rock. It looks like metal--gold metal. If you are just starting and have not yet had the opportunity to see much gold in its natural form, there are three easy tests which will validate your discoveries one way or the other:

Glitter Test: Gold does not glitter. It shines--sometimes it's bright, sometimes it's dull, but very seldom does it glitter. The thing about fool's gold (pyrites or mica), is that because of its crystalline structure, it tends to always be glittery. Take the sample and turn it in your hand in the sunlight. If it is gold, the metal will continue to shine regularly as the specimen is turned. A piece of fool's gold will glitter as the different sides of its crystal-like structure reflect light differently.

Hardness Test: Gold is soft metal, like lead, and will dent or bend when a small amount of force is applied to it. Pyrites, mica and shiny rocks are generally hard and brittle. Just a little amount of pounding will shatter them. Gold almost never shatters!

It is said that the old-timer used to put a specimen in his mouth and bite on it to test if it was gold. This is another way of testing the larger sized specimens. However, keep in mind that a larger sized piece of gold is worth a great deal and the resulting tooth marks could lessen its value. If you find a true piece of gold big enough for you to bite on, I can assure you that you will have no doubt that it is the real thing, simply because of its nature and its weight. But if you are still uncertain of your find, you might try using the sharp edge of a knife and gently press in on the specimen in a place which is less likely to be noticed. If it's gold, an indentation can easily be made into the metal with the blade of your knife. If it is a rock or iron pyrites, you will not dent its surface.

Acid Test: Nitric acid will not affect gold (other than to clean it); whereas, it will dissolve most of the other metals found in a streambed. Nitric acid can be purchased from some drug stores or prescription counters, and can sometimes be found where gold mining equipment is sold. If you question whether your specimen is some metal other than gold, immerse it in a solution of nitric acid. If your specimen is gold, it will remain rather unaffected. If it is most any other kind of metal, it will dissolve in the acid. Nitric acid will not affect iron pyrites or mica (fool's gold), but they are brittle and will not pass the hardness test.

CAUTION: Nitric acid can be dangerous to work with, and certain precautions must be taken to prevent harm to yourself and your equipment when working with it. These, and mixing instructions, are covered in Chapter 7 of this volume.

PLATINUM

Platinum is an industrial metal, one of a family of six separate metals: platinum, palladium, iridium, osmium, rhodium, and ruthenium. These metals are frequently naturally alloyed among each other so that they are seldom found separately.

Platinum is a valuable metal, its value ranging in the same neighborhood as that of gold. It also has a high specific gravity, sometimes even heavier than gold--depending on how much iridium is present.

Sometimes, platinum will be present in placer gold deposits, and so will become trapped in the same recovery systems which are used to recover gold. Sometimes platinum will be present in enough quantity that it could be worth a great deal of money to you to know what it looks like--so you don't discard it along with the waste materials.

Platinum is usually a dull silvery-colored metal, much like steel--only different, in that it is non-magnetic and is not affected by nitric acid like steel is, and platinum does not rust. Platinum usually comes in the form of large and small flakes, just like gold, and sometimes in the form of nuggets.

Platinum does not have as great an affinity for mercury (quick-silver) as do some other shiny metals. However, it can be made to have affinity for mercury by the use of certain involved chemical processes, or by putting a negative electrical charge into the mercury.

Russia has been the world's number one producer of the platinum metals for the last hundred years or so.

If at first you have difficulty in telling the difference between platinum and lead, remember that platinum usually takes on a dull shiny color, whereas lead does not shine at all--unless it is polished or covered with a coat of mercury. Lead (and mercury) are also easily dissolved by nitric acid, whereas platinum remains unaffected. Most other shiny, silvery-colored metals which will be found in the recovery system will be magnetic. Platinum is not.

SILVER

Most silver mining is done by the hardrock method of extracting ore from a lode and processing out the silver by chemical and mechanical procedures. The silver recovered out of most placer deposits is still alloyed with the gold and the black sands being taken out of the deposit.

Native silver has a specific gravity of about 10 or 11, so it is heavy. It will be trapped in most gold recovery systems if present as itself in a placer deposit--which is uncommon. Generally, silver in its native form does not look like shiny silver--like in silverware. It looks more like a silver-colored rock which is uncommonly heavy. Sometimes, native silver is so tarnished that it cannot be distinguished by color at all.

PLACER GEOLOGY

CHAPTER II

Thorndike/Barnhart's **Advanced Dictionary** defines "*placer*" as "*A deposit of sand, gravel or earth in the bed of a stream containing particles of gold or other valuable mineral.*" The word "*geology*" in the same dictionary is defined as "*The features of the earth's crust in a place or region, rocks or rock formations of a particular area.*" So in putting these two words together we have "*placer geology*" as the nature and features of the formation of deposits of gold and other valuable minerals within a streambed.

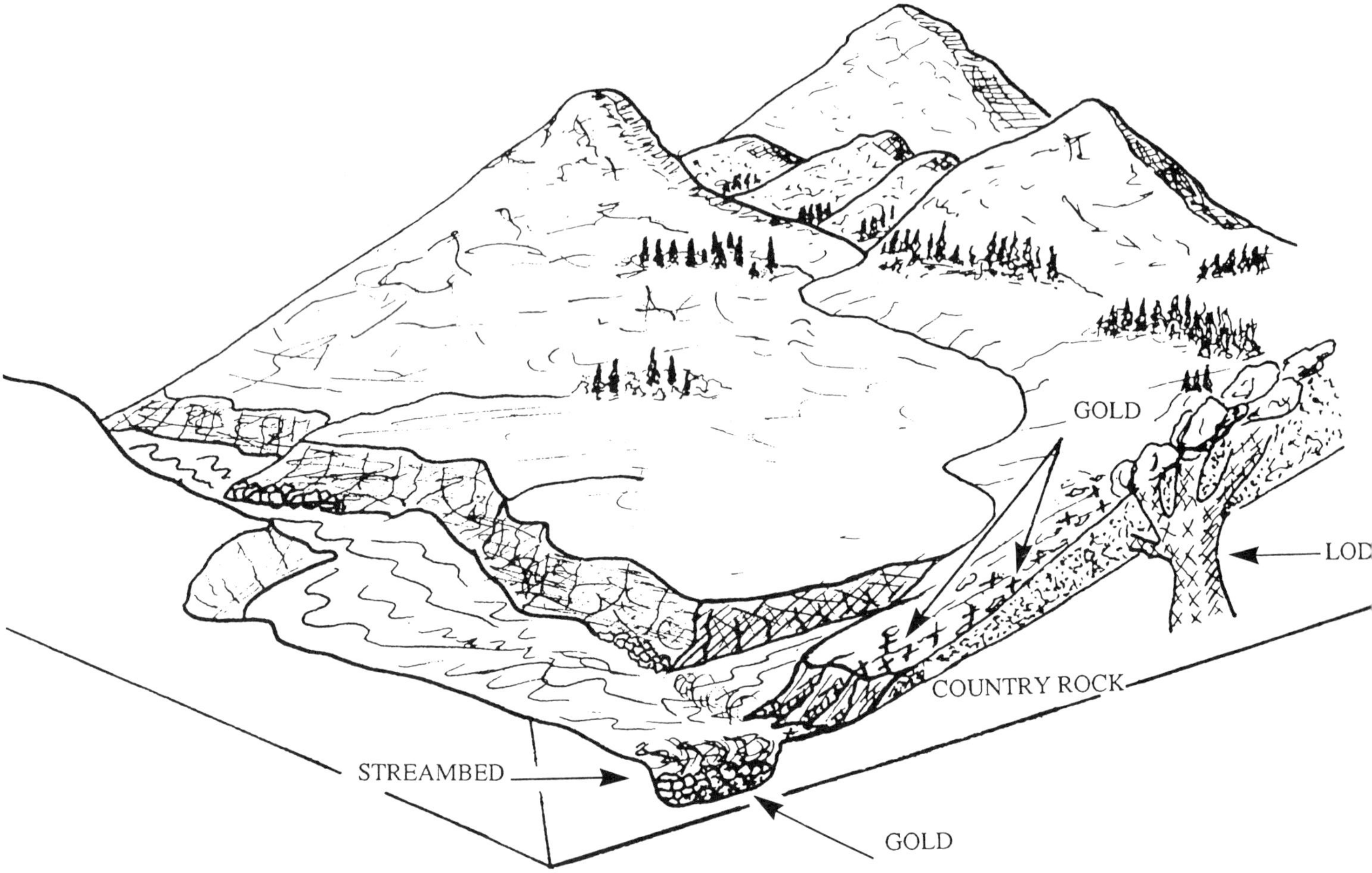

Fig. 2-1. Gold erodes from its natural lode and eventually is washed into a river or streambed.

The main factor causing gold to become deposited in the locations where it does is its superior weight over the majority of other materials which end up in a streambed. By superior weight, I mean that a piece of gold will be heavier than most any other material which displaces an equal amount of space or volume. For example, a large boulder will weigh more than a half-ounce gold nugget; but if you chip off a piece of the boulder which displaces the exact same volume or mass as the gold nugget, the nugget will weigh about six or seven times more than the chip of rock.

As gold is eroded from its original lode, gravity, wind, water and the other forces of nature may move it away and downwards until it eventually arrives in a streambed (see Figure 2-1).

There are several different types of gold deposits a prospector should know about, because they have different characteristics and are dealt with in different ways. They are as follows:

LODE: Any valuable mineral deposit, still in hardrock form, that fills a crack or fissure extending through the general country rock is a *"lode."* Lodes are the originating source of placer deposits.

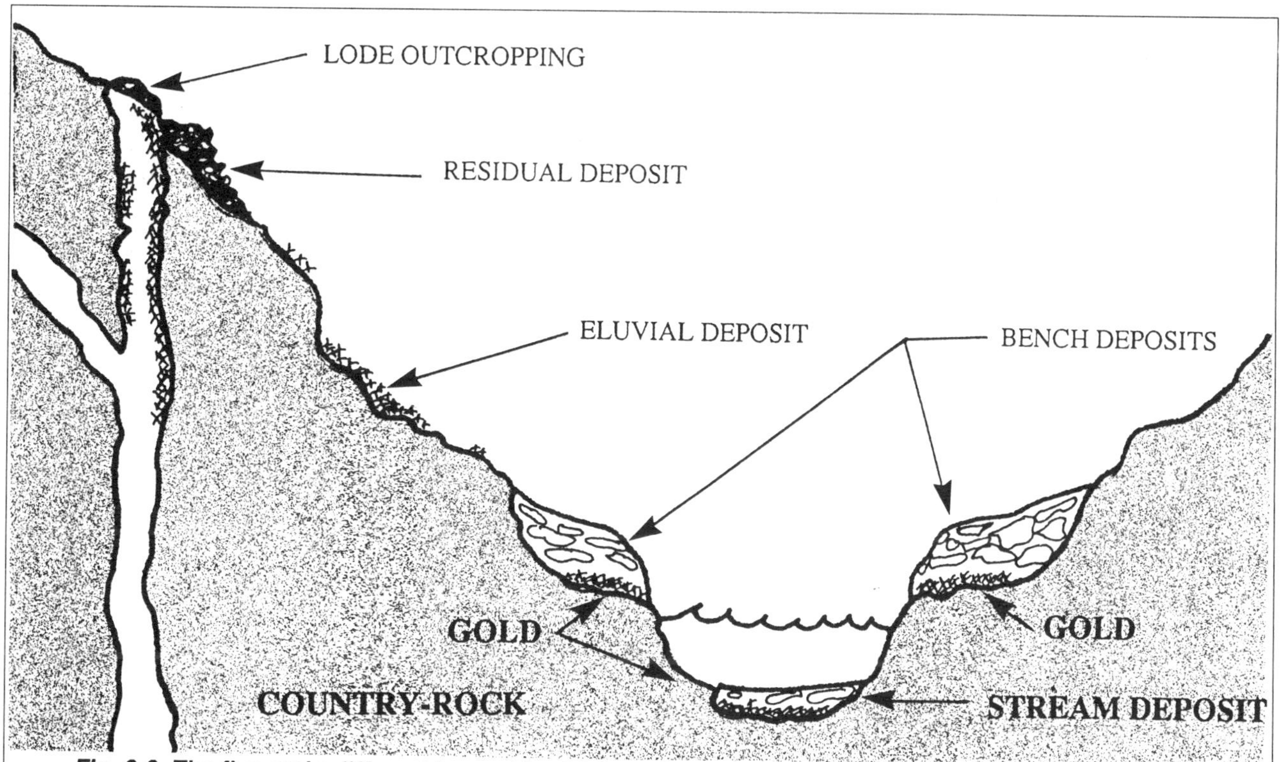

Fig. 2-2. The five main different types of gold deposit.

RESIDUAL DEPOSIT: A *"residual deposit"* consists of those pieces of the lode which have broken away from the outcropping of the vein due to the chemical and physical weathering, but have not yet been moved or washed away from the near vicinity of the lode. A residual deposit usually lies directly at the site of its lode, as shown in Figure 2-2.

ELUVIAL DEPOSIT: An *"eluvial deposit"* is composed of those pieces of ore and free gold which have eroded from a lode and have been moved away by the forces of nature, but have not yet been washed into a streambed (see Figure 2-2). The fragments of an eluvial deposit are often spread out thinly down along the mountainside below the original lode. Usually the various forces of nature cause an eluvial deposit to spread out more as its segments are washed further away from the lode deposit, as shown in Figure 2-3. The individual pieces of an eluvial deposit are popularly known as *"float."*

BENCH DEPOSIT: (Also *"terrace deposit"*) Once gold reaches a streambed, it will be deposited in different standard ways by the effects of running water. Most of the remainder of this chapter will cover these ways. During an extended period of time, a stream of water tends to cut deeper into the earth. This leaves many of the older sections of streambed high and dry. Old streambeds which now rest above the present streams of water are referred to as *"benches."* The accumulations of gold and other valuable minerals contained in an old, high streambed are called *"bench deposits."* Figure 2-2 shows bench deposits as they are. However, the diagram can be misleading in that at first it might appear that the gold out of an eluvial deposit can drop directly down to become the gold in the bench deposit. This is not the case. In actuality, an eluvial deposit might

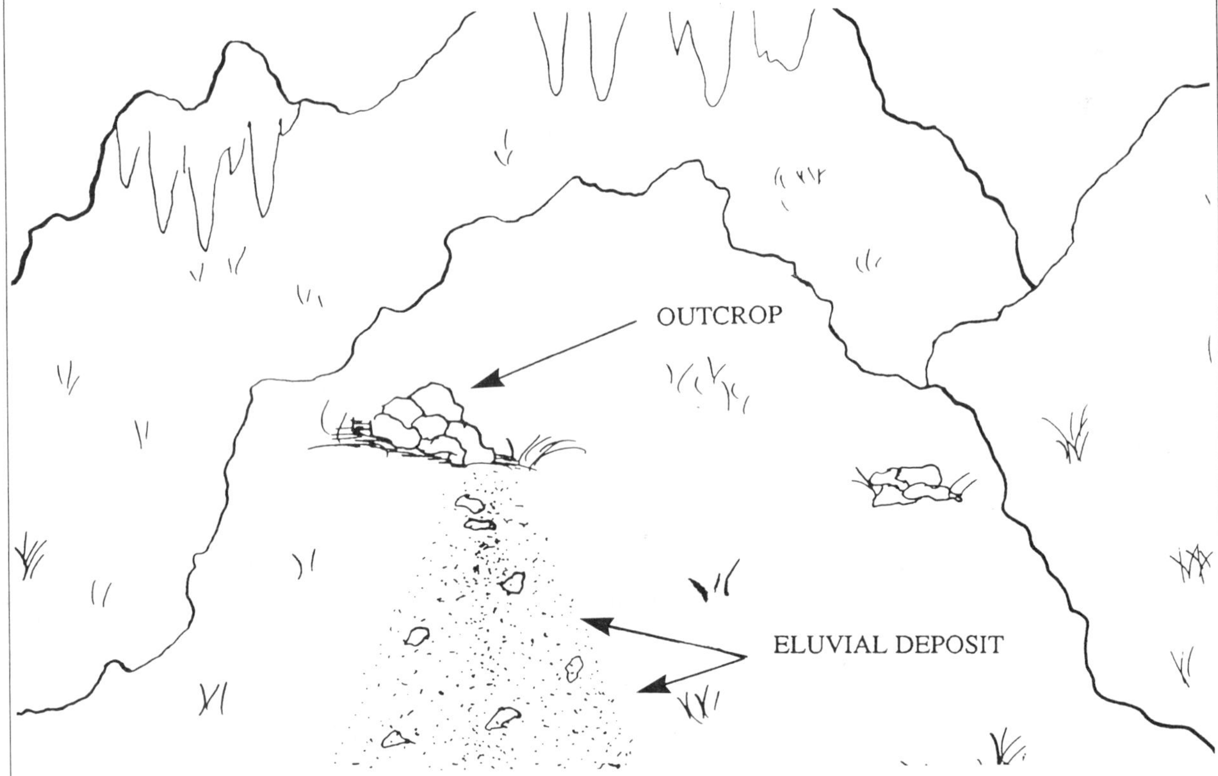

Fig. 2-3. An eluvial deposit contains those pieces of ore that have been swept away from the lode which have not yet been deposited by running water.

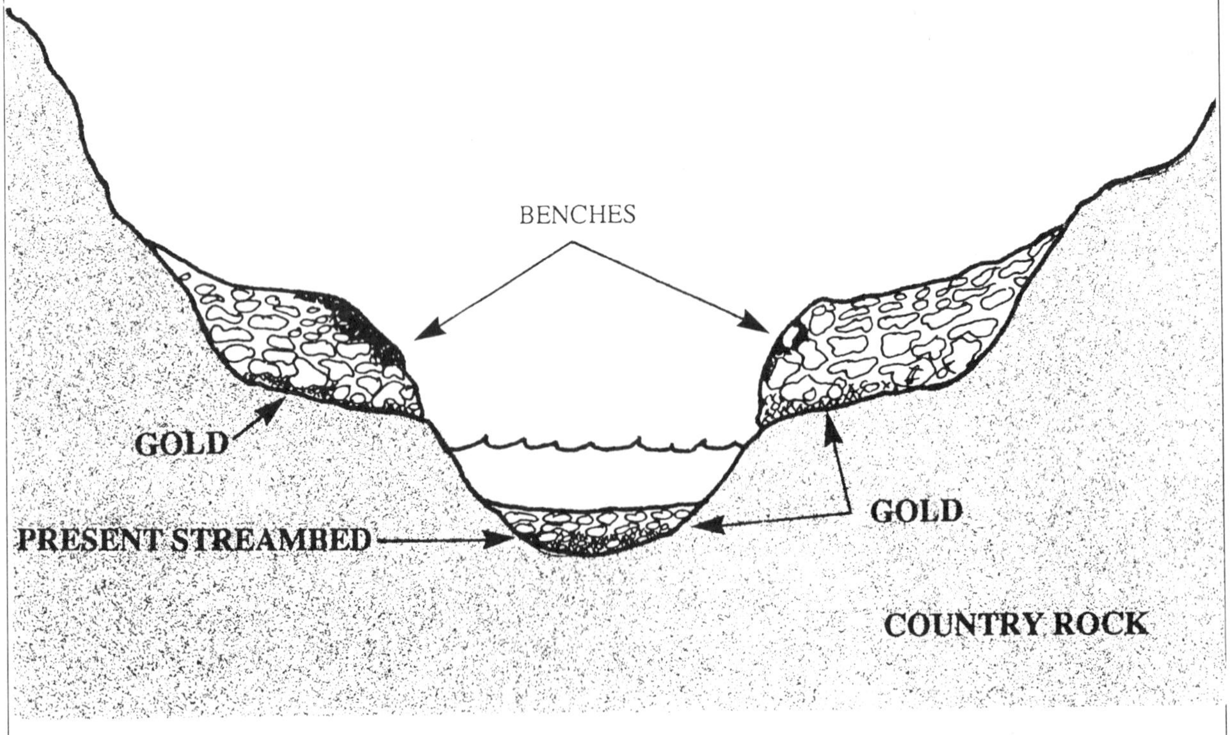

Fig. 2-4. Benches are the remains of older streambeds which once ran in that location.

be swept down to rest on top of an old streambed (bench), but it will still remain as an eluvial deposit until it is washed into a stream of water. A bench placer deposit contains the gold deposited in that streambed before it was left high and dry.

Many benches are lying close to the present streams of water, and are actually the remains of the present stream as it ran years and years ago, as shown in Figure 2-4.

Some dry streambeds (benches) are situated far away from any present stream of water. These are sometimes the remains of ancient rivers which ran before the present river systems were formed. Ancient stream benches are sometimes on top of mountains, far out into the deserts, or can be found near some of today's streams and rivers. The ancient streambeds, wherever found, are known for their rich deposits of gold.

Most surface gold mining operations today direct their activities at bench deposits. The reason for this is that the presence of an old streambed is evidence that it has never been mined before. Any gold once deposited there will still be in place.

STREAM PLACER: In order to discuss what happens to gold when it enters a stream of water, it is first necessary to understand the two terms: *''bedrock''* and *''sediments.''* Many millions of years ago, when the outer perimeter of the earth cooled, it hardened into a solid rock surface--called *''bedrock''* (or *''country rock''* when discussing the subject of lodes). All of the loose dirt, rocks, sand, gravel and boulders which lie on top of the earth's outer hardrock surface (bedrock) are called

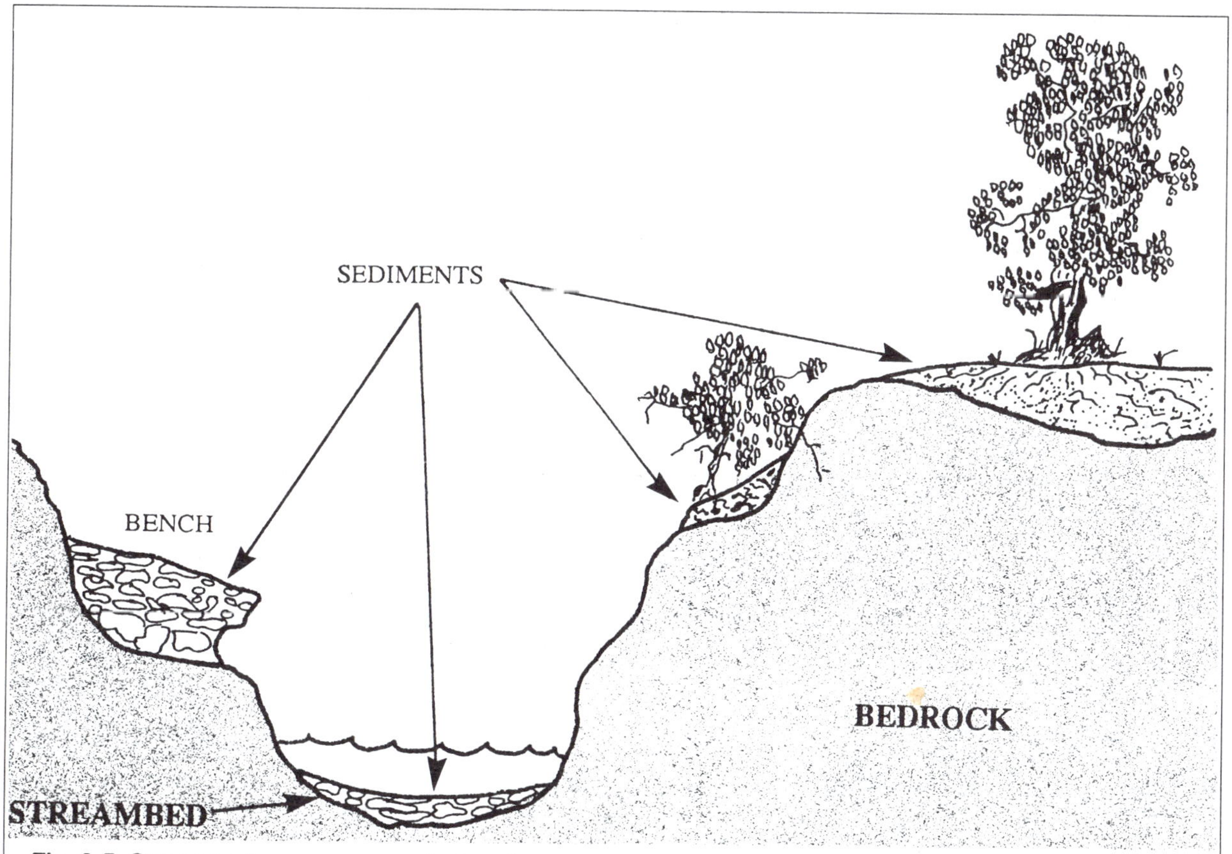

Fig. 2-5. Streambeds are composed of sediments which lie on top of bedrock.

''*sediments.*'' In some areas, the sedimentary material lies hundreds of feet deep. In other areas, especially in mountainous country and at the seashore, the earth's outer crust (bedrock) is completely exposed. Bedrock can usually be observed by driving down any highway and looking at where cuts have been made through the hard rock in order to make the highway straight and level.

Streambeds are composed of rocks, sand, gravel, clay and boulders (sediments) and always lie on top of the bedrock foundation (see Figure 2-5). Bedrock and country rock are the same thing.

A large storm in mountainous country will usually cause the streams and rivers within the area to run deeper and faster than they normally do. This additional volume of water increases the amount of force and turbulence that flows over the top of the stream beds lying at the bottom of these waterways. Sometimes, in a very large storm, the increased force of water is enough to sweep the entire streambed down the surface of its underlying bedrock foundation. It is this action which causes a streambed to cut deeper into the earth over an extended period of time. A storm of this size can also erode a significant amount of new gold into the streambeds where it will mix with the other materials.

Gold, being heavier than the other materials which are being swept downstream during a large storm, will work its way quickly to the bottom of these materials. The reason for this is that gold has a much higher specific gravity than the other streambed materials and so will exert a downward force against them. As the streambed is being vibrated and tossed around and pushed along by the tremendous torrent of water caused by the storm, gold will be vibrated downward through the other materials until it reaches something which will stop its descent--like bedrock.

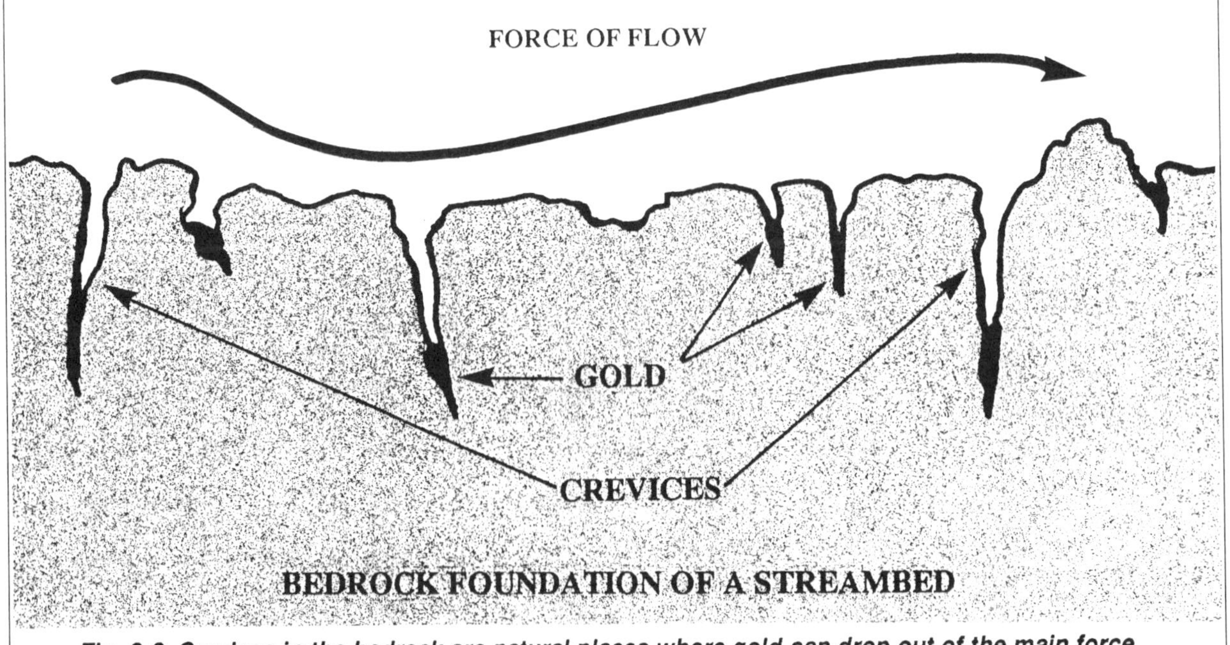

Fig. 2-6. Crevices in the bedrock are natural places where gold can drop out of the main force of the stream and become trapped.

With the exception of the finer sized pieces, it takes a lot of force to move gold. Since gold is about 6 or 7 times heavier than the average of other materials which commonly make up a streambed, it takes a lot more force to move gold down along the bedrock than it does to move the other streambed materials.

So there is the possibility of having enough force in a section of river, because of a storm, to sweep part of the streambed away, yet perhaps not enough force to move much of the gold lying on bedrock.

When there is enough force to move gold along the bottom of a riverbed, that gold can then become deposited in a new location wherever the force of the flow is lessened.

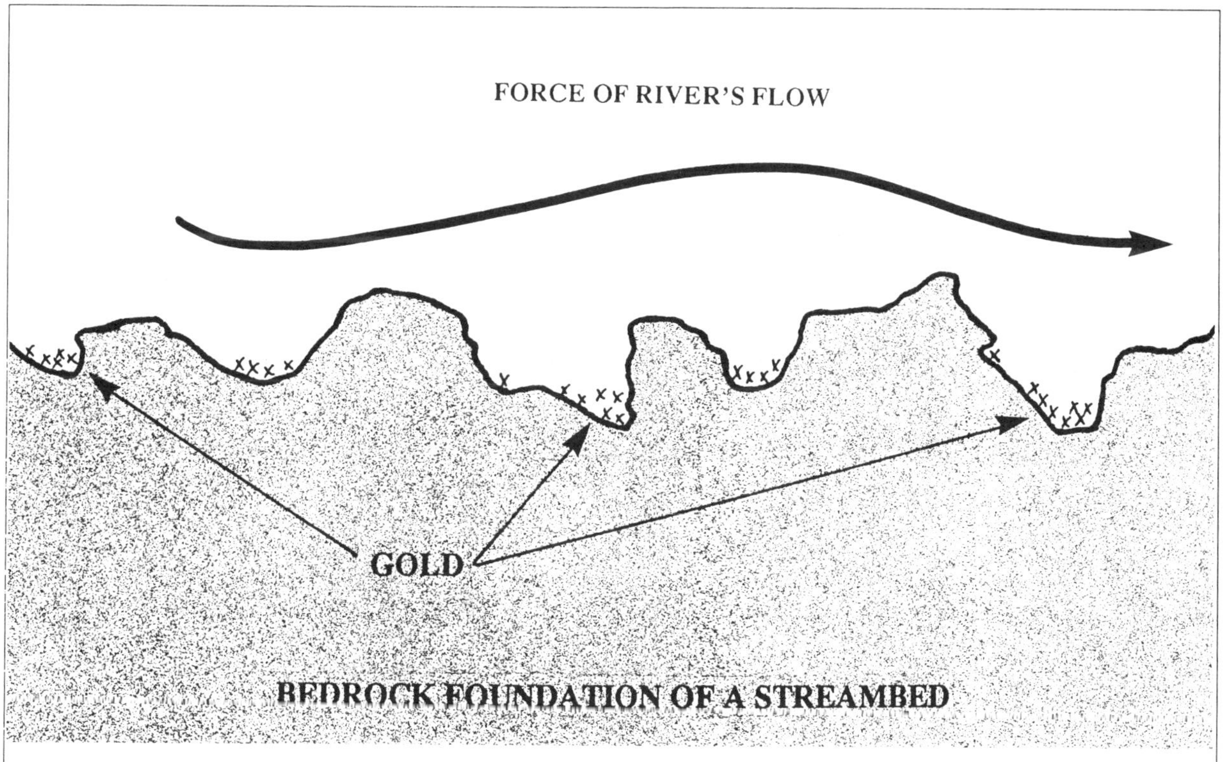

Fig. 2-7. Some bedrock surfaces are very rough and irregular--which allows for many gold traps.

BEDROCK GOLD TRAPS

Bedrock irregularities at the bottom of a streambed play a large role in determining where gold will become trapped. One good example of a bedrock gold trap is of a crack or crevice in the bedrock surface, as shown in Figure 2-6.

Many bedrock gold traps are situated so that the main force of water, being enough to move gold, will sweep the traps clean of lighter streambed materials. This leaves a hole for the gold to drop into and become shielded from the main force of water and material which is moving across the bedrock. And there the gold will remain, until some fluke of turbulence boils it out of the hole and back into the main force of water again, where it can then become trapped in some other such hole, and so on.

Some types of bedrock are very rough and irregular, which allows for many, many gold traps

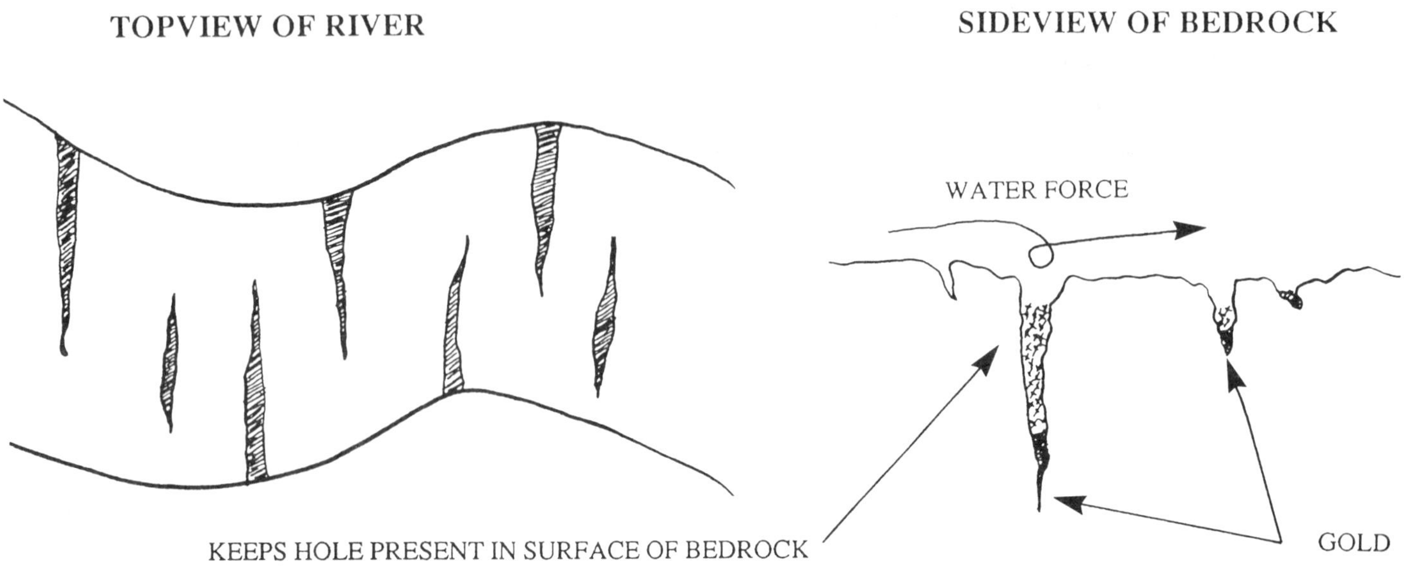

Fig. 2-8. Horizontal crevices often make excellent gold traps.

along its surface, as shown in Figure 2-7.

How well a crevice will trap gold depends greatly on the shape of the crevice itself and its direction in relation to the flow of water. Crevices extending out horizontally into a riverbed are often very effective gold catchers, because the force of water can be enough to keep the upper part of the crevice clean of material, yet the shape and depth of the crevice will prevent gold from being swept or boiled out once it's inside, as shown in Figure 2-8.

Crevices running lengthwise with the flow of the stream, or in a diagonal direction across the bed, can be good gold traps or poor ones, depending on the shape of the crevice and the set of circumstances covering each separate situation. For the most part, water force can get into a lengthwise crevice and prevent a great deal of gold from being trapped inside. However, this mostly depends on the characteristics of the bedrock surface, and there are so many possible variables that it is no use trying to cover them all--such as the possibility of a large rock becoming lodged inside a lengthwise crevice, making its entire length a gold trap of bonanza dimensions. There's really no need to say much more about lengthwise-type crevices, because if you are mining along and uncover one, you're going to clean it out to see what lies inside anyway.

Potholes in the bedrock foundation of a streambed have a tendency to trap gold very well. These usually occur where the bedrock surface is deteriorating and some portions are coming apart faster than others, leaving holes which gold can drop into and thereafter be protected from the main force of water (see Figure 2-9).

Bedrock dikes (upcroppings of a harder type of bedrock) protruding up through the floor of a streambed can make excellent gold traps in different ways, depending on the direction of the dike.

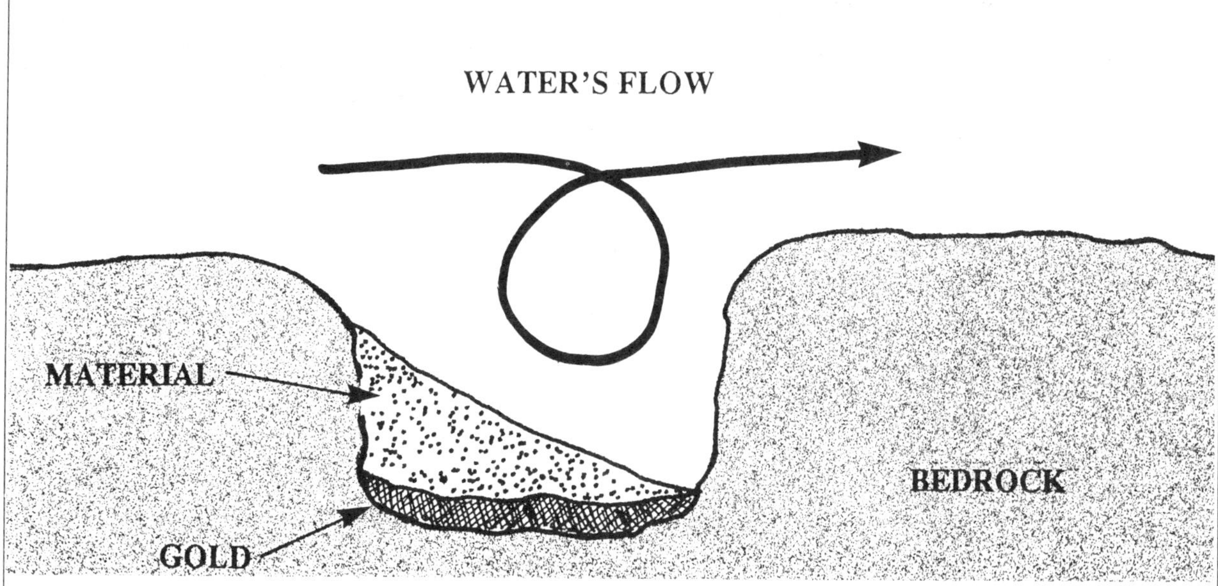

Fig. 2-9. Potholes in deteriorating bedrock surfaces often make for good gold traps.

For example, if a dike protrudes up through the floor of a streambed and is slanted in a downstream direction (see part A of Figure 2-10), gold will generally become trapped behind the dike where it becomes shielded from the main force of the flow. A dike slanting in an upstream direction is likely to trap gold in a little pocket just up in front, as shown in part B of Figure 2-10.

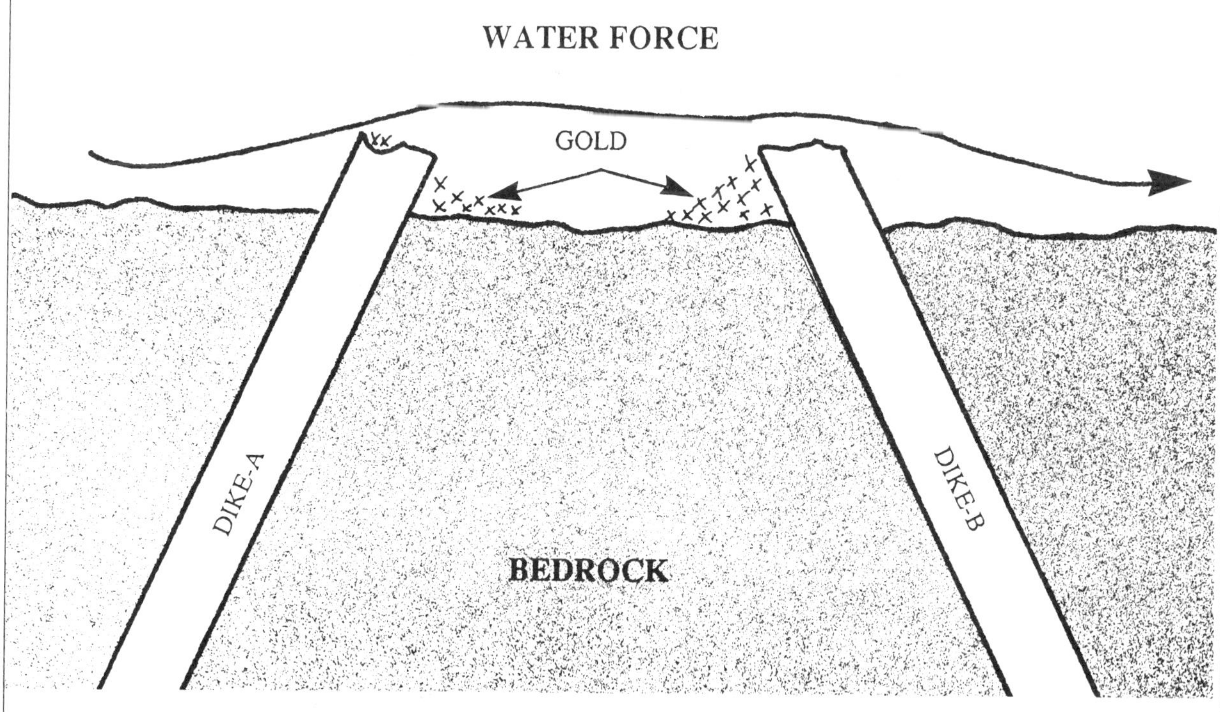

Fig. 2-10. Bedrock dikes trap gold in different ways depending on their direction to the stream's flow.

Hairline cracks in the bedrock surface of a streambed often contain surprising amounts of gold. Sometimes you can take out pieces of gold that seem to be too large for the cracks you find them in, and it leaves you wondering how they got there. Once in a while a hairline crack will open up into a space which holds a nice little pocket of gold, as shown in Figure 2-11.

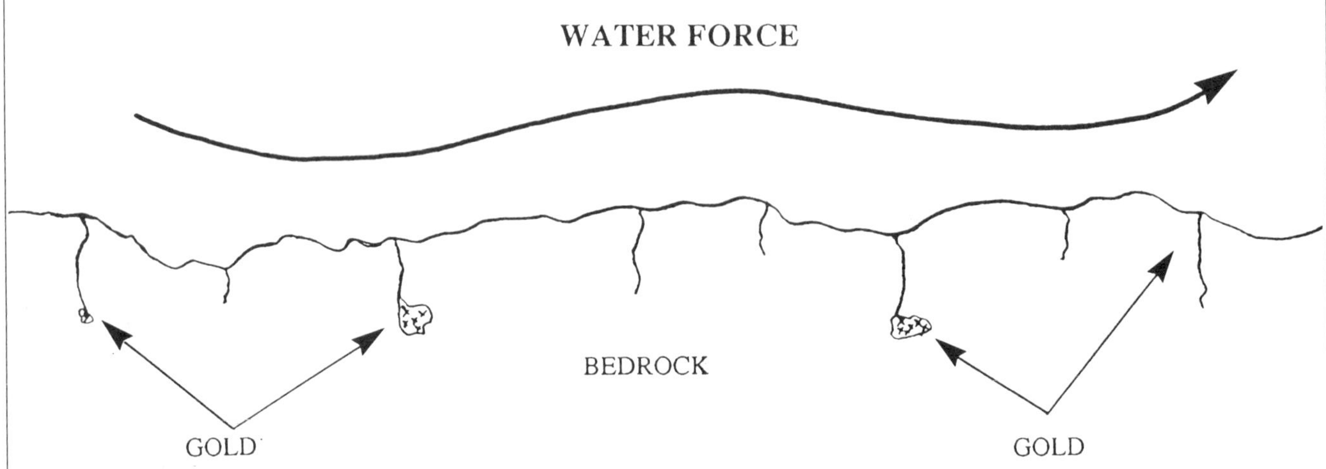

Fig. 2-11. Hairline cracks can hold plenty of gold, and sometimes open up into small pockets.

How smooth the bedrock surface is has a great deal to do with how well its various irregularities will catch gold. Some types of bedrock, like granite for example, are extremely hard and tend to become pounded into a smooth and polished surface (see Figure 2-12).

Polished bedrock surfaces like this do not trap particles of gold nearly as well as the rough types

Fig. 2-12. Smooth and polished bedrock surfaces do not retain gold very well.

Fig. 2-13. Rough bedrock surfaces tend to trap gold very well.

of bedrock surfaces do. Also, polished bedrock, which sometimes contains large, deep ''*boil holes*'' (holes which have been bored into the bedrock by enormous amounts of water turbulence), is often an indication of too much turbulent water force to allow very much gold to settle there.

Rougher types of bedrock, as shown in Figure 2-13, often being full of irregularities, both large and small, have the kind of surface where most paying placer deposits are found. This kind of bedrock can be very hard, and still maintain its roughness. Or it can be semi-decomposed. Either way, it will trap gold very well.

Basically, anywhere rough bedrock is situated so that its irregularities can slacken the force of water, in a location where gold will travel, is a likely place to find gold trapped.

Obstructions in a streambed can also cause the flow of water to slow down and can be the cause of a gold deposit, sometimes in front and sometimes to the rear of the obstruction. An outcropping of bedrock jutting out into a stream or river from one side can trap gold in various ways, depending on the shape of the outcropping and the direction it protrudes out into the stream. An outcropping extending out into the river in an upstream direction is most likely to trap gold in front of the outcropping where there is a lull in the water force, as shown in part A of Figure 2-14. Part B of the same diagram shows the gold deposit is most likely to be found on the downstream side of an outcropping which juts out into the river in a downstream direction, because that's where the force of the flow lets up.

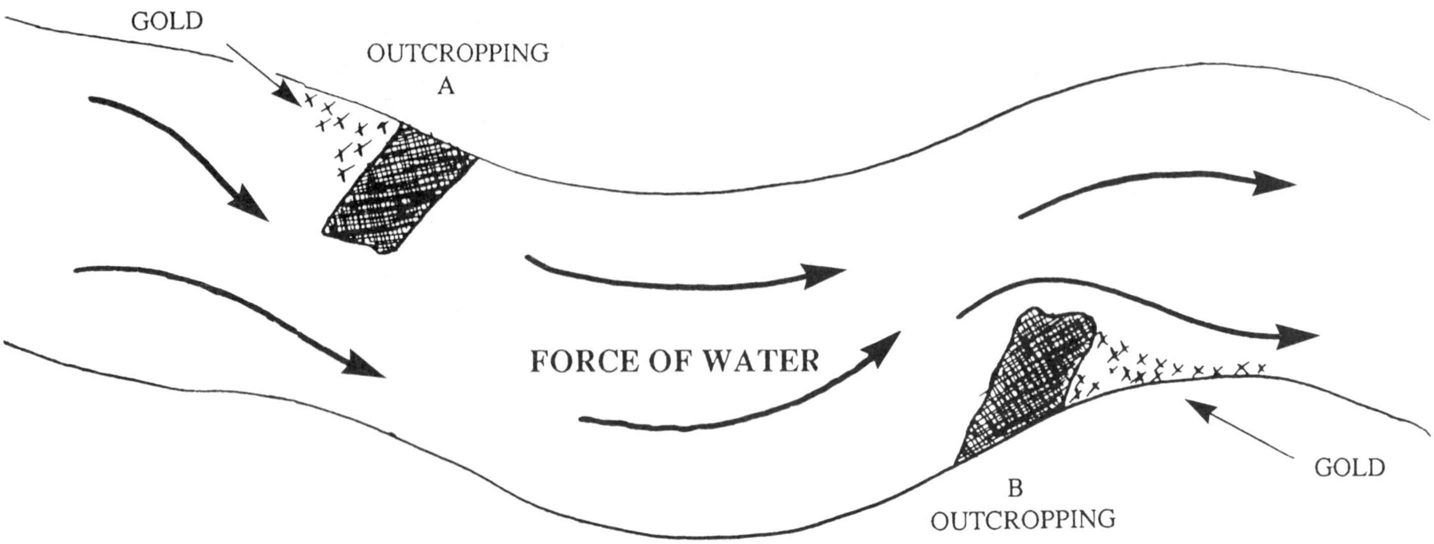

Fig. 2-14. An outcropping can be the source of a gold deposit in various ways--depending on the direction in which it juts out into the streambed.

LARGER GOLD TRAPS - PAYSTREAK AREAS

One of the most common locations in a stream or river to find a gold deposit is where the bedrock drops off suddenly to form a deep water pool. Anyplace where a fixed volume of water suddenly flows into a much larger volume of water is a place where the flow may slow down. Wherever the flow of water in a streambed slows down during a major floodstorm is a good place for gold to be dropped. And so it's not uncommon to find a good sized gold deposit in a streambed where there is a sudden drop-off into deeper water, as shown in Figure 2-15.

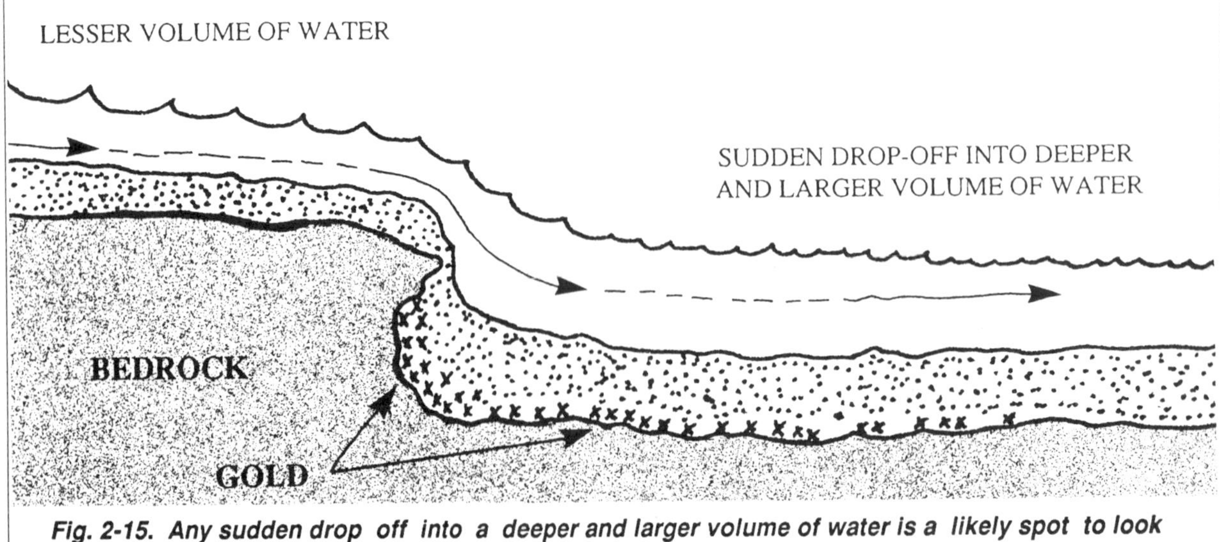

Fig. 2-15. Any sudden drop off into a deeper and larger volume of water is a likely spot to look for a deposit of gold.

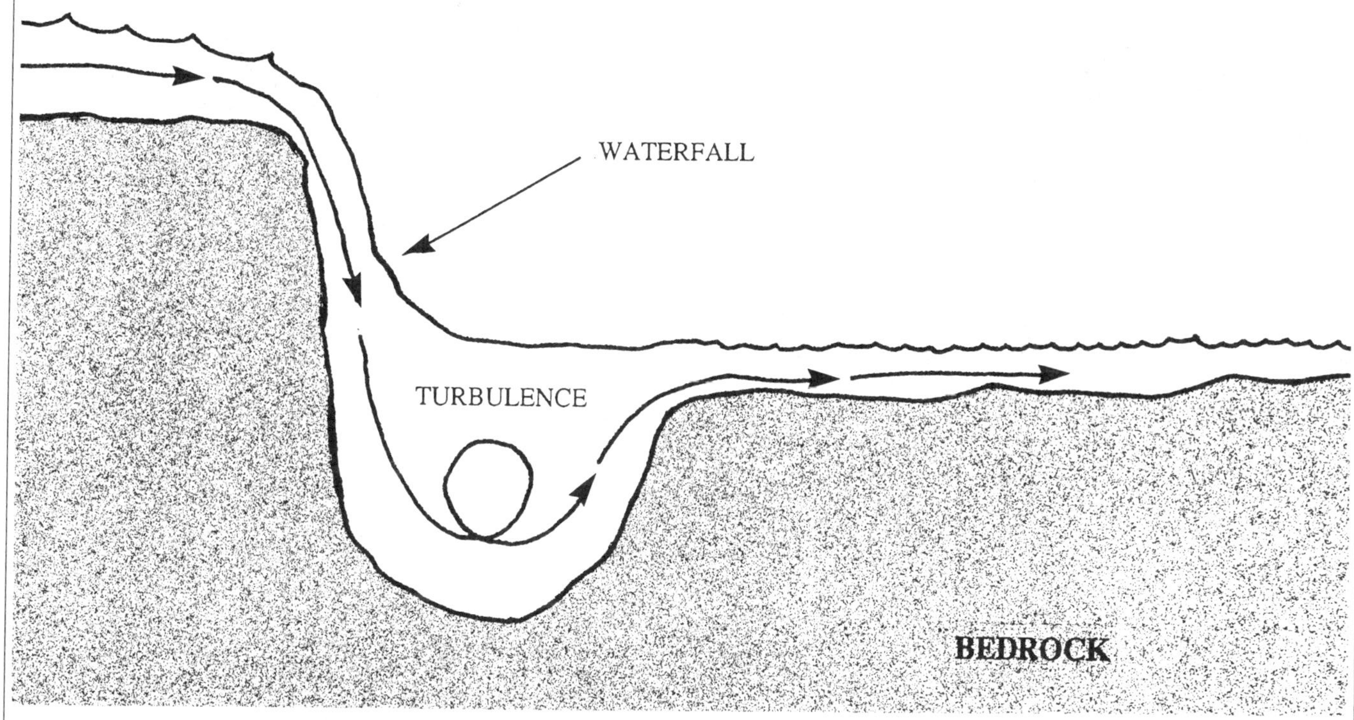

Fig. 2-16. Sometimes the amount of turbulence at the base of a waterfall is so great that everything is boiled out during high water.

A waterfall is the extreme case of a sudden bedrock drop-off and can sometimes have a large deposit of gold at its base--but not always. Sometimes the water will plunge down into the hole of the falls and create so much turbulence that any gold dropped into the hole during a storm will become ground up or boiled out (see Figure 2-16). This is also potentially true of any other lesser sudden drop-off locations inside a waterway.

On the other hand, sometimes large boulders can become trapped at the base of a falls and protect the gold from becoming ground up or boiled out by the turbulence. In this case the falls can become a bonanza, as shown in Figure 2-17.

In some waterfalls (or lesser sudden drop-off locations) the gold that has been boiled out will drop just outside of the hole, where the current has not yet had enough runway to pick up speed again--at least not enough to carry off much of the gold which arrives there, as shown in Figure 2-18.

Waterfalls are often the territory of the suction dredger, because this type of gold trap usually deposits the gold underwater. Yet this is not always the case. Sometimes during the low water periods of the season, some of the area below a falls is exposed, and often there is very little streambed to move in order to reach bedrock--where most of the gold is likely to be. The only dependable way to determine if gold will be present below a falls, or any other sudden drop-off location in a waterway, is to sample around and find out. Usually this is rather easy (unless you run

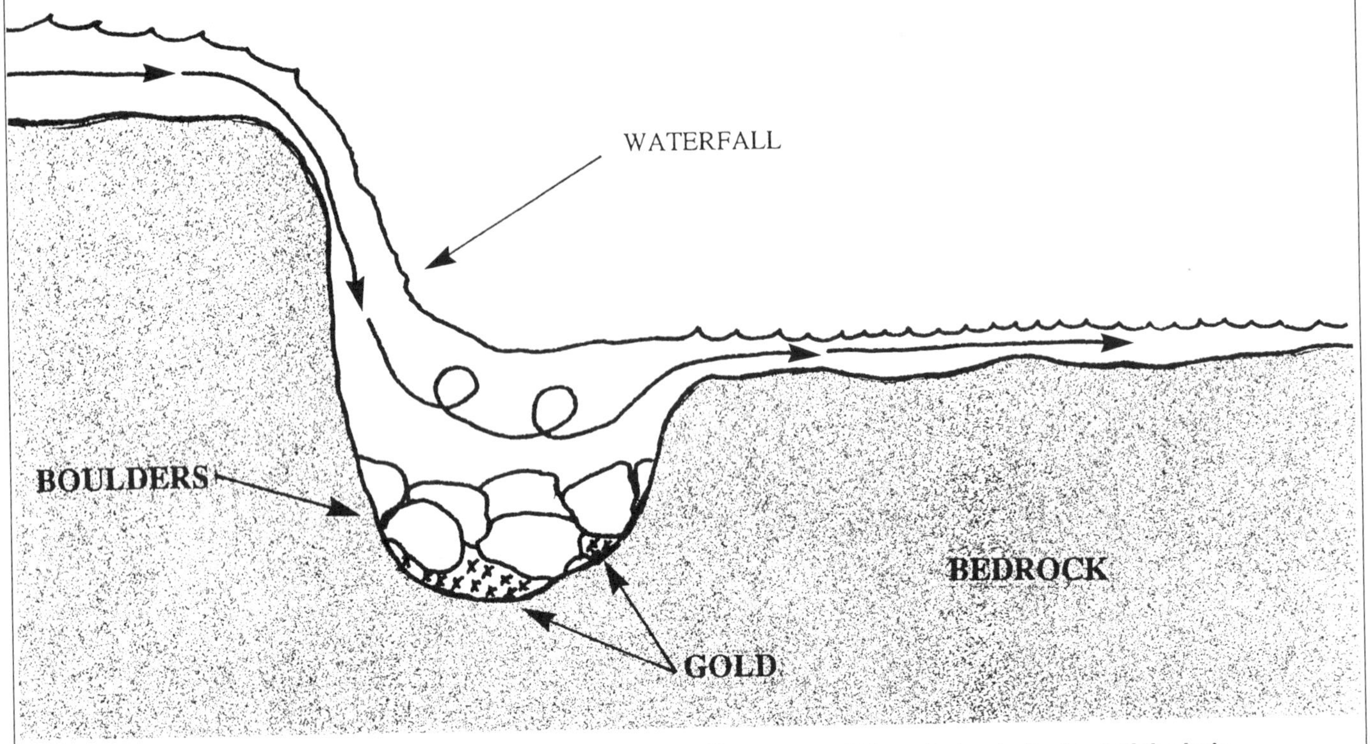

Fig. 2-17. Sometimes boulders at the base of waterfall will protect gold from being boiled out of the hole.

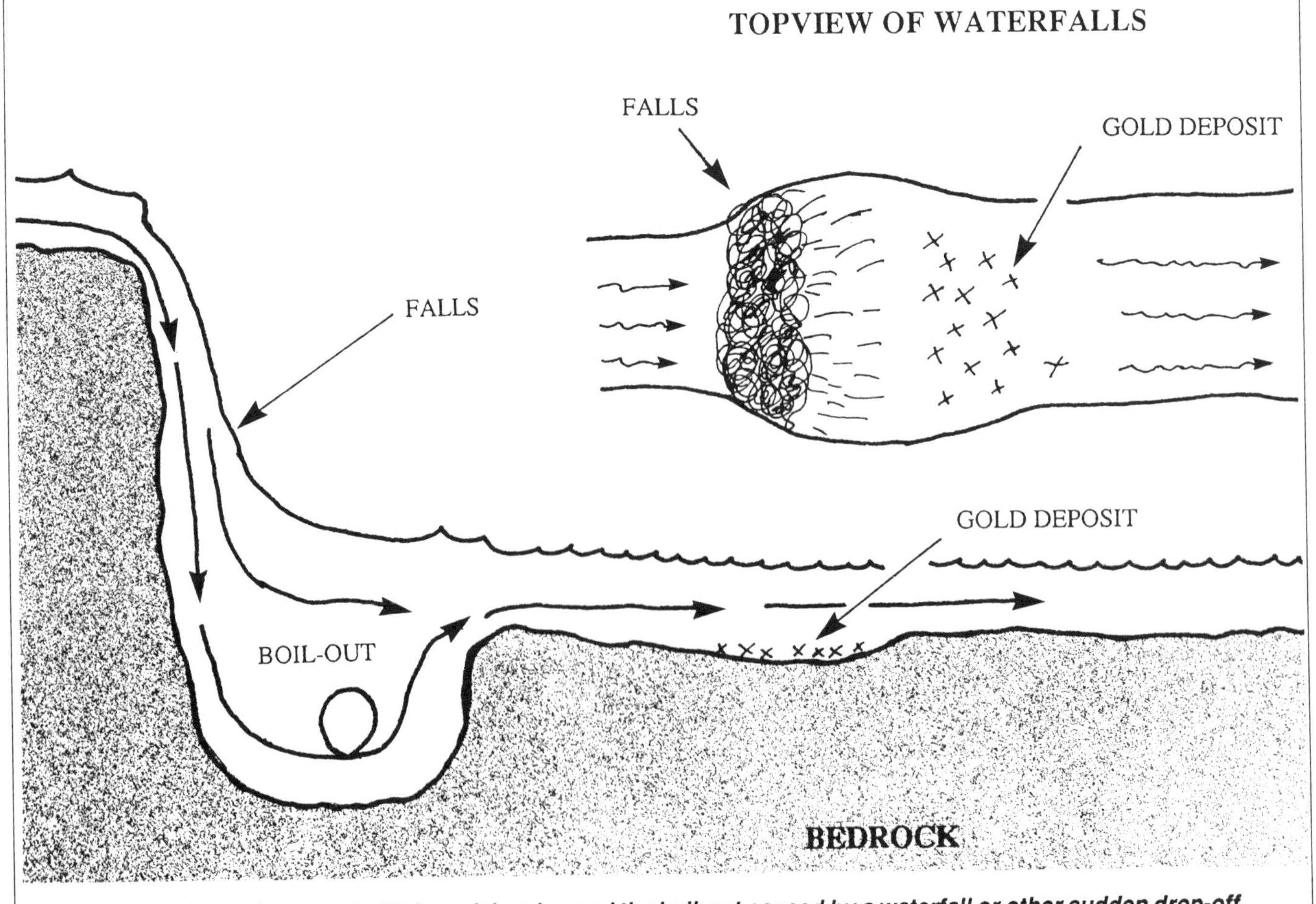

Fig. 2-18. Sometimes gold will deposit just beyond the boil-out caused by a waterfall or other sudden drop-off.

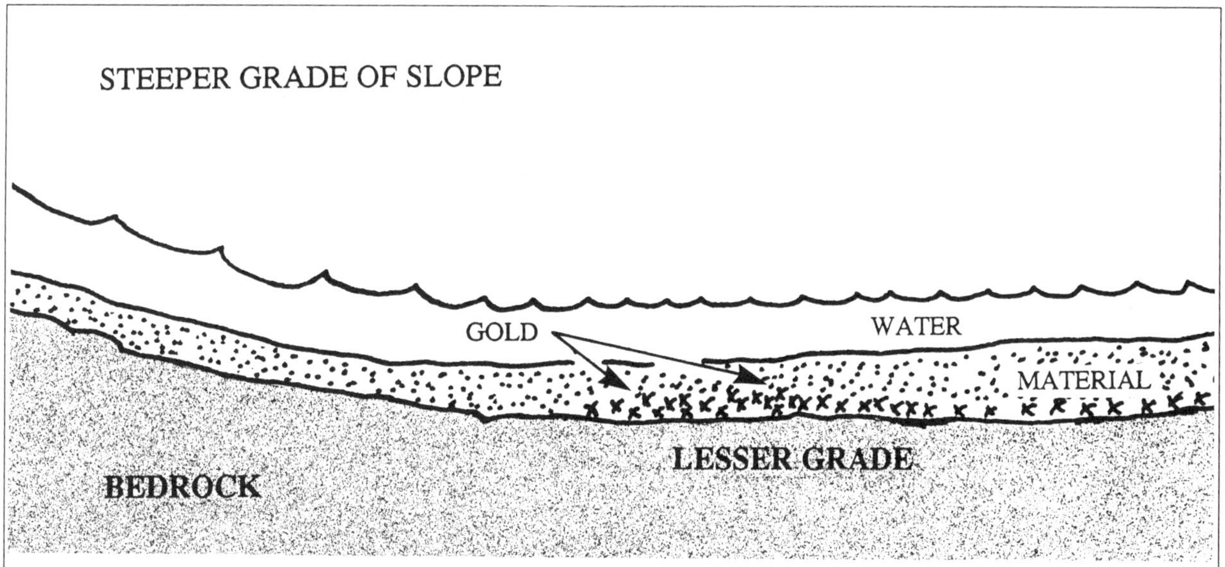

Fig. 2-19. The area just below where a streambed's slope lessens often contains a good sized gold deposit.

into huge boulders); because if the area has been boiled out and swept clean of gold, often the bedrock will be exposed or have a layer of light sand and gravel on top. Again, this is not always true. Each falls has its own individual set of circumstances.

Another common location where a good sized gold deposit is likely to be found is where the layout of the countryside causes the stream to run downhill at a rather steep grade for some distance and then suddenly it levels off. It's just below where the slope of the streambed levels off that the water flow will suddenly slow down. This is where you are likely to find gold (see Figure 2-19). Areas like this are known for their very large deposits (paystreaks).

Boulders are another type of obstruction which can be in a riverbed and cause gold to drop out of a fast flow of water. Boulders are similar to gold in that the larger they are, the more water force it takes to move them. Sometimes in a storm the force of water can pick up enough to sweep large amounts of streambed material and gold down across the bedrock. When this happens, the force may or may not be great enough to move the large boulders. A large boulder which is at rest in a stream, while a torrent of water and material is being swept past it during a large storm, will slow down the flow of the stream just in front, below, and just behind the boulder. This being the case, if the storm's torrent happens to sweep gold near the boulder, some of the gold is likely to drop where the slackening of current is, as shown in Figure 2-20.

One thing to know about boulders is that they do not always have gold trapped around them. Whether or not a specific boulder will have a deposit of gold along with it depends greatly on whether or not that boulder is in the direct path the gold would have taken when it traveled down that section of streambed during earlier major floodstorm periods.

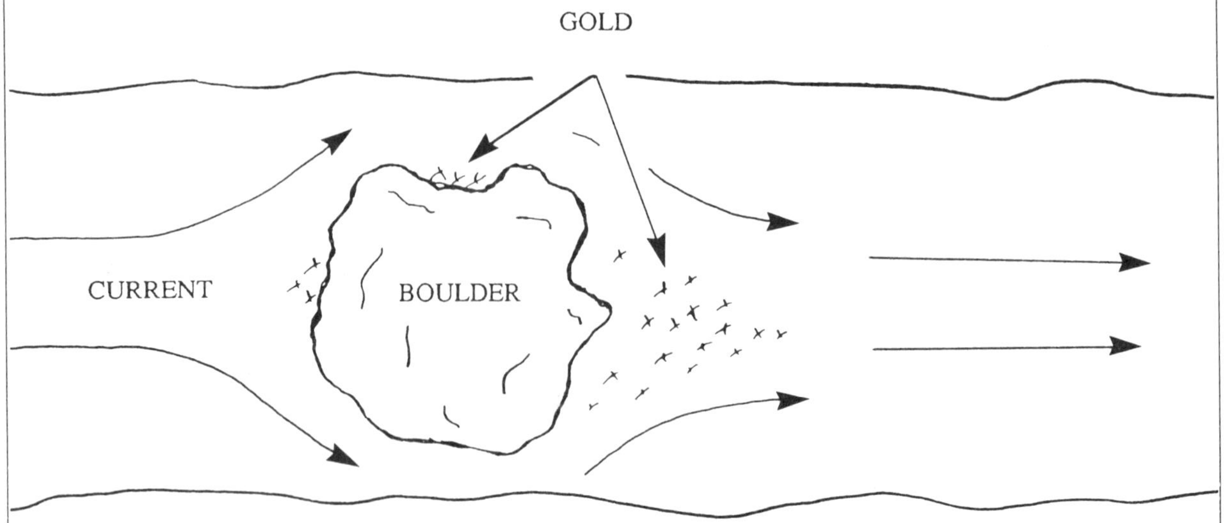

Fig. 2-20. A boulder at rest in a streambed during a large storm might trap gold wherever it slows down the water force.

THE PATH THAT GOLD FOLLOWS

Because of its weight, gold tends to travel down along a streambed taking the path of least resistance. For the most part this seems to be the shortest route possible between any major bends in the stream (see Figure 2-21).

Take note in Figure 2-21 that the route the gold is taking rounds each curve towards the inside of the bend. While this might not be the route gold always takes in a streambed, it is true that when it comes to curves, the majority of gold deposits are found towards the inside of bends. Very few are found towards the outside in comparison. Perhaps the reason for this is that the centrifugal force causes a much greater energy of flow to the outside of the bend. This creates less force towards the inside, which allows for gold to drop there.

It's important for you as a prospector to grasp the concept that under most conditions, gold tends to travel the shortest distance between the bends of a stream or river, and it also seems to follow the inside of the bends. Your best bet in prospecting is to direct your sampling activities towards areas which lie in the path that gold would most likely follow in its route downstream. This requires an understanding of what effects the various changes in bedrock and the numerous obstructions will have on changing and directing the path of gold as it is pushed downstream during extreme high water periods. For example, if you are looking for concentrations of gold around and behind boulders, you are much better off to check out the boulders lying in the path that gold should, in theory, take. This is likely to be more productive than just sampling boulders randomly in the streambed, no matter where they are located (see Figure 2-22).

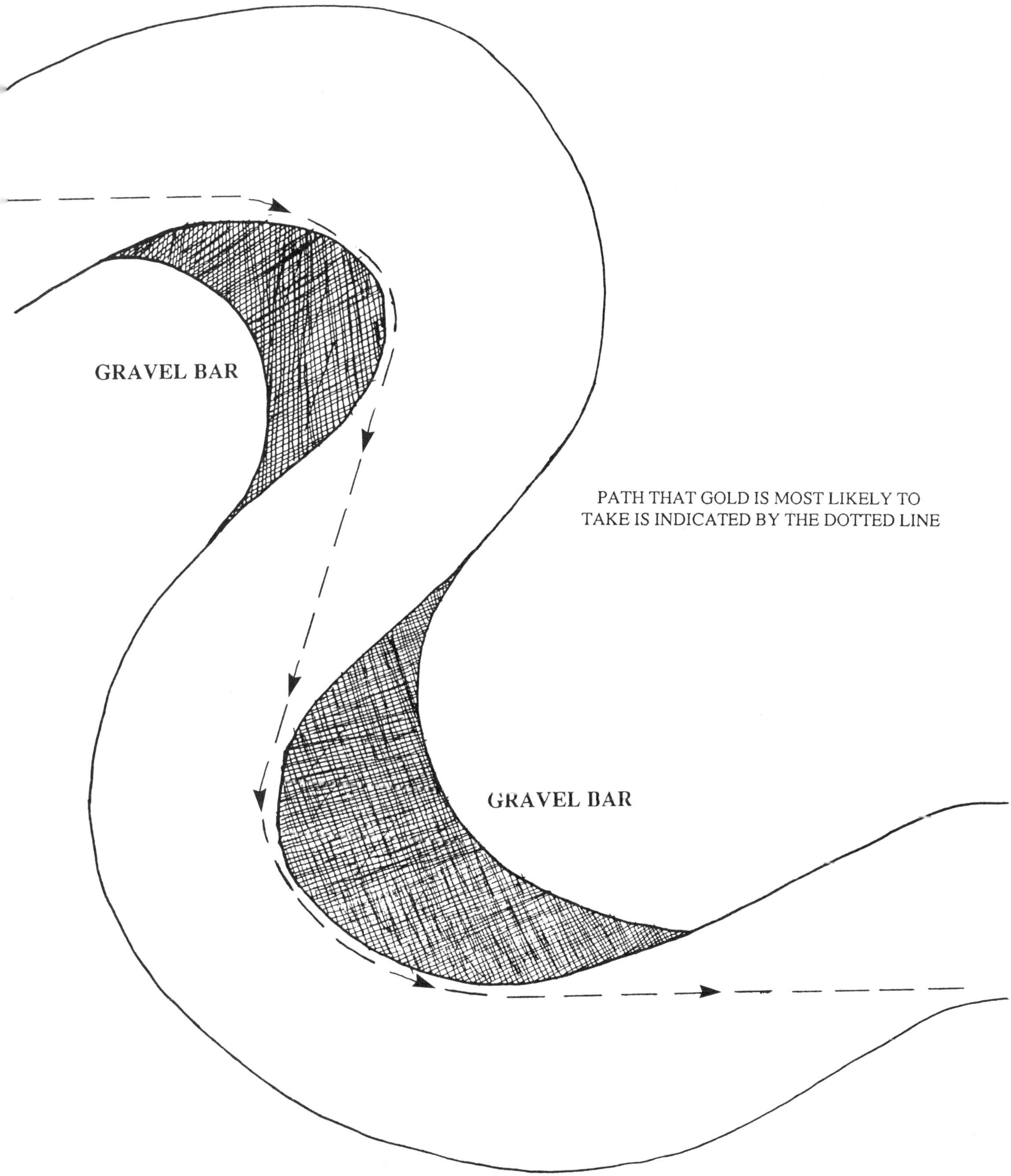

Fig. 2-21. Gold tends usually to follow the shortest route possible between any major changes in the direction of the stream or river.

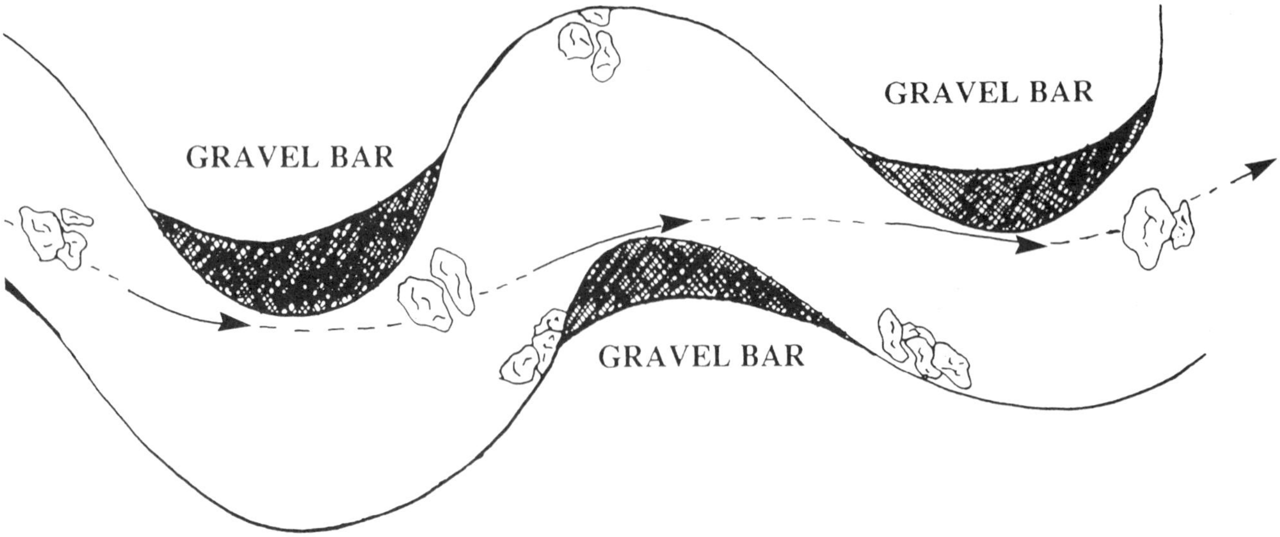

Fig. 2-22. The general path that gold should follow is indicated by the dotted line. Which boulders be would most likely to have gold concentrations along with them?

STREAMBED SUDDENLY WIDENS

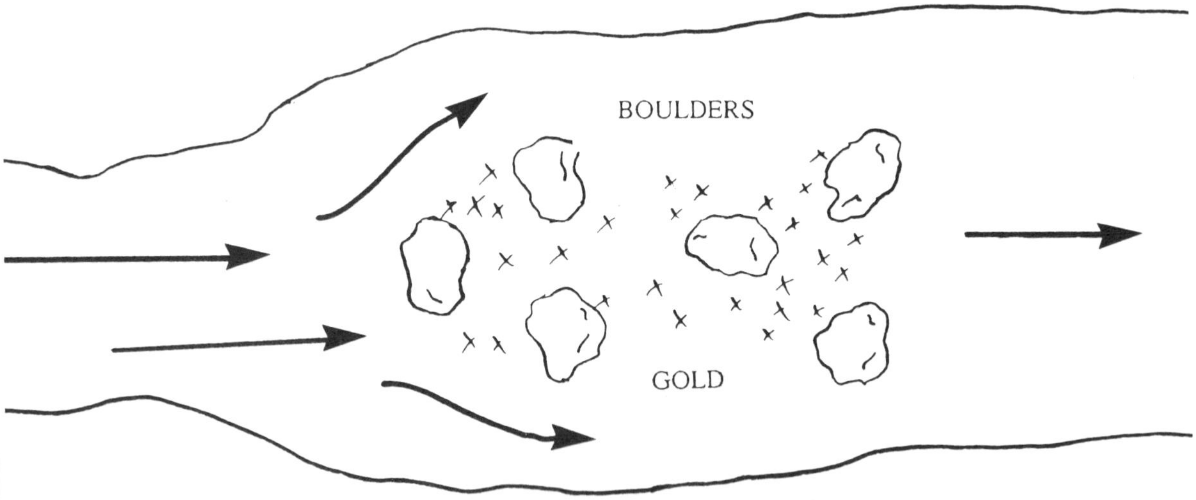

CAUSES WATER TO SUDDENLY SLOW DOWN
AND DROP BOULDERS AND GOLD

Fig. 2-23. Anywhere that a streambed suddenly widens enough to slow the force of the stream during highwater periods is a likely place to find a deposit of gold.

WHERE THE STREAMBED WIDENS

Another situation in a stream or river where there is likely to be a large deposit of gold (paystreak) is where the stream runs narrow or at a certain general width for a distance and then suddenly opens up into a wider portion of streambed. Where the streambed widens, the water flow will slow down because the streambed allows for a larger volume of water in such a location--especially during extreme high water periods. Where water force slows down is a likely place for gold to drop, as shown in Figure 2-23.

Notice in Figure 2-23 that some boulders have also dropped where the water force is suddenly slowed down. Boulders are similar to gold in that it takes a tremendous amount of force to push them along; and wherever that force lets up enough, the boulders will drop. So boulders are often found in the same areas where large amounts of gold are deposited. But gold is not always found where boulders are dropped. This is because there are so many boulders and the majority of them do not continuously follow the same route that gold generally takes. Nevertheless, it's well to take note that those places where boulders do get hung up, which are on the same route that gold should follow, are generally good places to direct some of your sampling activities.

ANCIENT RIVERS

About two million years ago, towards the end of the Tertiary geological time period, the mountain systems in the western United States underwent a tremendous amount of faulting and twisting, changing the character of the mountains into much of how they look today. It was during the same period when the present drainage system of streams, creeks and rivers were formed, which run pretty much in the westerly direction.

Prior to that, there was a vastly different river system, which generally ran in a southerly direction. These were the old streambeds which ran throughout much of the Tertiary period, and so are called *"tertiary channels"* or *"ancient rivers."* The ancient rivers ran for millions of years, during which time an enormous amount of erosion took place, washing enormous amounts of gold from the exposed rich lode deposits into the rivers.

The major changes occurring towards the end of that period, which rearranged the mountains and formed the present drainage systems, left portions of the ancient channels strewn about. Some portions were placed on top of the present mountains. Some were left out in the desert areas. And some portions were left close to, and crossed by, the present drainage system.

Geologists have argued that most of the gold in today's river systems is not gold that has eroded more recently from lode deposits, but gold that was eroded out of the old ancient riverbeds where they have been crossed by the present river systems.

The ancient channels, where they have been discovered and mined, have often proven to be extremely rich in gold deposits. In fact, many of the richest bonanzas that have been found in today's river systems have been discovered directly beneath where it has crossed the ancient streambed gravels. Other areas which have proven to be very rich in today's river systems have been found close to the old channels, where a few million years of erosion has caused some of the channel and its gold to be deposited into the present streambeds.

The ancient channels *(benches)* are notorious for their very rich bottom stratum. This strata is usually of a deep blue color; and indeed the rich blue color, when encountered, is one of the most certain indicators that ancient gravels are present. This bottom strata of the ancient gravels was referred to by the old-timers as the *"blue lead,"* probably because they followed its path all over the west wherever it led them.

Ancient blue gravels usually oxidize and turn a rusty reddish brown color after being dug up and exposed to the atmosphere. Often the ancient blue gravels are very hard and compacted--but not always.

Running into blue gravel at the bottom of a streambed does not necessarily mean that you have located an ancient channel. But it is possible that you have located some ancient gravels which might have a rich paystreak along with them.

Most of the high benches that you will find up alongside today's rivers and streams, and sometimes a fair distance away, but which travel generally in the same directio, are not Tertiary channels. They are more likely the earlier remnants of the present rivers and streams. These old streambeds are referred to as *"Pleistocene channels."* They were formed and ran some time between about a million and a half and 10,000 years ago, during the earlier part of the *Quaternary Period*, known as the *Pleistocene epoch.* Some high benches that rest alongside the present streams and rivers were formed since the passing of the Pleistocene epoch. These are referred to as *"Recent benches,"* having been formed during the *"Recent epoch"* (present epoch).

Some of these benches, either Pleistocene or Recent, are quite extensive in size. Dry streambeds are scattered about all over gold country, some which have already been mined, but many are still

Fig. 2-24. Usually all that is left of a high bench after it has been mined are piles of the larger sized streambed rocks and boulders.

untouched.

Usually, all that is left of a bench after it has been mined are rock piles, as shown in Figure 2-24. Notice in the picture that part of the unmined streambed is in the background, behind the trees.

Most of the hydraulic mining operations which operated during the early to mid-1900's were directed at high benches. *''Hydraulic mining''* was done by directing a large volume of water, under great pressure, at a streambed to erode its gravels out of the bed and through recovery systems, where the gold would be trapped (see Figure 2-25).

So some bench gravels have been mined, but many of them still remain intact. While most of the

Fig. 2-25. A hydraulic mining operation. Photo courtesy of Trinity County Historical Museum.

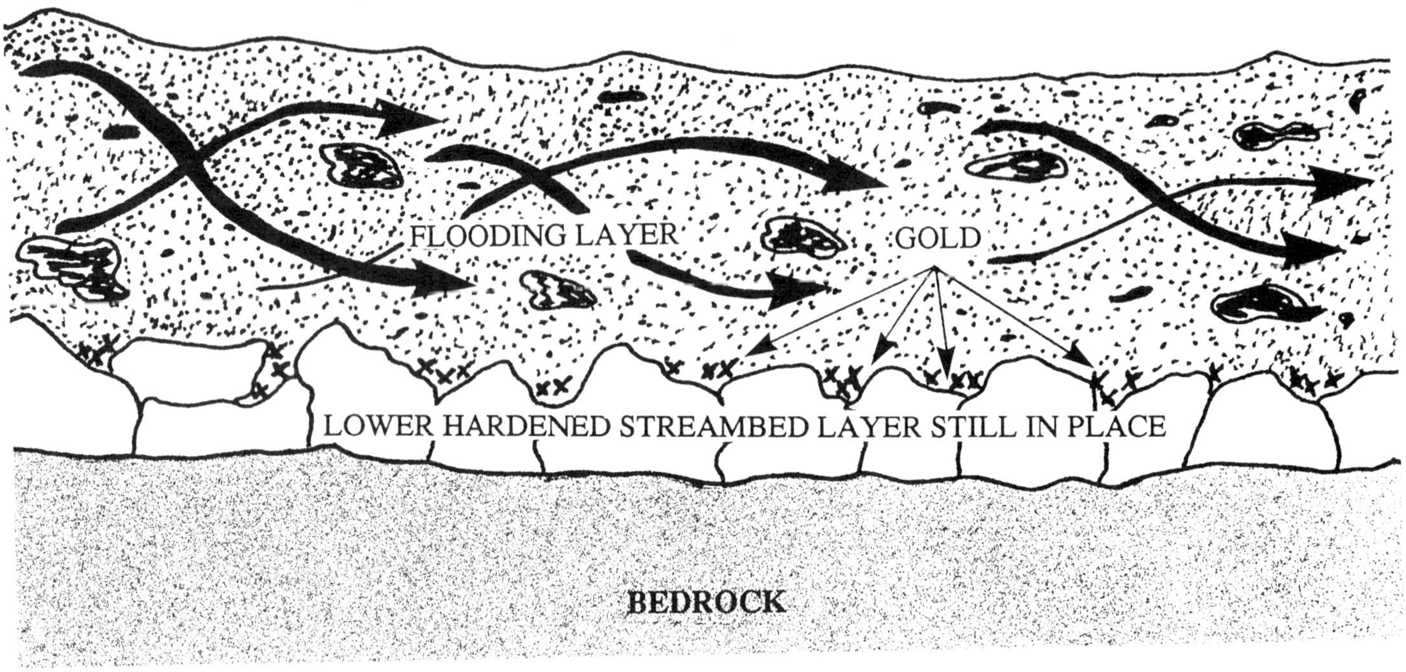

Fig. 2-26. Flood gold is that gold which rests inside and at the bottom of a flooding layer.

Pleistocene and Recent benches are not as rich in gold content as were the Tertiaries, it still remains true that an enormous amount of gold washed down into these old channels when they were running. They are pretty darn rich in some areas, and pay rather consistently in others. Also, any gold that ever washed down into any old bench which has yet to be mined, still remains there today.

FLOOD GOLD

A large percentage of the gold found in today's creeks and rivers has been washed down into them out of the higher bench deposits by the erosive effects of storms and time. A certain amount of gold is being washed downstream in any river located in gold country at all times--even if only microscopic in size.

The larger a piece of gold is, the more water force it takes to move it downstream in a riverbed. The amount of water force it takes to move a significant amount of gold in a riverbed is usually enough force to also move the streambed too. This would allow the gold to work its way quickly down to the bedrock, where it can become trapped in the various irregularities.

Some streambeds contain a high mineral content and grow very hard after being in place for a number of years.

Sometimes a storm will have enough force to move large amounts of gold, but will only move a portion of the entire streambed, leaving a lower stratum in place in some locations. When this happens, the gold moving along at the bottom of the flooding layer can become trapped by the irregularities of the unmoving *(false bedrock)* streambed layer lying underneath. The rocks in a stable lower stratum can act as natural gold traps, as shown in Figure 2-26.

Different streambed layers, caused by flooding storms, are referred to as *"flood layers."* The flood layers within a streambed, if present, are easily distinguished because they are usually of a different color, consistency, and hardness from the other layers of material within the streambed. That gold found at the bottom of, or throughout, a flood layer, is called *"flood gold."* Sometimes the bottom of a flood layer will contain far more gold than is present on bedrock. Sometimes, when more than one flood layer is present in a streambed, there will be more than one layer of flood gold present, too.

The larger a piece of gold is, the faster it will work its way down towards the bottom of a flooding layer of material as it's being washed downstream during a storm. Finer-sized pieces of gold might not work their way down through a flooding layer at all, but might remain up in the material.

So you can run across a flood layer which has a line of the heavier pieces of gold along its bottom edge, or a flood layer which contains a large amount of fine gold dispersed throughout the entire layer. You can also run across a flood layer which contains a lot of fine gold dispersed throughout, in addition to a line of heavier gold along the bottom edge.

Not all flood layers contain gold in paying quantities for the small-sized mining operation. But in gold country all flood layers do seem to contain gold in some quantity--even if only microscopic in size.

Some of the best areas to test for paying quantities of flood gold are where the stream or river widens out, or levels out, or changes the direction of its flow. Such places cause the flow of water in a storm to slow up. This can allow concentrations of flood gold to collect.

GRAVEL BAR PLACERS

Gravel bars located in streambeds flowing through gold country, especially the ones located towards the inside of bends, tend to collect a lot of flood gold, and sometimes in paying quantities even for the smaller sized operations. The flood gold in bar placers is sometimes consistently distributed throughout the entire gravel bar. Often the lower end of a gravel bar is not as rich as the head of the bar, but the gold there can be more uniformly dispersed throughout the material.

FALSE BEDROCK

Once in a while a prospector will uncover an extremely hard layer of material located just above the bedrock and mistake the layer for bedrock because of its hardness. A hard layer is often called *"false bedrock."* Such a layer can consist of streambed material, or of volcanic flows which have laid down and hardened on top of bedrock, or it can consist of any kind of mineral deposit which has hardened on top of the true bedrock.

There can be a good paying gold deposit underneath a false bedrock layer. But when there is, it is usually rather difficult to get at.

The top of every different storm layer in a streambed could be considered a "false bedrock" and be a surface area for gold to become trapped out of the flood layer which layed down on top.

WHERE TO FIND GOLD

CHAPTER III

THE LEGAL ASPECTS

The 1872 Mining Law says that any citizen of the United States (or person who has declared his intention of becoming one) can claim portions of the public lands in his own name for the purpose of developing the mineral resources on that land. This is generally true of any lands managed by the federal government which have not already been set aside for some other purpose.

The bulk of National Forest and Bureau of Land Management (BLM) land falls under the category of land which can be used for the purpose of mineral development. Most of the gold bearing country within the Western United States is located on the public lands and so is open for the gold prospector--whether as a hobby, adventure, or as a profession.

Some of the specific areas which are closed to mineral entry (meaning that you cannot file a mining claim on the land) are as follows: national park land, military land, Indian reservations, reservoirs, permanent lake beds, city owned land, state owned land, private land, state monuments, and some sections of forest lands which are set aside by the Department of Fish and Game for study or development reasons. Wilderness and primitive areas are also withdrawn.

Just because a section of land is withdrawn from mineral entry does not always mean that you can't prospect in the area. It does mean that you can't claim the mineral rights within that section of land for yourself. In some of these places, like Fish and Game Department withdrawals, it's possible that prospecting can be done on the land. It's a matter of contacting whoever is in charge to find out. This is also true of private and Indian-owned lands.

There are maps available which outline the areas that are primitive, National Parks, Indian reservations and private land, as well as the areas that are just plain National Forest land and open to mineral development by the individual. These maps can usually be obtained by contacting the U.S. Forest Service (or ranger district) having jurisdiction over the area you are interested in.

Some sections of rivers within the national forests have been entirely withdrawn from mineral entry, yet are left open to be prospected by anyone who is interested. Some Forestry Departments have withdrawn sections of river for recreational purposes so outsiders can have a place to mine recreationally, without having to worry about filing claims and such. Information on places open to the public, along with data concerning the local rules and regulations, can also be obtained by contacting the BLM and U.S. Forest Service within the areas of your general interest.

Actually, there are just a few rules and regulations the small time operator has to worry about. He usually does not effect enough environmental change to warrant the necessity of having an approved plan of operation with the Forest Service, at least not at the time of this writing.

In the state of California, it is necessary to apply to the Fish and Game Department for a permit before operating the suction nozzle of a dredge in any stream, creek or river. At the time of this writing, the permit costs $27 for California residents and around $110 for non-residents, and is just a matter of filling out a simple form at any Fish and Game Department within the state. The permit lasts until the beginning of the new year--at which time a new permit has to be filled out and paid for--just like a fishing license. The Fish and Game Department in California has certain dredge size limitations for some of its rivers and streams, and also has time limits for dredging operations in some of the waterways. Their reason for this is to prevent any damaging effects to the fish movements. When you fill out and pay for your yearly permit, the Fish and Game Department will also give you a list of the various dredge size limitations and the dredging periods which apply to the streams, rivers, and creeks they are interested in protecting. If you should want to use a dredge of a larger size, or operate during the off-time period in such a waterway, it is necessary to apply for a special permit with the Department of Fish and Game. Sometimes the permit will be directly approved. Other times, the Department will want to send someone out to the spot and have him do an inspection. Special permits cost extra.

One other rule, which is strictly enforced, is that a U.S. Forest Service-approved spark arrester is required on any internal combustion engine which is operated in the National Forest. This is true of chain saws, motorcycles, and also of the small engines used to pump water to the various types of mining equipment. These spark arresters are almost always available wherever gold mining equipment is sold.

The vast majority of land contained inside the boundaries of national forests is open to mineral entry. It's not uncommon to go into an area which has all the signs of being a good gold-bearing location and find that much of that specific area appears to be already claimed up by the individuals who got there first. In this case, if you are interested in finding an open spot to file on, it is necessary to go to the County Clerk's office and learn how to research the area that you want to file on. This is usually not hard to do, but it takes time and patience--how much, depends entirely upon how many claims have been filed in the area of your interest. The County Clerk will help you around and show you what you need to go about your research.

You do not need to file a mining claim in order to mine or prospect in an area. The purpose of a mining claim is to secure the rights to mine the minerals on that section of land for yourself.

If the mineral rights on the section of ground you would like to mine are already legally claimed by someone else, you can sometimes work out some kind of a deal with the owner. Some claim owners will allow you to prospect on a small scale for no other commitment other than to let them know how much gold you find and where. After all, a 20-acre claim covers a lot of ground, and any smart owner wants it sampled thoroughly to get an idea of its value and where the gold lies.

Sometimes the owner of a mining claim will agree to a percentage deal. Ten percent to the owner is the most widely accepted deal at this writing, unless the ground that the claim owner is going to allow you to mine is already tested and has proven to be rich paying ground. If already proven, his percentage can be more--how much more usually depending upon the highest bidder who is present.

Some claim owners prefer to not have anyone else mining on their claims. It's not uncommon in gold country to see claims which are paying medium sized dredges (one man operation) one to two ounces of gold per day, consistently. Some claims are paying far better amounts. So it's understandable that the owner of a rich claim may not want a bunch of bystanders hanging around the mining operation. The main thing to remember is that the owner of a valid mining claim does have the total rights over the minerals on that section of ground. If he says "No !," you might as well go find another piece of ground because he does have that right, as will you, if and when you claim a section of mineral rights for yourself.

LOCATING GOLD-BEARING GROUND

Deposits of gold have been located in all 50 states. Gold can often be found in many plants, animals, in sea water, and even in your own back yard. If you are seriously interested in mining gold, the thing you will want to do is to locate it in large enough amounts to make it worthwhile to expend the energy required to recover it.

When starting out in gold prospecting, your best bet is to begin in a location which is an already proven gold producing area. It's a proven fact that a geological location which has trapped gold in the past is likely to do so again. It's easy to get the idea that an area which has already been heavily mined is all played out (no more gold left). This would probably be so if it were not for the large flood storms that have occurred since earlier mining activity was done. In our case, we have about 120 years of erosion that has taken place since most of the rivers and streams were heavily mined. Some very large storms have occurred during this period. Many of the river and streambeds which once produced well are producing well again. If you are just beginning and are trying to figure out where to start your operation, one of your best bets is to locate already proven gold bearing ground and start there.

GEOLOGICAL REPORTS AND MAPS

As already mentioned, during the earlier days of mining, it was mandatory by law for a miner to sell his gold to the government. The government, in turn, kept records of how much gold was bought from the various mines within the various counties. These records are open to public scrutiny. As a result, books have been published which give an account of how much recorded gold has been mined out of the various rivers, streams, creeks, and other locations within the various gold producing areas. These books are usually available in places where gold mining equipment is sold. The books are valuable to a prospector who wants to place himself in an already proven gold producing area. For example, in looking over one such book, I found that Trinity County in Northern California produced better than 2 1/2 million ounces of recorded gold. That makes it a high producer when compared to Modoc County, which has produced no more than 15 thousand ounces according to the same book. In looking this over, we can determine that of the two counties, Trinity County would probably be the better direction to go in an effort to search out paying quantities of gold. The book goes on to mention that the Trinity River, which flows through Trinity County, has produced over a million ounces of recorded gold. This information is valuable, because it tells us that the Trinity River has been a good gold producing river in the past and so is likely to be a good gold producer at the present.

There are specific county reports available from the Bureau of Mines which give a wealth of data about the different kinds of minerals which have been located within the various counties, and where they have been found. These reports often include several maps, including a geological-type map showing the many different geological rock formations within the county, and sometimes also the placement of the Tertiary gravels. There is also usually a map included which shows the locations of the mines and mineral resources within the specified county. The report usually includes a large section that gives a listing of all or most of the mines which were selling gold to the government prior to or during the time when the report was being written. This listing usually includes data concerning each separate mine, such as: its name, location, name of owners at the

GOLD, PLACER—Continued

Map no.	Name of claim, mine, or group	Location	Owner (Name, address)	Geology	Remarks and references
	Trinity Gold and Mining Company				See La Grange, herein.
	Trinity Gold Mining Company				See Lower Buckeye.
	Trinity Gold Placer Mining Syndicate, Ltd.				See Blythe.
	Trinity River Hydraulic Mine				See Hawkins Bar.
	Trinity River Mining Company				See English Tom.
	Turney (Wickline)	Sec. 30, T 35 N, R 10 W, MDB&M		10-ft. gravel bank.	Small-scale operation at Dedrick on Canyon Creek. In 1946 L. L. Turney and Joe Wickline mined with bulldozer. Gold recovered in sluice box 20 ft. long. Small producer.
	Two Sisters and Wonder	Sec. 25, T 35 N, R 8 W, MDB&M	Undetermined		Placer claim about 3½ mi. NE of Minersville. No published description. Idle. (Averill 41:88.)
	Tyson	Sec. 30, T 35 N, R 10 W, MDB&M	Undetermined		Placer claim S of Dedrick. No published description. Idle. (Averill 41:88.)
	Union Gulch Placer	Sec. 9, T 33 N, R 9 W, MDB&M	Undetermined		About 2 mi. E of Weaverville. No published description. Idle. (Averill 41:88.)
164	Union Hill	Sec. 6, T 32 N, R 9 W, MDB&M		River terrace gravel.	175 ft. above Trinity River about 1 mi. NE of Douglas City. First worked 1862, again 1906-14. Leased during early 1920's. Idle. See McMurry and Hupp also. (Crawford 94:314; 96:465; Brown 16:915; Tucker 22:97, 207; 23:58; Haley 23:94; Logan 26:49.)
	Unity				See Nugget Bar.
165	Up Grade (Bonus, Good Enough)	Sec. 18, T 6 N, R 7 E, HB&M			Placer deposit about 3 mi. SW of Denny at confluence of New River and Panther Creek. Sampling by open cuts, adits, and shafts in 1939 said to have indicated large deposit of gravel suitable for hydraulicking. (Averill 41:64.)
166	Upham (Pine Tree)	Sec. 29, 30, T 32 N, R 8 W, MDB&M		Terrace gravel 15 to 50 ft. higher than present bed of North Fork Indian Creek; gravel about 24 ft. deep above hard hornblende schist bedrock.	Hydraulic mine 8 mi. SE of Douglas City. Water brought from Indian Creek through 3,800-ft. flume. (Averill 33:72; 41:64.)
167	Uphill Mining Company (Hornet Bar)	Sec. 5, 8, T 32 N, R 9 W, MDB&M. Sec. 1, T 34 N, R 11 W, MDB&M. Sec. 29, T 34 N, R 10 W, MDB&M		Gravel over greenstone bedrock.	Short-lived dredge operation. Gold averaged 4½ cents per cubic yard. See also Indian Creek (Bennett) dredge, herein.

Fig. 3-1. Example of a single page in a geological report, under the "Tabulation of Mines" section. Note: Monetary figures based on $35 an ounce.

time, geology concerning the deposit, how it was being mined, and sometimes how rich the material was and how much recorded gold was produced by the mine (see Figure 3-1).

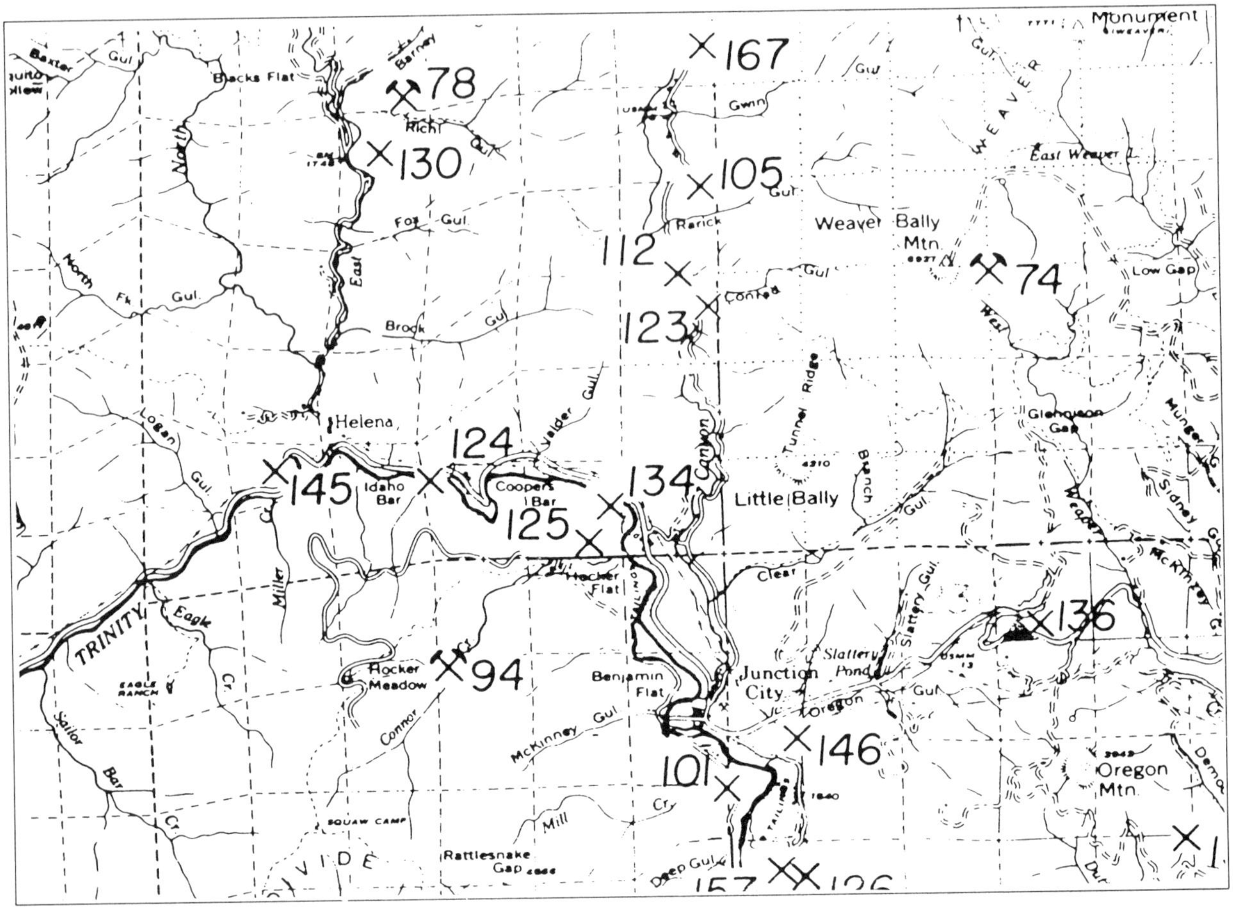

Fig. 3-2. The criss-crosses show where earlier placer mining operations occurred. The crossed picks show where earlier lode mining took place. Each mine is numbered.

There is probably no need to go into how valuable such a report can be to a prospector. Basically, it gives you an individual breakdown of a specific county, showing where gold has already been located in paying quantities and the specific geological information concerning those areas.

In California, for information concerning where to buy specific county geological reports, write to: California Division of Mines, P.O. Box 2980, Sacramento, CA 95812.

Sometimes a specific county report will be out of print and can't be purchased. In this case, the Division of Mines should be able to tell you where to locate one for study purposes. Usually the larger libraries and museums within a county have copies of the specific county geological reports which have been published about their own areas.

If you are seriously considering the idea of prospecting for paying quantities of gold, and do not already have a good general location picked out, or you are looking for a better area, you should consider the idea of looking over the available *"Where To Find Gold"* books, and then the geological reports which lay out the counties of interest to you. In following this procedure you are utilizing the accumulated data from most of the mining which has already occurred. If you do a good job of it, you are almost certain to place yourself into a good gold-bearing area.

Another thing to keep an eye out for when looking for good present-day placer ground, is the presence of high benches in the vicinity--especially near the stream or river that you are interested in. Mainly, it is these benches that are eroded away and washed into the present streams during storms that put new deposits of gold into the streams and rivers. One hundred and twenty years of natural erosion and storms has been enough to deposit a large amount of new gold into some of the present rivers and streams. Perhaps some of them are not as rich as they once were, but some of them pay pretty well and consistently to those who know how to find the deposits.

Rivers in California, which were supposedly mined by the 49'ers, are paying regularly in 15-, 30- and 40-ounce concentrated gold deposits, with 80-ounce deposits being turned up here and there. Its also not uncommon to hear about a 100-, 200- or even 400-ounce catch being turned up by a small to medium-sized dredging operation and being mined out during the course of a mining season--but this is more rare. It is the 5 to 20-ounce concentrations that are rather consistent and not too difficult to find, if you are willing to do the work of sampling to locate them. By concentrations, I don't mean finding all of the gold in a single crevice or hole. The concentrations (paystreaks) are usually spread out over a small area--say 15-feet wide by 30-or 40-feet long. They can be smaller--or larger--depending on each individual situation. The gold can be found to be inside the bedrock traps, or up in the material itself, or may be found in both areas.

These concentrations are usually found in a position in a stream or river where the water force is made to slow down because of a sudden major change in the bedrock, a group of obstructions in the stream flow, where the river makes a turn, or other such places as covered earlier. In concentrations such as these you usually don't get all of the gold in a single day, but instead are more likely to be into excellent wages for a week or a month or two or maybe more. It depends on the size of the deposit and how long it takes you to mine it out once you have found it.

Deposits such as these are being turned up rather consistently in the already proven gold country which was gone through by the earlier miners. It was rich when they went through it, too.

Sometimes these present-day deposits are discovered directly below where an old high bench is *bleeding off* (eroding) into the present stream or river. I know of one case where an 80-ounce deposit was taken in about six days out of a small concentrated area just below where a bench had been eroding into the river. The gold was all lying on top of hard packed streambed layer, under about six inches of loose sand and gravel--and all right up along the stream bank. Not all concentrations are found just below old benches--not by a long shot. But it's not too uncommon to hear of concentrations just below high benches, and so it is worth a bit of your sampling energy to test such areas.

Maps showing the locations of mines and mineral resources, which often come along with a county geological report, can be very helpful in telling you where good paying high bench deposits were located in relation to the streams or rivers you might be interested in. The locations of placer mines are shown on such maps, and when these are indicated some distance away from the present body of water, you know that the mine was processing high bench gravels (see Figure 3-2). On many of these maps, as shown in this figure, each mine is numbered. To be sure that the mine was processing old bench gravels, you can look up its number within the *''tabulation of mines''* contained in the geological report and read the information concerning that individual mine, as shown earlier in Figure 3-1.

The evidence of high bench mining having once occurred near a stream or river is an indication of potential rich paying gravels which may still be present to some extent within that area. The last 120 years of erosion may have redeposited a large amount of gold out of the high deposits and into the present stream or river. This is not a sure thing; it's just a possibility. But, when researching where gold is likely to be, it is good procedure to put as many possibilities in your favor as you can. If there are high bench deposits which contain(ed) paying quantities of gold near a stream or river, or the eroded remnants of old such beds are still present, it's almost a certainty that the last 120 years of erosion has washed some of its gold into the present stream--how much, remains to be seen. The indication of an old mine being present shows that the deposits were there and that they were probably relatively rich, too.

USING TOPOGRAPHIC MAPS

Topographical maps are put out by the Geological Survey of the Department of Interior. They can be purchased at most any sports equipment shop within the area you are interested in. The purpose of the topographic map (topo) is to show the surface features of the area that it covers, including the mountains, valleys, hills, lakes, rivers and streams, bridges, roads and trails, etc. This is accomplished with the use of ''*contour lines,*'' showing the elevations in feet above or below sea level. All points on a single contour line have the same elevation. To make the elevations easy to follow on a topo, usually every 5th line is drawn more boldly than the other lines, with its elevation numbered, as shown in Figure 3-3. On most, each separate contour line indicates a change of 50 feet in altitude.

In looking over a topo, you can see the different grades of slope on the various hills, mountains, streams and rivers within the areas you are interested in. Where the contour lines are closer together, it indicates the grade of drop in that location is steeper than in an area where the contour lines are further apart. So, in this way, a topo can give you a top view of the general lay of the land.

Topos also include the various roads and trails within the area. By looking these over, and the elevation changes, you can get a good idea of the accessibility of a location.

Topo maps also show where the large tailings piles lie from the earlier large scale mining operations. These are an index of where large-scale operations were mining--such as the old bucket line dredges which sometimes processed thousands of yards of material per day (see Figure 3-4).

There are two specific geological situations which often cause large concentrations of gold to deposit--both of which are easily spotted on topographical maps. The first of these is where a gold

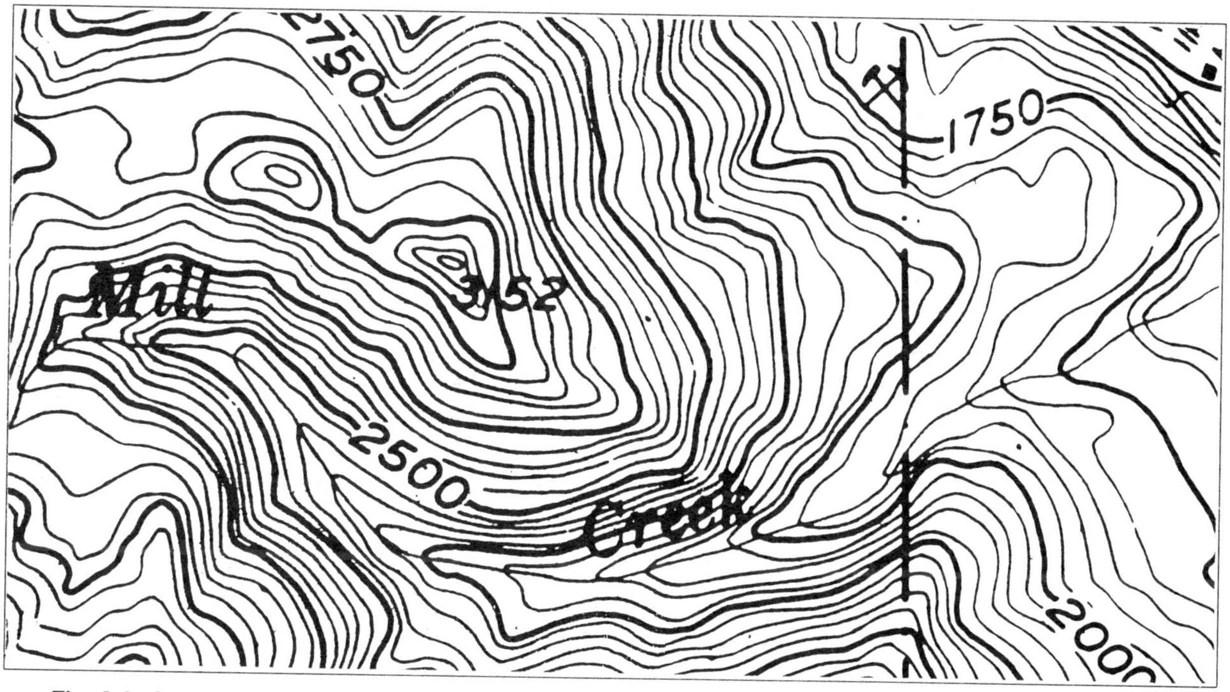

Fig. 3-3. Contour lines.

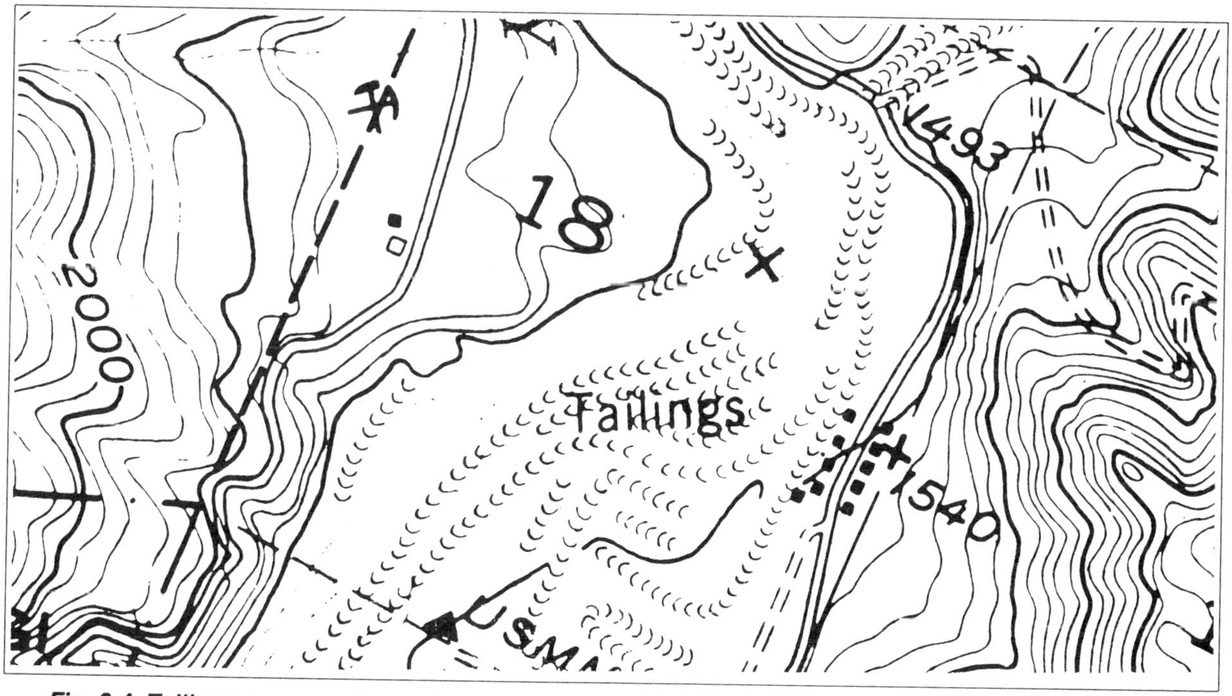

Fig. 3-4. Tailings are sometimes shown on topographical maps, also.

bearing stream, river or creek drops off steadily at a rather steep and steady grade, and then suddenly levels off to a degree for some distance. As mentioned earlier, the reason locations like this pay so well is because where the grade of slope in a waterway is lessened, the water force of the entire stream of water slows down--which can be the cause of a major dropping area for gold. This can be true of any major change of slope in any gold-bearing waterway. It doesn't have to be from steep slope to level. It can be a steady dropping off to just a little bit less dropping off. Anywhere on a map where the slope of grade in a stream or river is lessened in any amount, the water force is likely to be less below that general area, and a deposit of gold might be present.

On a topographical map, the grade of a river or stream can be accurately observed by following the contour lines and noting their distance apart. Where the contour lines along the streambed are closer together, it indicates that the downward grade of slope in that location is more steep, and the water will generally be moving faster there. Where the contour lines in the stream suddenly become further apart, it indicates that the slope of the stream is less steep in that area, and the water force is likely to slow down there (see Figure 3-5). The best areas for finding gold are where the grade lets up and remains that way for some distance.

The other geological situation which pays rather consistently, and is easily found on a topo, is where the channel of a riverbed remains rather narrow or at a certain width for a distance, and then suddenly opens up into a larger and wider streambed. During extreme high water periods, the larger

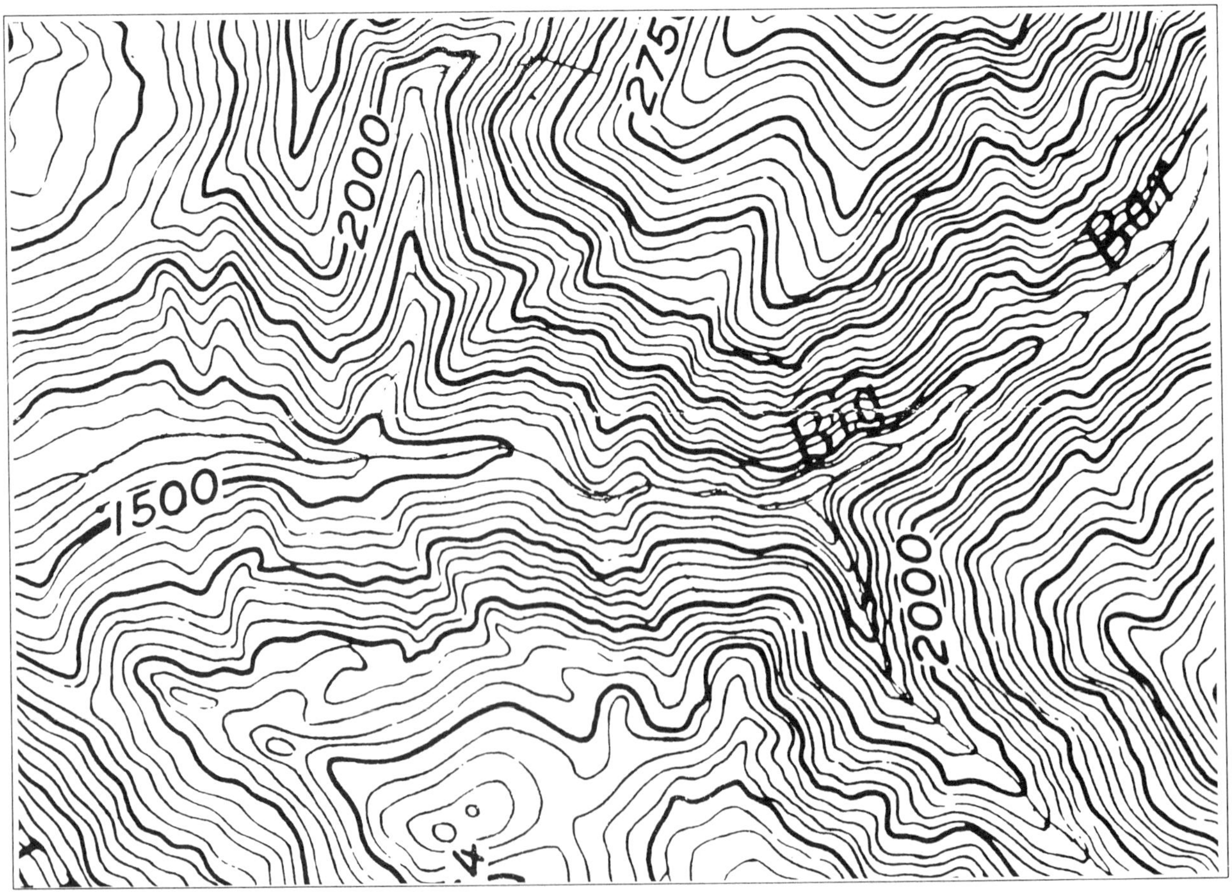

Fig. 3-5. Follow Big Bar Creek from right to left, and notice how the contour lines are further and further apart. This indicates a gradual lessening of slope.

area in such locations will cause the water force of the river to spread out and slow down. Figure 3-6 shows an example of this happening and how it looks on a topo.

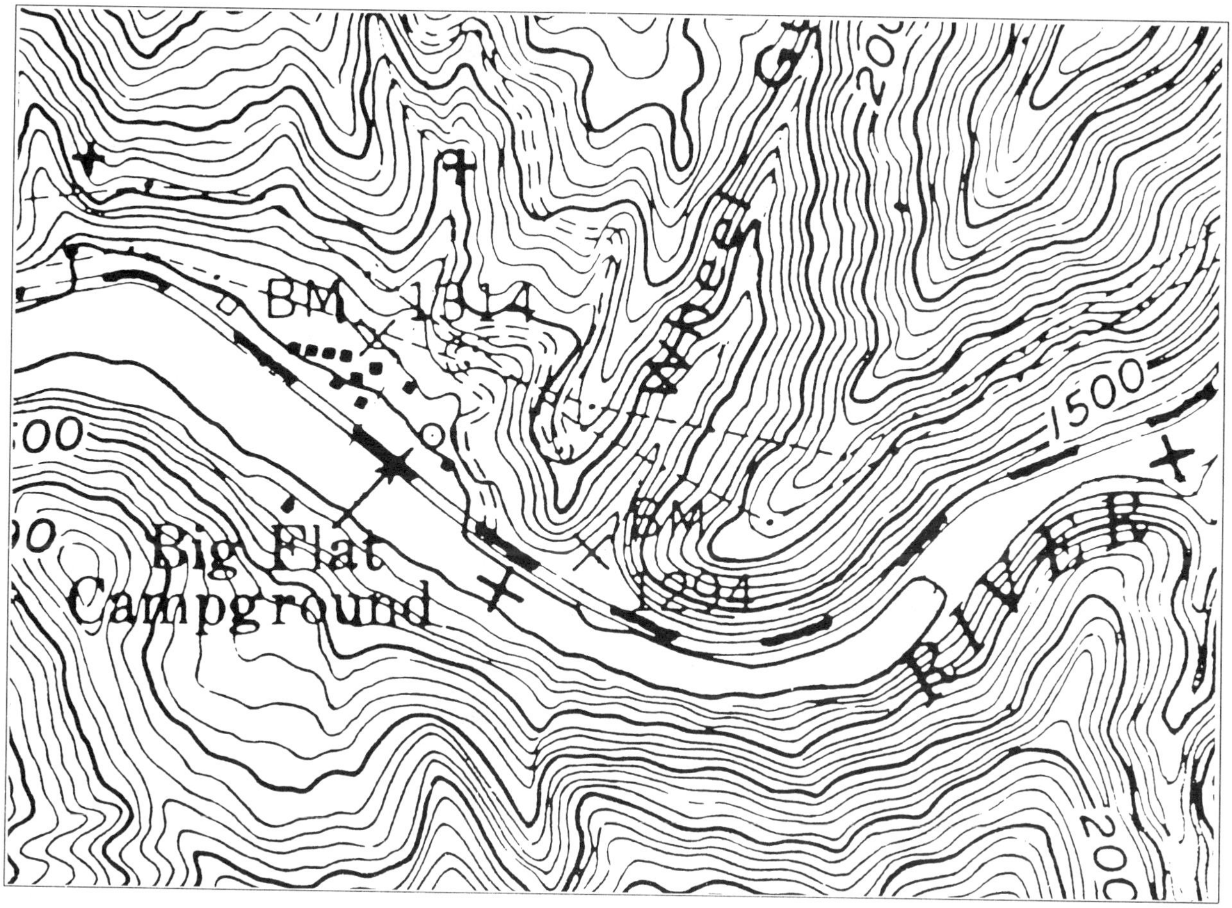

Fig. 3-6. Notice where the basic riverbed suddenly opens up much larger in size.

The best gold deposits in this situation are found where the riverbed opens up and remains open for some distance. Sometimes the deposit is found some distance downstream from where the bed first opens up. This is because it might take some distance for the force of water to slow down enough to drop gold. This is also true of major changes in slope, as covered earlier. Sometimes more than one deposit can be found in situations like this, depending on the size of the change. In one case I know of, where the river remained rather narrow for some miles and then opened up (see Figure 3-6) and remained opened up for some miles, several deposits have been turned up--with thousands of ounces of gold being taken out by suction gold dredgers within just a few years. The whole area was rich, and by all reports on its history, the whole area was thoroughly mined during the late 1800's, with it being very rich then, too.

APPLICATION OF GOLD FINDING INFORMATION

It is not uncommon to be sampling out a location, which by all visible indications should have a gold deposit present, and yet not be able to find one. There are numerous reasons why a gold deposit might not be where it should be--the main reason being that someone else got to it first. Once in a while gold deposits are located in places where all the visible signs show that they should not be. Geological conditions do change over an extended period of time. Bleedoffs from ancient gravels and other high bench deposits can sometimes lay gold in places where the present geological conditions might not warrant investigation.

It is for these reasons--all the discrepancies which occur--that there is a common saying amongst miners that ***"Gold is Where you Find it."*** No statement more true about the yellow metal has ever been made. However, there is always a reason for gold to be in the location where it is found, as well as a reason for it not being there when the present signs indicate that it should be. The multitude of environmental and man-made factors which have come into play over the years which have placed gold in deposits, are not all apparent or in effect at the present. An incident or two of not finding gold where it appears that it ought to be, and perhaps an incident of someone else finding gold where it appears that it ought not be, can sometimes leave an inexperienced person believing that **"Gold is where you find it, and it doesn't really matter where you look!"** This is not a healthy viewpoint from the standpoint of the author.

In my experience, the most successful prospectors have been the ones who have learned to look for gold where it ought to be and understand that sometimes it won't be there for reasons which cannot be presently seen. It is also true that the really experienced prospector can almost always see the reasons why it is or isn't there.

While it is true that gold is going to be where you find it, there are certain factors about gold which are also invariably true. You can use these factors to determine where to look for it. As mentioned earlier, gold is about six times heavier than the average of other materials contained in a streambed. It generally works its way quickly down to the bottom of the other materials that it is moving along with during a flood storm. It takes much more force to push gold along than it does to move normal streambed materials. Wherever that force is lessened, gold can become deposited. That change can be because of irregularities in the bedrock, a major change in the size or slope of the streambed, a bend in the river, or obstructions in the path of the water's flow. Gold tends to follow a common path down a streambed--that path usually being the shortest possible route between the directional changes. Places where boulders tend to get hung up in the river, along the line of travel which the gold should follow, are places where gold is likely to get hung up, as well. All these factors are true. They follow the natural laws which make up our physical universe. And, it is the experienced and successful prospector who uses them in his (or her) search, putting as many of them in his favor as possible in his effort to find gold deposits.

One example of the use of the data covered in this chapter is as follows. In looking for a place to dredge large quantities of gold, we open up a ***"Where To Find Gold"*** book and pick out Trinity County, California as a good gold producing area. There are many other good gold producing areas--some of them better than Trinity County in the book (like Siskiyou County). But for our purposes, we choose Trinity County because it's nearby. Plus, the book mentions the million ounces taken out of the river. This interests us, because we are going to dredge.

We break out our Fish and Game Department permit lists and find out that the Trinity River is open year round for dredging from the confluence of the North Fork of the Trinity on downward until it reaches the confluence of the South Fork. Again, this interests us because we don't want to be limited by time when we get into a large gold deposit. We look over the Forest Service map and see that this entire stretch of river runs through National Forest, with just a few scattered pieces of private property here and there. This is good. So we get hold of the Trinity County geological report, break out the maps, and look over that section of river to see where the various successful placer mining activities took place. In doing so, we find that the large bucketline dredges operated down to just above the confluence of the North Fork and several successful high bench operations worked just above and below the confluence of the North Fork (see no. 124 and no. 145 in Figure 3-2).

In tracing downriver from the North Fork on our topographic map, we see that the highway follows the river for its entire length from this point. We like this, because it gives us excellent accessibility. We also see that the river drops off at more or less a steady grade and runs rather narrow for some miles until the streambed suddenly opens up into about twice that width in the small town of Big Flat (see Figure 3-6 for exact location of example). The topo shows us that the river runs wide for some miles afterwards.

We know that many of the small towns in gold country were settled by mining communities during the early days where large deposits of gold were located, and we take the presence of the small town of Big Flat in that location as a positive sign.

All signs so far indicate that this general area should be pretty good, so we drive there to have a closer look. In doing so, we notice that the exposed bedrock in the area all seems to be rough and semi-decomposed. We also notice lots of old mining tailings on the banks where the old benches used to lie, and we notice some bench deposits still in place. An obvious good sign is that there seems to be a lot of present gold dredging activity.

One short trip to the local Forest Service office and we find out that there is a long stretch of river in that same location which has been withdrawn from mineral entry so that anyone, including us, can mine there. The rangers show us the boundaries of the free dredging area and give us a permit to dredge there--all in about 10 minutes' time.

We go down to the site to look it over and run into the local gold buyer who lets us know that the area is hot-hot-hot and swears up and down that he has bought over 300 ounces of gold out of that one section of river within the past two seasons alone. And he mentions that most of the gold taken out of the area was not even sold to him! Some inquiries about the other claims within the area indicate that they too are paying very well and that percentage deals can be worked out with the owners.

''Big Flat'' is a legitimate location and this story is true. Gold is still being taken out in paying quantities at the time of this writing, but much of the area has been worked out during the last few seasons (at least until the next big storm).

The purpose of the story is not to start a major gold rush towards Big Flat, but to show how the ***"Where To Find Gold"*** books, geological reports, topos, forest service maps and a general knowledge of placer geology can be used to pinpoint locations which are likely to have large paying placer deposits. Granted, you don't have all these maps and reports on hand. However they are not hard to come by and the ***"Where To Find Gold"*** book is really all that you need to get started. There are many such books available. Once you have one in your possession, you can study it and decide which county or counties you are interested in and write to the various agencies for the available reports and maps. Another way is to go to the county itself; get the maps and reports, and have a look at the area firsthand while you study them. A few phone calls will almost always find you the geological reports of a county within any of its larger sized towns.

The idea is to put yourself into a geological location which has as many factors in your favor as possible. Finding rough bedrock is good. Locating it on the inside of a bend in the stream is better. If there are boulders present, in the right places, the chances of a gold deposit being there are increased. If there is a steeper grade which flattens out not too far up stream, and if there are signs of lots of earlier mining activity--such as rock piles upon the streambanks, all the better for a good gold deposit to be present. Get the idea? And finally, if there is a road or trail present, it will make it much easier for you to get in and sample that section of ground quickly. We will talk about sampling in a later chapter.

ACCESSIBILITY

It is sometimes said that the more difficult it is to get into an area, the less likely it is that it has been mined earlier. This is true only to a degree. You have to remember that during the early rush days, most of the western United States was rather inaccessible. That didn't stop the 49'ers. Today, highways, roads, and trails go through a lot of ground that has not been prospected since the late 1800's, and some which was never touched at all. So it's not uncommon to see paying quantities of gold being recovered directly off the roadside.

You also hear of tremendous finds being made up in the more inaccessible areas too. At the time of this writing it doesn't seem necessary to HAVE to get up into the remote locations in order to locate good gold finds (even though they are there), because there seems to be a lot of acceptable paying accessible ground which has yet to be thoroughly mined. So, accessibility is not necessarily a factor as to where you will find gold today, although it is a factor in where you will be able to haul your equipment without a lot of additional effort. This is something to take into consideration during the planning stages. Topos can be a big help in this.

BELOW OLD HYDRAULIC MINES

Some of the larger-sized hydraulic mining operations moved so much material that entire mountainsides were washed away in the process. Experts have agreed that as much as a third to one half of the gold moved by the hydraulic operations was lost because of the tremendous volumes of material washed through their recovery systems all at once. The lost gold would have been washed away along with the tailings. Locations where hydraulic tailings were washed into a present-day stream or river are always good places to do some sampling. These areas have proven to be rich time and time again. A third or half of all the gold from an operation which moved hundreds of thousands, or even millions, of cubic yards of gold-bearing material could cause quite a large deposit, or series of deposits, downstream.

In one known area, two personal friends of mine were prospecting around in a wash below the old dumping ground of a medium-sized hydraulic operation that ran sometime during the early 1900's. One of the fellows happened to notice that the culvert passing underneath the highway was made of corrugated steel, which seemed to him would act much the same as the riffles in a sluice box. There were also railroad ties placed along the bottom edge of the culvert to protect it from the tremendous pounding of rocks and materials that would wash through during large storms. Since he was sampling the area anyway, he decided to test some of the material lying in the bottom of the culvert. I guess the combination of railroad irons and corrugated steel acted as a good gold trap. They ended up taking better than 13 ounces out of the culvert within the next two days!

There's a similar example of a professional dredger I know. He owns a claim located on a river where a medium-sized hydraulic operation (operated in the 1930's) dumped its tailings into the river. For some time, it has been rumored that this guy's claim is so rich that he consistently pulls about a pound of gold each day that he dredges, and has been doing so for years. Now one thing about gold mining is that the stories tend to grow as they are passed along from one person to the next. I do alright myself when I am into a good deposit. But a pound a day, consistently, for a matter of years, seems a bit much to swallow. So, one day I happened to be present on the scene while he was doing his final cleanup after a full day of running. By my conservative estimate he had no less than 6 ounces of gold in his pan--with probably 60% of it being of jewelry grade (nuggets). And for whatever it's worth, he was obviously disappointed with his production for the day!

There are many other such confirmed stories--some of them even better--about the gold deposits which have been located below where the tailings of an old hydraulic operation dumped into a present-day stream, creek or river. Keep this in mind when you are searching out likely spots.

CONSTRUCTION CUTS

Wherever construction operations have cut into or through old dry streambeds and exposed the bedrock, and/or the different strata of gravels, is a good place to sample for deposits of gold. It is only in the more recent past that many of the present roads and highways were built through gold country. Some of these cut directly through old Recent and Pleistocene benches in places, and sometimes you will run across the ancient Tertiary gravels in this way, too. Usually, these construction cuts are made directly through the channel, leaving the bottom stratum and bedrock entirely exposed to the scrutiny of any prospector who happens along (see Figure 3-7).

Fig. 3-7.

While the Highway Department is not likely to be happy about you starting up a gold mining operation alongside the road (it's probably against the law), you can easily take a few samples of the exposed lower strata of such cuts to see if the ground is rich enough to pursue further action. If the ground proves to be rich, perhaps the streambed can be traced to a point further away from the road and an operation can be started up there.

Foundations for houses and buildings are often dug into an old streambed. Sampling could be done before the concrete is poured. It might be worth your while. Some very rich deposits have been found in just this way. The same thing holds true when digging pools and wells. Did you ever hear the story about the lady who found gold in her back yard when she decided to take samples out of her well? It turned out that her house was built upon a very rich section of riverbed and the gold that she recovered was worth in excess of the value of her property--house and all!

NATURAL EROSIVE CUTS

Anywhere that natural erosion has cut a path through an old streambed is a place worth doing a bit of sampling. This is especially true when the streambed is cut in all the way down to bedrock or the lower strata of material. In this case, the erosion has already done the hardest part of the job in uncovering the lower strata for you--where the paydirt is most likely to be found. Sampling locations such as this is usually rather quick and easy, and sometimes very rewarding.

LAKEBED DEPOSITS

Sometimes the spot where an old river or stream reached a lake was a major dumping ground for gold, because that was where the force of the river dissipated to nothing. This is true of old dried up lakes and of the present ones--manmade or otherwise. The mouth of the stream, and sometimes out into the lake itself, can be found to be very, very rich, and is a good place to sample. In the case of the present lakes, I recommend that you check with the local authorities before starting, to make sure it's not illegal. It could be against the law to mine in some lakes.

BEACH DEPOSITS

Millions of years of erosive activity have caused an enormous amount of gold to wash down through the rivers and into the ocean. Experts have said that the vast majority of eroded gold has washed into the oceans and has yet to be recovered. Much of this gold is of microscopic size and remains in a state of suspension in the seawater itself. So far, no means of recovering this ultra-fine gold has yet been developed which has proven to be economically viable in a large-scale operation.

Some of the gold which has washed down into the ocean is also of a larger size and can be recovered by the standard recovery methods. This gold can sometimes be found on the beaches nearby the gold bearing rivers, where the fine particles have been washed up by storms and tides.

Northern California, Oregon, Washington state and Alaska are all well known for their gold beaches. These are usually covered in the various ***"Where To Find Gold"*** books.

Occasionally, certain tidal changes cause entire beaches to be covered with a layer of exposed fine gold, which causes many of the locals and tourists to rush down and try to recover it. Sometimes the next change in the tide will wash it all away, or cover it up. The main point here is that the gold beaches should not be discounted as possible locations to prospect for and recover gold.

ANCIENT BEACHES

If you have spent much time traveling near the sea shores, you might have noticed what appears to be lots of old dry streambeds scattered about. Sometimes they are located close to the ocean, and in other cases, miles and miles away. Some of these are quite massive and it's hard to believe that they were once streambeds because they are so large. Yet they are laid out just like a streambed. It's ancient ocean bed! And, in some places, as in northern California, Oregon, Washington state and in Alaska, they can be moderately rich in fine gold.

Some major operations were run on ancient ocean beds during the early 1900's with good results. Sometimes these old beds are quite extensive in height and size, and so are out of the range of the small-scale operator. However it is not uncommon to find streams and creeks which cut directly through old ocean bed, and these might be found to be rich--along with any other places where the natural erosion could cause a concentration of gold to take place.

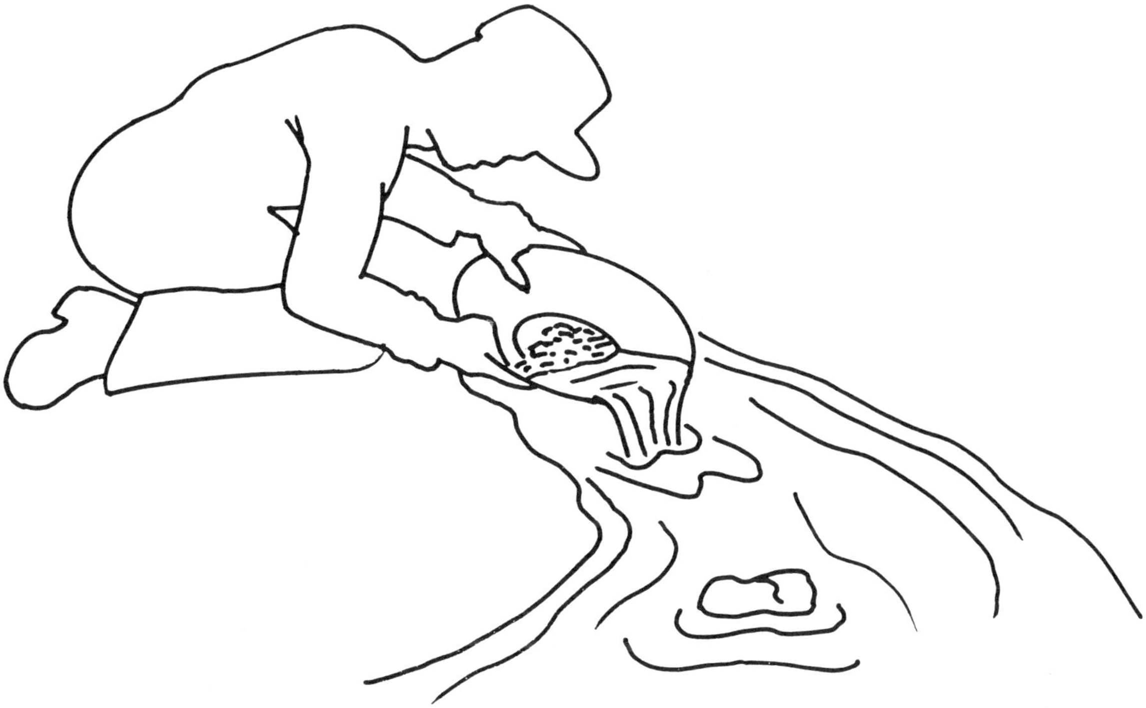

PANNING FOR GOLD

CHAPTER IV

Processing streambed material with the use of a gold pan is probably the oldest method of recovering placer gold in existence. Some of the first gold pans consisted of Indian woven baskets and wooden bowls--called *"beteas."* In the 1800's, Americans developed gold pans made out of tin, which have evolved into today's ultra modern space age plastic and steel spun gold pans.

The gold pan still remains one of the most valuable tools used by today's modern prospector, whether he is running a small-time operation or one of large scale. Sometimes the recovery systems on the larger pieces of mining equipment, like highbankers and dredges, are cleaned up with the use of a gold pan. *"Cleanup"* is the procedure of separating the gold from the other heavy materials which get trapped in a gold recovery system.

The fundamental principle of most of today's recovery systems is that if you have a device which will recover (trap) the heavier materials out of the streambed material that is being processed, then you will catch the gold--because gold is the heaviest of all. These recovery systems very seldom catch only gold. They also trap other heavy materials out of the streambed, such as pieces of lead and iron, old nails and coins, and the heavier iron rocks and sand--called *"black sand."* All these heavy materials which collect in a recovery system, gold included, are called *"concentrates."*

Towards the end of the day or operating period, when it is time to collect the gold which has been trapped in a recovery system, it is done by cleaning all of the concentrated materials out of the system, and then separating the gold from the other heavy materials. This process is often done with the use of a gold pan.

At first, if you are rather new to the subject, it might strike you that gold mining is still in the dark ages. Not true! There simply is no quicker and more effective way of separating gold from the other heavily concentrated material that comes out of a recovery system when operating on a small scale. There are specialized concentration machines available which will do most of the separation for you. But even these do not accomplish complete separation. So, many cleanups, or at least the final stages, are done with the use of a gold pan.

One of the other main uses of a gold pan by today's modern prospector is in sampling out different locations. This is done in an effort to find new and better deposits to mine.

One reason the gold pan is such an asset as a sampling tool is that it is light and can be carried anywhere with little additional effort. It's also very accurate and effective as a gold recovering device. The gold pan is quick and easy to use. It takes no time at all to set one up for operation, and it only takes a little water to work one properly. To put it simply, the gold pan is one of the most effective gold recovery tools ever developed, and has a wider range of operational ability than any other gold recovering device.

On an amateur basis, the gold pan can be profitable to play with and provide hours of recreational enjoyment.

By using a gold pan, a beginner can learn the basics of gold recovery and of mining in general. For when one has learned how to use a gold pan effectively, he or she understands the basics of mining gold--whether it is realized or not.

The gold pan is very simple to learn how to use, and recovers gold so well that even a beginner would have difficulty losing a piece of gold the size of a small pellet--even on the first few attempts.

TYPES OF PANS

There are a host of different kinds of gold pans on today's market. Rather than go into each one of them individually, I'll go over the main characteristics and their advantages and disadvantages, and let you decide which pan you will be most comfortable in using.

Plastic pans are molded from space age plastics, and for the most part are unbreakable--although there are a few on the market which break rather easily. The way to tell if a plastic pan is unbreakable is by taking it in both hands and putting a strain on it, as shown if Figure 4-1. In doing so, and by watching and feeling, you'll get the idea if the pan will continue to bend or will break when additional force is applied. Be careful not to apply so much pressure that you break a breakable one. I hereby refuse to accept financial responsibility for all the gold pans which are going to be destroyed by this! (The shop owners are going to love me.)

Fig. 4-1. Gold pans tend to take a lot of banging out in the field. It's well to have an unbreakable one.

Steel pans will bend, but it takes a tremendous amount of force to bend one, like dropping it over a cliff or running over it with your car. For the most part, with everyday normal use, there is no need to worry about the durability of either the unbreakable plastic pan or the steel one.

Plastic pans are lighter than the steel ones. If you are planning on an extra long and difficult hike for sampling purposes, it could make a difference. However, those who prefer the steel pan over the plastic kind say that the difference in the weight factor is not enough to make them change over.

Plastic gold pans generally float, whereas a steel one will quickly sink to the bottom--unless it is dropped into the water in such a position that it floats like a boat. These factors can either be an advantage or disadvantage, depending on the circumstances in which you are using the gold pan. For example, if you are prospecting near deep water and your steel pan slips off the streambank and sinks, it means that you will probably have to go underwater to retrieve it--or abandon it there. Most of the plastic pans will float and you can fish them out with a stick or wade out a bit and grab them. Yet in the case of shallower water which is running fast, like in a set of rapids, if you drop your plastic floater, you have to be very fast indeed to catch it. The steel gold pan will usually sink and anchor itself to the bottom. Plastic pans will not rust. Steel ones will--especially if a set of wet concentrates out of a recovery system are allowed to sit in one for an extended period of time. Yet, some miners insist that the rust in their gold pan helps them to recover the fine gold. Other prospectors prefer a smooth surface at the bottom of their gold pans. It's all a matter of preference.

By the way, the best way to prevent a steel pan from rusting is to empty it out and place it upside down when not in use.

One advantage to a plastic pan is that nitric acid can be used to clean gold directly in the pan if you should decide to do so during the final cleanup (Chapter 7). Nitric acid will attack steel. So a steel pan cannot be used in conjunction with acid during the final cleanup stages.

Another advantage to a plastic pan is that you can use a magnet to separate the magnetic black sands (iron) out of the final concentrates directly in the pan during cleanup (Chapter 7), whereas a steel gold pan will react to the magnet and interfere with this process.

During the final cleanup steps back at camp (Chapter 7), there is always a need for a metal container to put your gold into, to heat it up and dry it out. A steel gold pan can come in handy when doing this part of the cleanup.

It's a good idea to have a plastic gold pan on hand if you are planning to use an electronic metal detector in your prospecting activities. In this way, when you get a signal from the ground you can shovel material into the pan and then scan the bottomside the pan to see if you have recovered the target which is causing the signal. A steel gold pan cannot be effectively used in this procedure, because it will cause the detector to read out falsely.

Many plastic gold pans are molded with a small set of *riffles* (gold traps) in one part of the pan. Different manufacturers have designed different types of riffles--each with its own unique way of assisting in the recovery of gold. Some of these riffles, when utilized correctly, do help in the recovery of gold, and are referred to as *''cheater riffles.''* This is because sometimes with the use of them, a novice can quickly learn to pan nearly as fast and as well as an old pro who has had years of experience.

One disadvantage that a plastic pan has is that it will melt with very little provocation. So keep yours well out of range of the campfire and also out of the rear window area of the car on a hot, sunny day.

Plastic pans are made in black, blue, green and other dark colors so that the small particles of gold will stand out better. This gives you more visible control over the fine particles of gold while working them in the pan.

It has been said that the steel gold pan has an advantage over the plastic one in that it can be used to cook with. While this may be true, it is not recommended that you use your steel pan for cooking. The cooking greases and oils are absorbed into the pan and can thereafter affect how well the pan will recover gold. Oil has a tendency to attach itself to pieces of gold--which are then susceptible to attaching themselves to air bubbles, and floating out of your pan while you are working it!

Steel pans usually come new with a thin layer of protective oil on them as a rust preventive before they are sold. It is generally recommended that this oil be cleaned out of a new steel gold pan before it is used out in the field.

There are several ways of cleaning the oil out of a steel pan. One is to rub it out with a rag which has been dampened with paint thinner or acetone. Perhaps the best and most often used method of cleaning the oil from a steel pan is to heat it slowly over a stove or an open fire until it reaches a dull red glow and then drop it into clean cool water. This process leaves the pan with a dark blue hue, which gives a good dark background to make the fine particles of gold stand out better. Care must be taken when cleaning a steel gold pan in this manner to prevent using excessive heat--which can cause the pan to warp.

Any gold pan that is used regularly has a tendency to collect a certain amount of body oils. Body oils can also affect your gold pan in the losing of gold particles. So it is good practice to clean your gold pan periodically, whether it is steel or plastic.

A plastic pan can be effectively cleaned by rubbing it with a clean rag which has been dipped in isopropyl alcohol.

Brand new plastic pans sometimes also have to be worked over before use. The test is whether or not the pan remains wet after dipping it into clean water. If water beads up on the surface of the pan, the oily, smooth surface still needs to be cleaned. Using a soapy Brillo Pad on a new plastic gold pan is one of the best methods of preparing it for use in the field.

One of the main differences between gold pans is the difference between the *"drop center"* pan and the *"straight-angle bottom"* pan. The difference is in the shape of the angle between the sides and the bottom of the pan (see Figure 4-2).

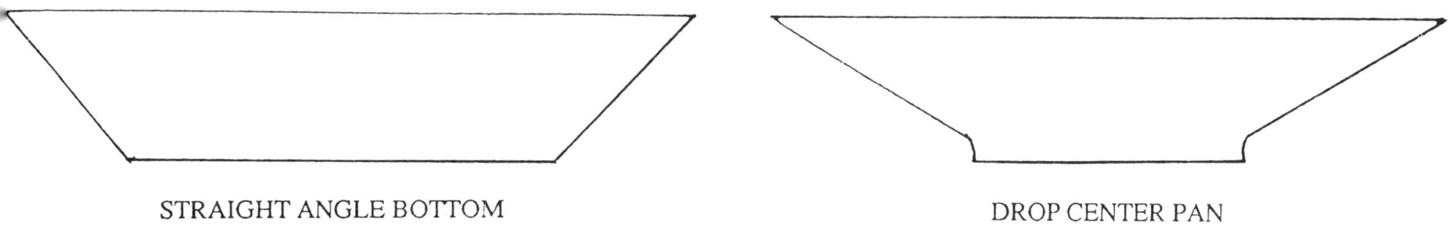

Fig. 4-2.

"Drop Center" gold pans are most commonly found in plastic, although some steel versions of this are available. The dropped center acts as a protective trap for gold which has worked its way to the pan's bottom, and prevents gold from sliding forward when the pan is tilted forward and worked (see Figure 4-3 on next page).

"Straight-angle bottom" gold pans can be either steel or plastic. Some prospectors prefer this type of pan because it allows them to separate ALL of the heavy black sands from the gold out in the field while panning. This way, it doesn't have to be done later during the final cleanup steps (Chapter 7). The trap in the drop-center pan has a tendency to hold onto some of the black sands too, along with the gold. This can make final separation more time-consuming.

When out in the field, it's not uncommon to hear arguments as to which pan is better. These arguments are silly, because the different pans--whether steel or plastic, drop center or straight

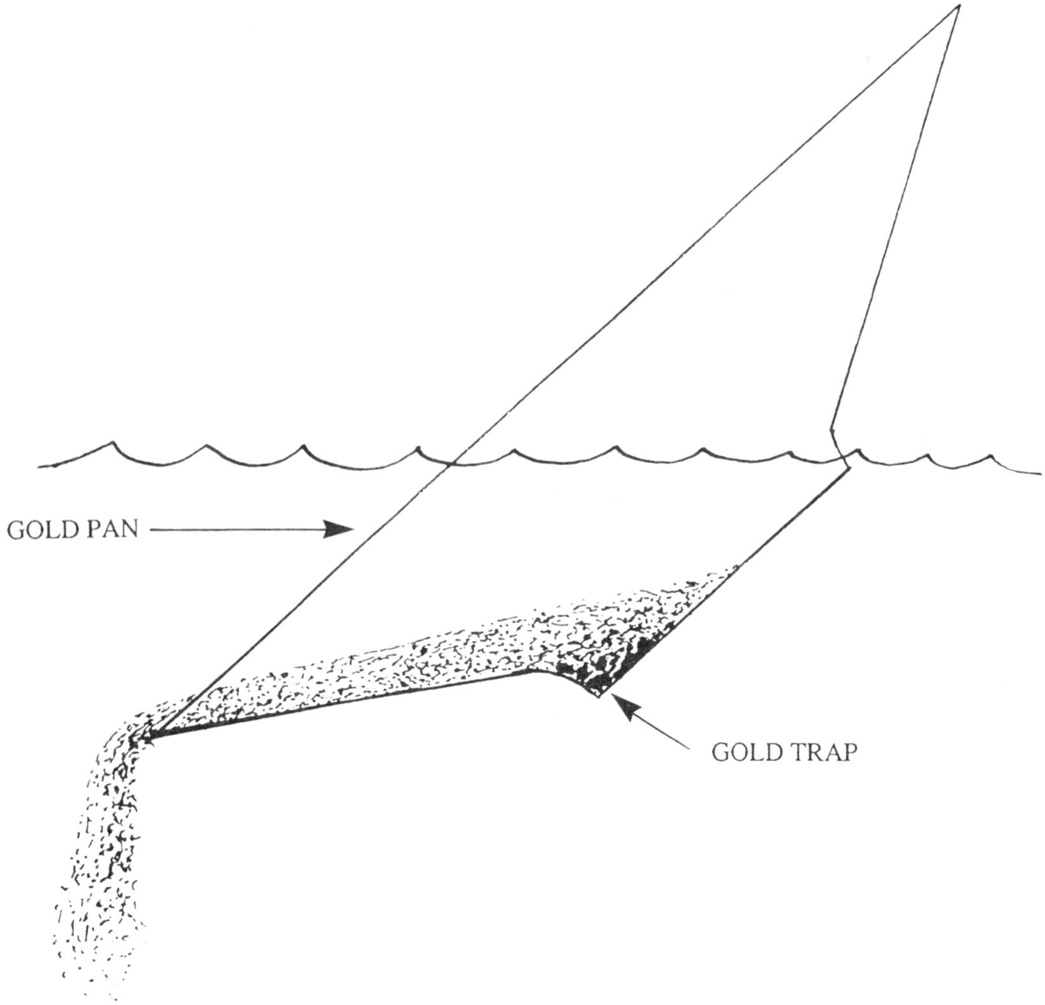

Fig. 4-3. A drop center-type gold pan tends to keep gold from sliding forward while the pan is tilted forward when it's being worked.

angle bottom, each have advantages for different uses. For this reason, many prospectors have several different kinds of gold pans, and use the one most suited to their needs.

The main pan which a prospector decides to use in his everyday panning is a matter of personal preference. If you are just beginning and don't feel qualified to choose the "best one" for you, don't worry about it too much. It's my opinion that once you understand the basics of panning gold, you can make any of these pans work well--and develop more skills as you do so. After all, some of the earliest miners made woven baskets work!

Gold pans come in a multitude of sizes, ranging from 6 to 18 inches in diameter, and possibly even larger (see Figure 4-4). The average size usually seen in use out in the field is about 14 to 16 inches in diameter.

SQUARE/RECTANGULAR GOLD PANS

One of the most innovative developments in gold mining equipment during the recent past is the new square/rectangular gold pans (see Figure 4-4). They work on the same basic principles as

Fig. 4-4. Gold pans are available in different sizes and shapes.

the conventional round pans, only they are easier to learn to handle. They are also more efficient, because water flow over the straight-edged face of the pan is consistent, whereas water flow velocity differs across the rounded edge of a conventional pan. This allows faster and more efficient panning with the new square/rectangular models--and I highly recommend them.

THE GOLD PAN AS A PRODUCTION TOOL

The main thing to remember about the use of a gold pan is that while it is effective as a gold-catching device, it can only process a limited amount of streambed material. For this reason, the gold pan is normally not used as a production tool in professional use, other than in the most remote locations where it would be very difficult to haul large pieces of equipment, and where there is only a limited amount of material present which is paying well enough to make the panning worthwhile.

The gold pan is most commonly used to locate richer paying ground (sampling), so that larger production equipment can be brought in to that spot in order to work the ground at greater profit.

There are stories in the old mining records about the ground being so rich during the days of the 1849 gold rush that as much as 96 ounces of gold were recovered from a single pan. That's $43,200 at today's rate of exchange, and must have been some very rich ground indeed!

Stories like that are rare and ground like that is not run across very often. However, it's not too uncommon to hear of prospectors today who are able to consistently produce better than an ounce per week with a gold pan in the high country, and have the gold to show for it. Some do better, but these guys have been at it for a while and have located hot spots. I personally know of two guys who support themselves with a gold pan, and one of them lives pretty well. As mentioned earlier, the gold pan gives you unlimited accessibility, and these guys concentrate on the pockets in the exposed bedrock along the edges of the creek beds in their areas, picking up a few pieces here, a few there, and a little pocket once in awhile. It adds up, and to them it's better than punching a time clock.

There is still plenty of rich ground to be found in gold country if you are willing to do the work involved in finding it.

GOLD PANNING PROCEDURE

Panning gold is basically simple, once you realize that you are doing the same thing that the river does when it causes gold to concentrate and deposit in various locations.

The process basically consists of placing the material that you want to process into your pan, and shaking it in a left to right motion underwater in order to cause the gold, which is heavy, to work its way down toward the bottom of the pan. At the same time, the lighter materials, which are worthless, are worked up to the surface where they can be swept off. The process of shaking and sweeping is done until only the heaviest of materials are left--namely the gold, silver, and platinum, if present.

Once you are out in the field, you will notice that no two people pan gold exactly alike. After you have been at it a while, you will develop your own little twists and shakes to accomplish the proper result.

Here is a basic gold panning procedure to start off with that works well and is easy to learn:

STEP 1: Once you have located some gravel that you want to sample, place it in your gold pan--filling it about 3/4 of the way to the top. After you've been at it a while, you can fill your pan to the top without losing any gold. While placing material in your pan, pick out the larger sized rocks, so that you can get more of the smaller material, and gold, into the pan (see Figure 4-5).

Fig. 4-5. Panning Step 1: Fill pan about 3/4 full of material

STEP 2: Choose a spot to do your panning. It's best to pick a location where the water is at least six inches deep and preferably moving slightly--just enough to sweep away the mud and silty water as it is washed from your pan. This way, you can see what you are doing better. You don't want the water moving so swiftly that it will upset your panning actions. A mild current will do, if available.

It's always best to try to find a spot where there is a rock or log or streambank or something that you can sit down on while panning. You can pan effectively while squatting, kneeling or bending over, but it does get tiresome. If you are planning to process more than just one or two pans, sitting down will make the job much more pleasant.

STEP 3: Carry the pan over to your determined spot and submerge it underwater (see Figure 4-6).

STEP 4: Use your fingers to knead the contents of the pan in order to break it up fully and cause all of the material to become saturated with water. This is the time to work apart all the clay, dirt,

Fig. 4-6. Panning Step 3; Submerge pan fully under water.

roots, moss and such with your fingers to ensure that all of the materials are fully broken up and in a liquid state of suspension in the pan.

The pan is underwater while doing this. Mud and silt will be seen to float up and out. Do not concern yourself about losing any gold when this happens. Remember: gold is heavy and will tend

Fig. 4-7. Panning Step 4; Use fingers to fully break up material in the pan.

to sink deeper in your pan while these lighter materials are floating out (see Figure 4-7).

STEP 5: After the entire contents of the pan have been thoroughly broken up, take the pan in your hands (with cheater riffles on the far side of the pan) and shake it, using a vigorous left and right motion just under the surface of the water. This action will help to break up the contents of the pan even more and will also start to work the heavier materials downwards in the pan while the lighter materials will start to surface.

Be careful not to get so vigorous in your shaking that you slosh material out of the pan during this step (see Figure 4-8). Depending on the consistency of the material that you are working, it may be necessary to alternate doing steps four and five over again a few times to get all of the pan's contents into a liquid state of suspension. It is this same liquid state of suspension that allows the heavier materials to sink in the pan while the lighter materials emerge to the surface.

Fig. 4-8. Panning Step 5; Shake pan just underwater in a vigorous left and right motion.

STEP 6: As the shaking action causes rocks to rise up to the surface, sweep them out of the pan using your fingers or the side of your hand. Just sweep off the top layer of rocks which have worked their way up to the pan's surface (see Figure 4-9).

Don't worry about losing any gold while doing this, because the same action which has brought the rocks to the surface will have worked the gold deeper down toward the bottom of the gold pan.

Rotating your pan in a circular motion underwater will help to bring more rocks to the surface where they can be swept off in the same way.

When picking the larger rocks out of the pan, make sure that they are clean of clay and other

Fig. 4-9. Panning Step 6; Sweep the larger sized rocks over the side as they emerge to the surface of the pan.

particles before you toss them out. Clay sometimes contains pieces of gold and also has a tendency to grab onto the gold in your pan.

STEP 7: Continue to do steps five and six, shaking the pan and sweeping out the rocks and pebbles, until most of the medium sized material is out of your pan.

STEP 8: Tilt the forward edge of your pan downward slightly to bring the forward bottom edge of the pan to a lower position, as shown in Figure 4-10. With the pan tilted forward, shake it back

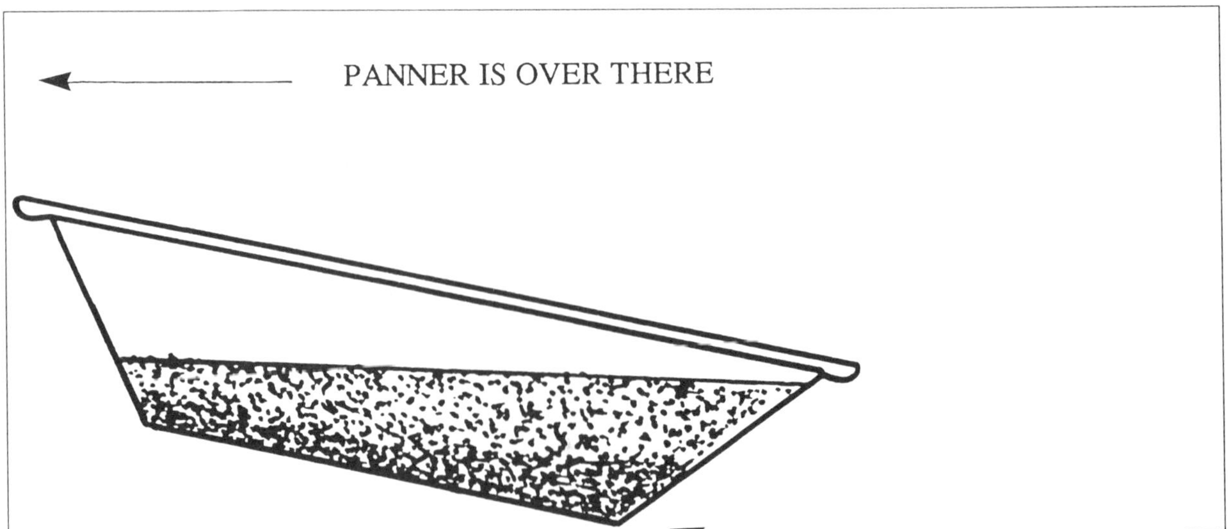

Fig. 4-10. Panning Step 8; Tilt the pan forward slightly and shake.

and forth using the same left and right motion. Be careful not to tilt the pan forward so much that any material is spilled over the forward edge while shaking.

This tilted shaking action causes the gold to start working its way down to the pan's forward bottom edge, and continues to work the lighter materials to the surface where they will be swept off.

Fig. 4-11. Panning Step 9; Move the pan back and forth to get the water to sweep the top layer of lighter material out of the pan.

STEP 9: Carefully, by using a forward and backward movement, or a slight circular motion, just below the surface of the water, allow the water to sweep the top layer of worthless, lighter materials out of your pan. Only allow the water to sweep out a little at a time, while watching closely for the heavier materials to be uncovered as the lighter materials are swept out (see Figure 4-11). It takes some judgment in this step to determine just how much material to sweep off before having to shake again so that no gold is lost. It will just take a little practice in panning gold before you will begin to see the difference between the lighter materials and the heavier materials in your pan, and get a feel for knowing exactly how much material that can be safely swept out before re-shaking is necessary. When you are first starting, it is best to re-shake as often as you feel that it is needed to prevent losing any gold. *When in doubt, shake!* There are a few factors which can be pointed out to help you with this. Heavier materials are usually darker in color than the lighter materials. You will notice while shaking the pan that it is the lighter colored materials that are vibrated to the surface. You will also notice that as the lighter materials are swept out of the pan, the darker colored materials are uncovered.

Materials tend to get darker (and heavier) as you work your way down toward the bottom of the pan, where the darkest and heaviest materials will be found, they being the purple and black sands, which are minerals of the iron family. The exception to this is gold, which is heaviest of all. Gold usually is of a bright and shiny metallic color and shows out well in contrast to the other heavier materials at the bottom of the gold pan.

One other factor to keep in mind is that the lighter materials sweep out of your pan more easily than do the heavier materials. As the heavier materials are uncovered, they are increasingly more resistant to being swept out of the pan, and will give you an indication of when it is time to re-shake.

As you work your way down through your pan, sometimes gold particles will show themselves as you get down to the heavier materials. When you see gold, you know that it's time to shake again.

There is another popular method of sweeping the lighter materials out of the top of your pan which you might prefer to use. It is done by dipping your pan under the water and lifting it up, while allowing the water to run off the forward edge of the pan, taking the top layer of material along with it (see Figure 4-12).

Fig. 4-12. Panning Step 9; Washing the top layer of material out of the pan by the dipping method.

STEP 10: Once the top layer of lighter material is washed out of your pan, re-shake to bring more lighter materials to the top. By "lighter materials," I mean in comparison to the other materials in the pan. If you continue to shake the "lighter materials" to the top and sweep them off, eventually you will be left with the heaviest material of all, which is the gold. It doesn't take much shaking to bring a new layer of lighter stuff to the surface. Maybe 8 or 10 seconds worth of shaking will do it, maybe less. It all depends on the consistency of the material and how much gold is present.

Continue to pluck out the larger-sized rocks and pebbles as they show themselves during the process.

STEP 11: Every few cycles of sweeping and re-shaking, tilt your pan back to the level position and re-shake. This keeps any gold from being allowed to work its way up the forward edge of your pan.

STEP 12: Continue the above steps of sweeping and re-shaking until you are down to the heaviest materials in your pan. These usually consist of old pieces of lead and other metal, coins, BB's, old bullets, buckshot, nails, garnets, small purple and black iron rocks, and the heavy black sand concentrates--which consist mainly or in part of the following: magnetite (magnetic black sands), hematite (nonmagnetic black sands), titanium, zircon, rhodolite, monazite, tungsten materials, and sometimes pyrites (fool's gold), plus any other items which might be present in that location which have a high specific gravity--like gold, silver, and platinum.

BLACK SANDS

Heavy black sands, the kind that end up at the bottom of your gold pan, are usually present to some degree anywhere in a streambed. When panning the lower stratum of streambed, just above bedrock, you will usually end up with at least some black sand at the bottom of your gold pan.

The presence of a large amount of heavy black sands in a particular streambed location is not a surefire index of the presence of gold being nearby. However, it does mean that some heavier materials have traveled over and concentrated to some degree in that particular location. There is a greater quantity of black sand in a streambed than of gold. Black sands, being much lighter than gold, do not necessarily follow the same path that gold does. Heavy black sands tend to concentrate in a much wider range of locations than gold does. However, heavy black sands do tend to always concentrate in the same locations that gold does. So, if there is no black sand present in a particular streambed location, it is less likely that you will find gold there. In this way, the presence of heavy black sands--or the lack of them, tells a story of its own. You will want to watch for them while prospecting new ground.

Gold has about three or four times the weight of the average black sands. It's about six or seven times heavier than the average of other materials generally found in a streambed. So, as you are panning, the materials become more and more heavy as you work toward the bottom of the pan. It's a bit more difficult to separate gold down through heavier material than it is to vibrate it down through lighter material while panning. This doesn't really amount to a problem when streambed materials are being panned, because usually there are not enough heavy black sands in a single pan to amount to much.

When panning down a set of concentrates that have been taken out of a sluice box or some other type of recovery system, much more black sand will be present, and it becomes more difficult to work the gold down through the heavier materials as one gets towards the bottom of the pan. Even so, the gold is heavier than the heavy black sands, and the process of shaking and sweeping remains basically the same as before, except it's necessary to be a little more careful to prevent losing any gold out of your pan during the process (see Figure 4-13).

Fig. 4-13. It's necessary to be a little more careful when sweeping out the black sands to to prevent loss of gold.

Once down to the heaviest black sands in your pan, you can get a quick look at the concentrates to see how much gold is present by allowing about a half cup of water into the pan, tilting the pan

Fig. 4-14. The final concentrates can be swirled away to see how much gold is present.

forward as before, and shaking from left to right to place the concentrates in the forward bottom section of your pan. Then level the pan off and swirl the water around in slow circles. This action will gradually uncover the concentrates, and you can get a look at any gold that is present. The amount of gold in your pan will give you an idea of how rich the raw material is that you are sampling (see Figure 4-14).

A magnet can be used at this point to help remove the magnetic black sands from the gold pan. Take care when doing this. While gold is not magnetic, sometimes particles of gold will become trapped in the magnetic net of iron particles which lump together and attach to the magnet. I prefer to drop the magnetic sands into a second plastic gold pan, swish them around, and pick them up again with the magnet. Depending on how much gold this leaves behind, I might do this step several times.

Many beginners like to stop panning at this point and pick out all the pieces of gold (colors) with tweezers. This is one way of recovering the gold from your pan, but it is a very slow method.

Most prospectors who have been at it for awhile will pan down through the black sands as far as they feel that they can go without losing any gold. Then they check the pan for any colors by swirling it, and pick out any of the larger-sized flakes and nuggets to place them in a gold sample bottle, which has been brought along for that purpose. Then the remaining concentrates are poured into a small coffee can and allowed to accumulate there until the end of the day, or week, or whenever enough concentrates have been collected to make it worthwhile to process them with mercury in order to recover all of the gold out of them (covered in Chapter 7). This is really the better method if you are interested in recovering more gold, because it allows you to get on with the job of panning and sampling without getting deeply involved with a pair of tweezers. Otherwise you can end up spending 25% of your time panning and 75% of your time picking.

It is possible to pan all the way down to the gold--with no black sands, lead, or other foreign materials left in the pan. This is often done among prospectors when cleaning up a set of concentrates which have been taken from the recovery system of a larger piece of mining equipment--like a sluice box or dredge.

Panning all the way down to gold is really not very difficult, once you get the hang of it. It's just a matter of a little practice and being a bit more careful. Most prospectors when doing so, prefer to use the smooth surface of gold pan as opposed to using the cheater riffles (see Figure 4-15).

When panning a set of concentrates all the way down to the gold--or nearly so, it's good to have a medium-sized funnel and a large-mouth gold sample bottle on hand. This way, once you have finished panning, it's just a matter of pouring the gold from your pan into the sample bottle via the funnel, as shown in Figure 4-16. Pill bottles and baby food jars often make good gold sample bottles for field use because they are usually made of thick glass and have wide mouths. Plastic ones are even safer.

Another method is with the use of a gold snifter bottle (see Figure 4-17). This is a small hand-

Fig. 4-15. Panning down through the black sands to the gold is not too difficult once you get the hang of it.

Fig. 4-16. A Funnel can come in handy when it's time to transfer your gold into a sample bottle.

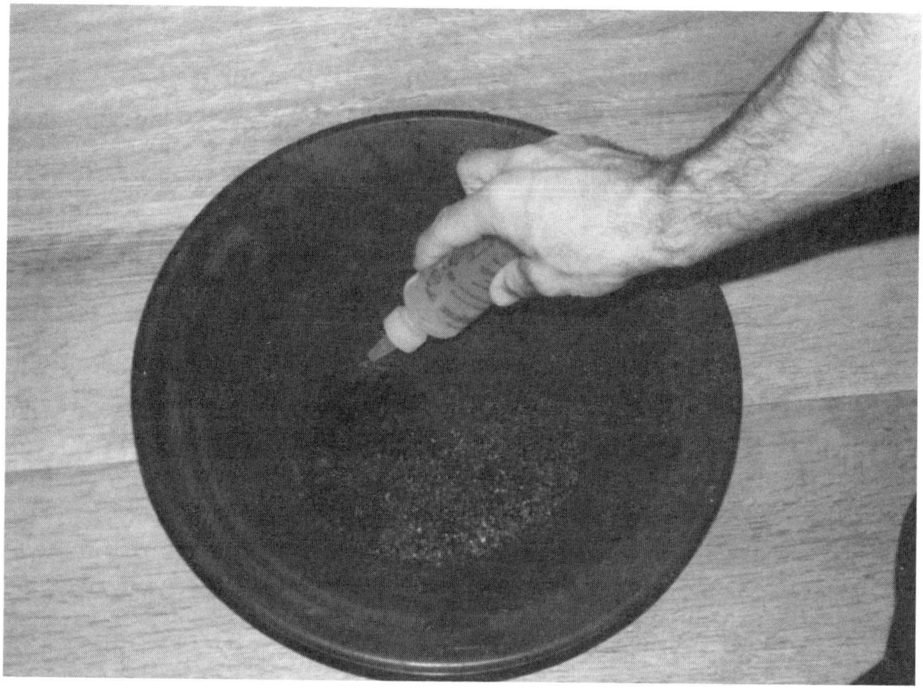

Fig. 4-17. Snifter bottle photo.

sized flexible plastic bottle with a small sucking tube attached to it. Squeezing the snifter bottle creates a vacuum inside, and submerged gold from the pan can consequently be sucked up through the tube.

If you do not have a snifter bottle or funnel on hand, try wetting your finger with saliva and fingering the gold into a container, which should be filled with water. The saliva will cause gold and concentrates to stick to your finger until it touches the water in the container. This works, but the funnel method is faster.

PRACTICE GOLD PANNING

If you are not in a known gold producing location, but want to do some practice panning to get accustomed to it and acquire some skills before going out into the field, you can practice in your own back yard. Use a washtub to pan into and some diggings from your garden, or wherever, to simulate streambed materials. I recommend that you throw in some rocks and gravel along with the dirt so that it takes on an actual streambed consistency. Take some pieces of lead, buckshot or small lead fishing weights, cut them up into various sizes ranging from pellet size down to pinhead size, and pound some of them flat with a hammer. This puts the pieces of lead in the same form as the majority of gold found in a streambed--flake form. They will act in much the same way as will flakes and grains of gold. Leave a few of the pieces of lead shot as they are so that gold nuggets can also be simulated.

When panning into the tub, place a few of these pieces of lead into your pan, starting off with the larger-sized pieces first. Keep track of how many pieces of lead you use each time so that you can see how well you are doing when you get down to the bottom of the pan. Practice panning in this manner can be very revealing to a beginner, especially when he or she continues to put smaller and smaller pieces of lead into the pan as progress is made.

If a person is able to pan small pieces of lead successfully, then he or she will have no difficulty whatsoever in panning gold (higher specific gravity) out of a riverbed. And who knows? You may end up with gold in your pan--right out of your own back yard! It wouldn't be the first time.

Out in the field, one sure sign that you are panning correctly is the accumulation of heavy black sands at the bottom of your pan. If you are recovering the magnetite--specific gravity 5.2--then you will recover the gold--specific gravity 19.3.

On the large gold-producing rivers of the western United States, there are actually very few places in which you can pan for gold and not turn up any. Direct your efforts toward the lowest stratum of streambed material and the bedrock irregularities. Many of these rivers, or sections of them, have been assayed to carry in excess of a million dollars in gold per mile (minimum figures). So there really is no lack of good gold-bearing ground near water in which to perfect your gold panning techniques.

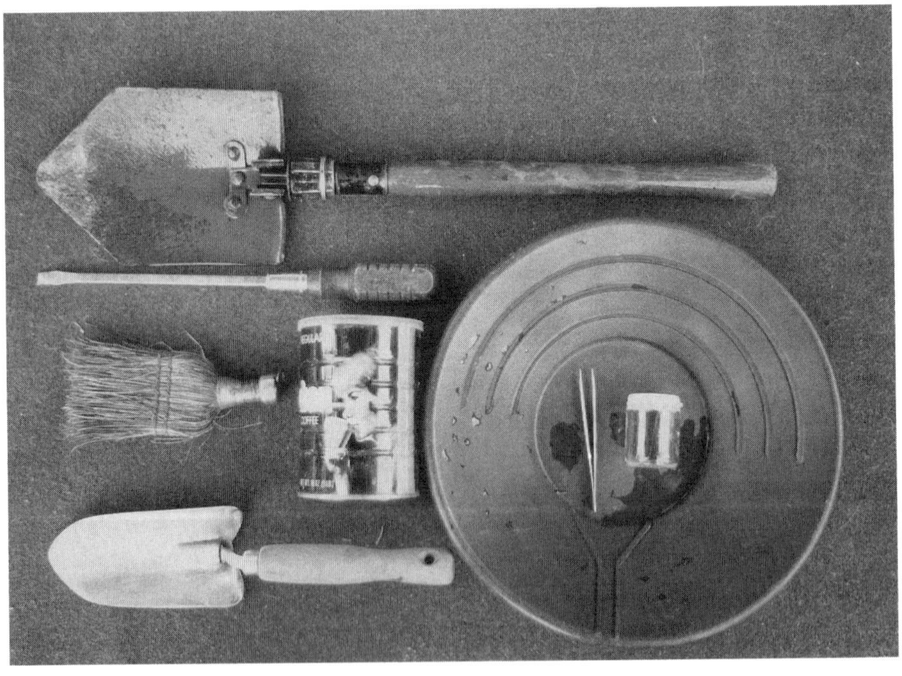

Fig. 4-18. Here are the tools that are normally used in a gold panning or sampling operation.

If you are just beginning and there is an experienced miner around, talk him into giving you a demonstration. This helps you to catch on faster.

The general equipment needed in order to conduct a gold panning or sampling operation is as follows: shovel--for digging, screwdriver--for breaking open bedrock irregularities and scraping them clean, whisk broom--for cleaning up dry bedrock, garden trowel--for digging out bedrock traps, small coffee can--for valuable black sand concentrates, gold pan--for processing the gold out of material, gold bottle and tweezers-- for the gold (see Figure 4-18).

One other tool that sometimes comes in handy when you are cleaning up bedrock is a *''suction gun''* (see Figure 4-19).

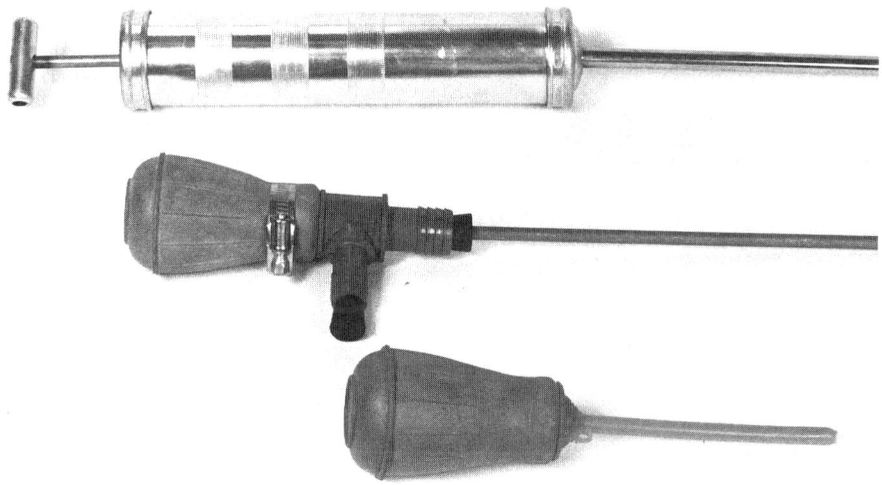

Fig. 4-19. "Suction gun" (top) and "gold snifters" are used to suck material from the bottom of bedrock crevices.

Once in a while, you will run into a bedrock crack, crevice, or some other kind of irregularity which you cannot easily get into and thoroughly clean out with the normal panning tools. In this case, you can fill the crevice with water and suck the contents out with the use of one of these suction tools--as shown in Figure 4-19--or you can use a dry-land dredge.

MACK-VACK

The Mack-Vack is a wonderful new innovation created for gold mining just within the past several years. It is actually a portable gas-powered shop-type vacuum cleaner which has enough power to suck up rocks, sand and gravel. It is an excellent tool for cleaning out crevices and bedrock irregularities. The manufacturer calls it a *''Dry Land Dredge.''* It can be used for cleaning up on bedrock, for sucking material out of various gold traps, or for mossing operations (covered later). The unit is becoming very popular--particularly among recreationalists who want to get more gold, but don't have the time or energy to put out much effort. Materials are sucked into the Mack-Vack's

Fig. 4-20. The Mack-Vack is used as a dry-land dredge.

catch bucket. Then the material can be carried to water and be panned off or run through a concentration device.

If you are panning in some location and are not turning up an acceptable amount of gold, the thing to do is move to another likely spot in which to find it. Use your knowledge of stream placer geology and try there, and keep on trying different spots until you locate some ground which is paying in gold values to your liking. This is the basic procedure of sampling. One luxury of using a gold pan as a sampling tool is that it's very easy to pick up and move around until you have found a place hot enough to work seriously.

CLEANING UP SLUICE BOX CONCENTRATES

There is really not much difference between panning down a set of concentrates taken from a sluice box and those materials taken from a streambed. The main difference is that sluice box materials are usually much more concentrated than the normal streambed materials are, and so you will be dealing with more black sands--and hopefully more gold, too.

Panning concentrates goes much more smoothly if you remove all of the larger stones and pebbles before you start. This can be done very quickly by pouring the concentrates through a set of mesh screens. Excellent results can be obtained by passing them through a 1/4 inch mesh screen first (hole openings around 1/4 inch).

Before discarding the larger rocks and pebbles that will not pass through a mesh screen, always inspect them first to make sure you don't throw away any large pieces of gold. These usually show

Fig. 4-21. A large handled kitchen-type wire strainer works well as a clean-up classification tool.

themselves quite readily--but it pays to make sure.

The second screening of your concentrates can be done using window screen mesh. A large kitchen wire screen strainer works very well in doing this final screening (see Figure 4-21). Separation of the different sized materials in this way is most easily done underwater. The concentrates which pass through this last screening are placed in a different container from the materials that will not pass through, and each is panned separately.

Interestingly enough, it's the black sands (smaller sized concentrates) that slow down the panning of the larger-sized concentrates. The larger concentrates get in the way of your being able to pan off the black sands effectively. So separating the two and panning them apart makes the process much faster.

When panning the larger-sized concentrates taken from this last screening, keep a close watch for flakes of gold while you are sweeping. Shaking does not cause them to sink down very far in the pan when concentrates of this size are being worked. This is especially true when you have a lot of gold in the material--which may be the case when cleaning up a sluice or dredge.

Because of the large amount of black sands that accumulate in a sluice box which need to be panned during cleanup, it's well to develop your panning skills to the point where you can pan all the way down to the gold, or nearly so.

If you are cleaning up a set of concentrates in which the sand will fill more than one pan, rather than take each pan all the way down to gold, it is faster to take each pan to the heavier concentrates and save them for a final pan. Then take it all down to the gold at one time.

Many experienced miners do the above process, panning each pan of their cleanup down to the point where they are afraid of losing some gold (panning quickly). Then the final concentrates are panned into a washtub, so the heaviest black sands can be saved (see Figure 4-22). The reason for this is that it takes a lot of time to pan a set of heavy concentrates all the way down to gold without losing some in the process--even if just a small amount. Also, some of the grains of black sands have gold inside them. These are called *''locked in values,''* and is the reason why some of the black sands appear to be heavier than they ought to be.

Fig. 4-22. The final concentrates can be panned into a wash tub without worry of losing any gold values.

The more fine gold that is in a set of concentrates, the more difficult it is to separate all of it from the black sands, and the longer it takes to do so. So, some experienced miners will do the final panning into a washtub--being careful to get as much of the fine gold as possible, but not bothering to take all day at it, either. It doesn't matter if some of the fines are lost out of the pan and drop into the tub; because afterwards, the concentrates are poured into a coffee can or some other similar container and accumulated for later processing (see Figure 4-23).

There are several things you can do with a set of concentrates in order to profit from them. Perhaps you'll want to make a winter project out of recovering the rest of the gold from them with the use of mercury (covered in Chapter 7). It's possible that you know, or will be able to find, someone who owns a professional vibrating table, who will be willing to run your concentrates over the table to extract the gold values. Or, maybe you will want to sell your concentrates. There are people around who are willing to pay cash for black sand concentrates--even after you have taken out all of the values that you are able to recover. Black sands can bring in just about any amount of

Fig. 4-23. The final concentrates, after being panned, can be saved for further processing at a later period.

money, depending on how much gold and other valuable minerals they contain. Sometimes, black sands contain a considerable amount of gold even when it is not visible (locked in values). As much as 100 dollars a pound (dry pound) is being paid for some concentrates at this very writing. One or two dollars per pound of concentrates is more common among the small time operations having a daily cleanup. Even so, that's nothing to laugh at when you're panning 10 or 20 pounds a day out of your sluice box. It pays expenses!

SLUICING FOR GOLD

CHAPTER V

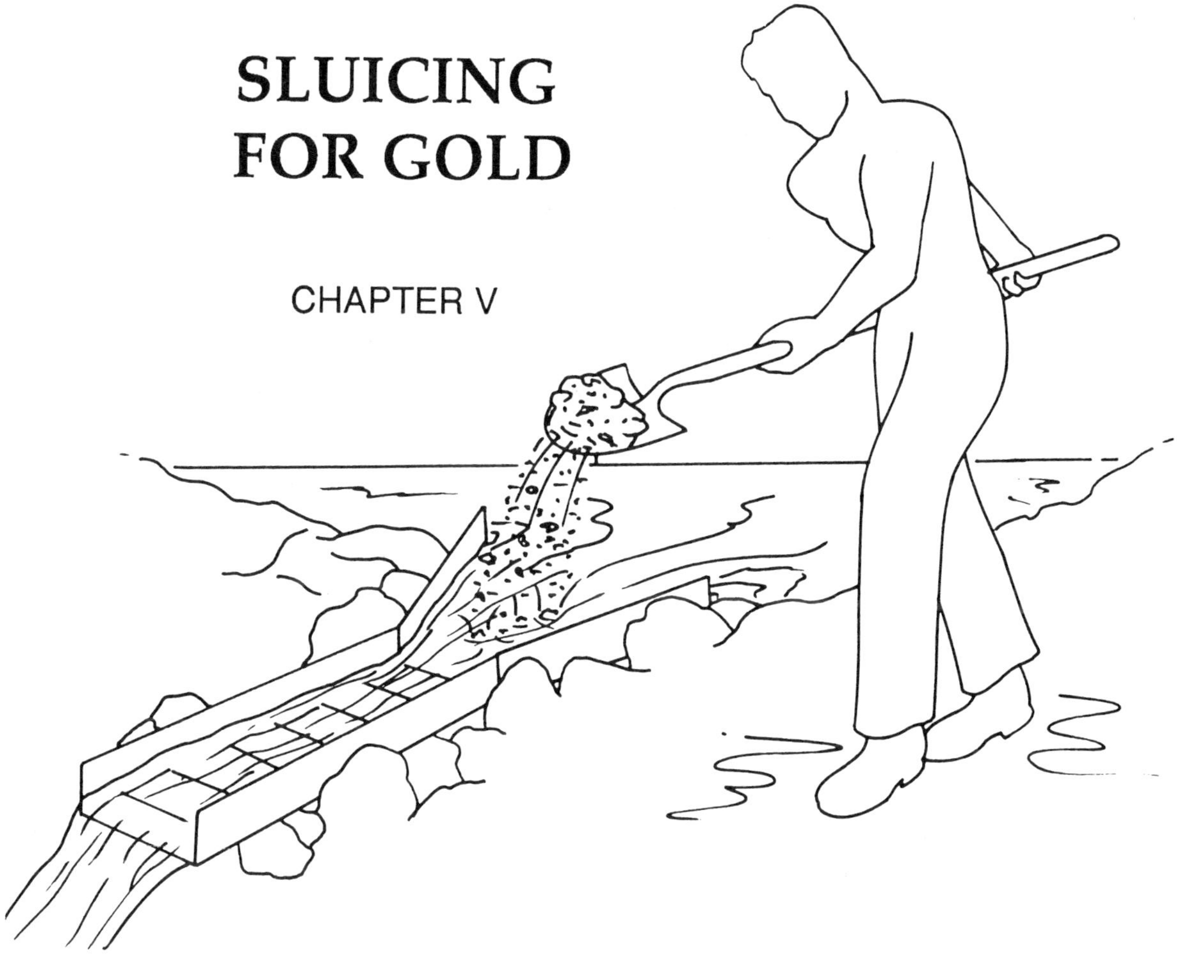

A *"sluice box"* is a trough-like gold recovering device which has a series of obstructions or baffles--called *"riffles"*-- along its bottom edge (see Figure 5-1).

While a steady stream of water is directed to pass through, streambed material is shoveled into the upper end of the box. The flow of water washes the streambed materials through the sluice and over the riffles, which trap the gold out of the material (see Figures 5-1 and 5-2).

The reason a sluice box works is that gold is extremely heavy and will work its way quickly down to the bottom of the materials being washed through the box. The gold will quickly be dropped behind the riffles and remain there, because there is not enough water force behind the riffles to sweep the gold out into the main force of water again.

A sluicing operation, when set up properly, can process the gold out of material just about as fast as it can be shoveled into the box. This can be anywhere from 10 to 200 times more material than a panning operation can handle, yet with similar efficiency in gold recovery. How much material can be shoveled into a sluice box greatly depends on the consistency and hardness of the material in the streambed itself, and how easily it can be broken away.

A sluice requires a steady flow of water through the box to operate at its best efficiency. Most

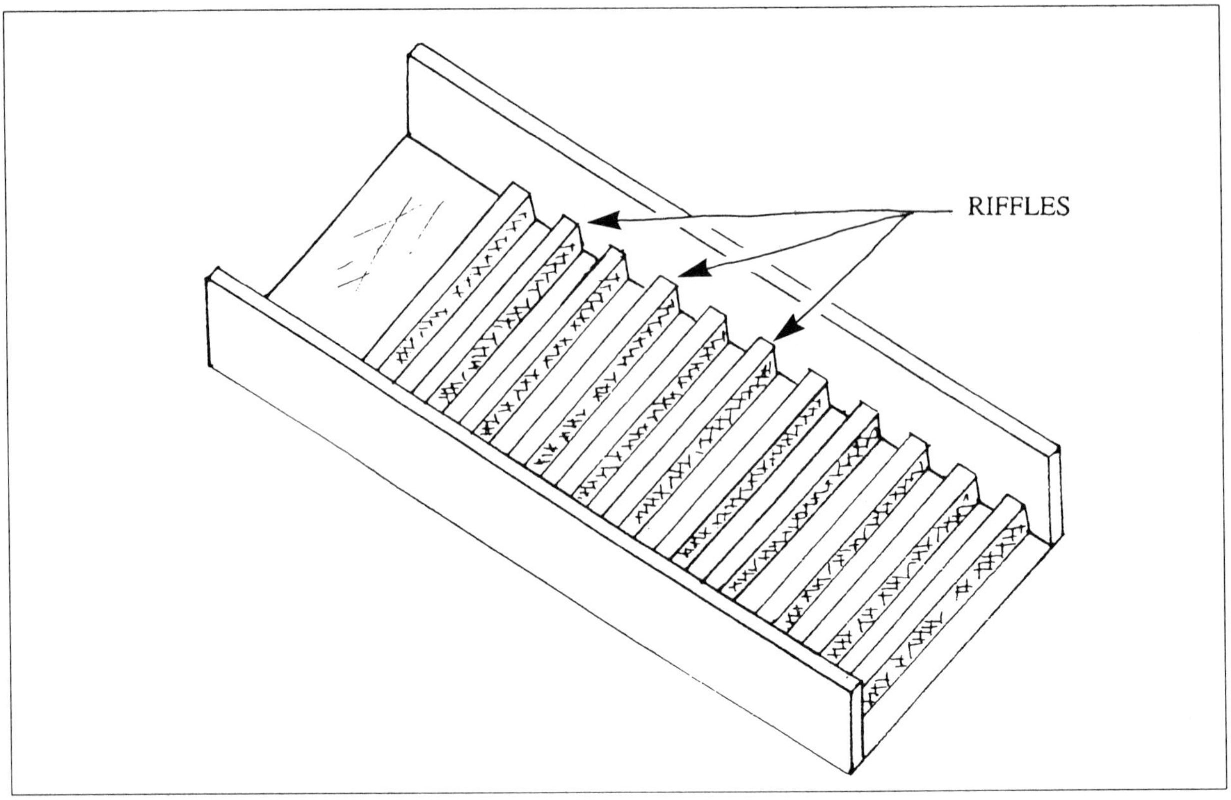

Fig. 5-1. Sluice box.

often, the box is placed in a stream or creek where water is moving rather swiftly, with the sluice being placed in such a way that a stream of water is directed through the box. In locations where water is available, but is not moving fast enough to be channeled through the box for sluicing purposes, the water can be pumped or siphoned to the box with excellent results (covered later). How much water is available, and whether or not it will need to be transported to your sluice box, is something that needs to be considered during the planning stages of a sluicing operation .

Because so much more material can be processed with a sluice than with a gold pan, streambed materials which contain far less gold values can be mined at a profit. If the streambed material had to pay a certain amount in gold values in order to be worked with a gold pan to your satisfaction, gravel containing perhaps 1/100th as many values can be worked with the same result with the use of a sluice box. This is an important factor to grasp, because it means the modern sluice box opens up a tremendous amount of ground that can be profitably mined by an individual.

If streambed material is located which can be profitably worked with the use of a gold pan, working the same material with a sluice greatly increases the amount of gold which will be recovered in the same amount of time.

Sluice boxes are the most broadly used recovery systems in the gold mining industry today. They come in a wide range of sizes, from miniature sluice boxes to the medium sized boxes -- which are often found on the suction dredges, to the larger sized sluices employed by the large-scale bench mining operations where the sluices are fed by heavy equipment, bulldozers, front-end

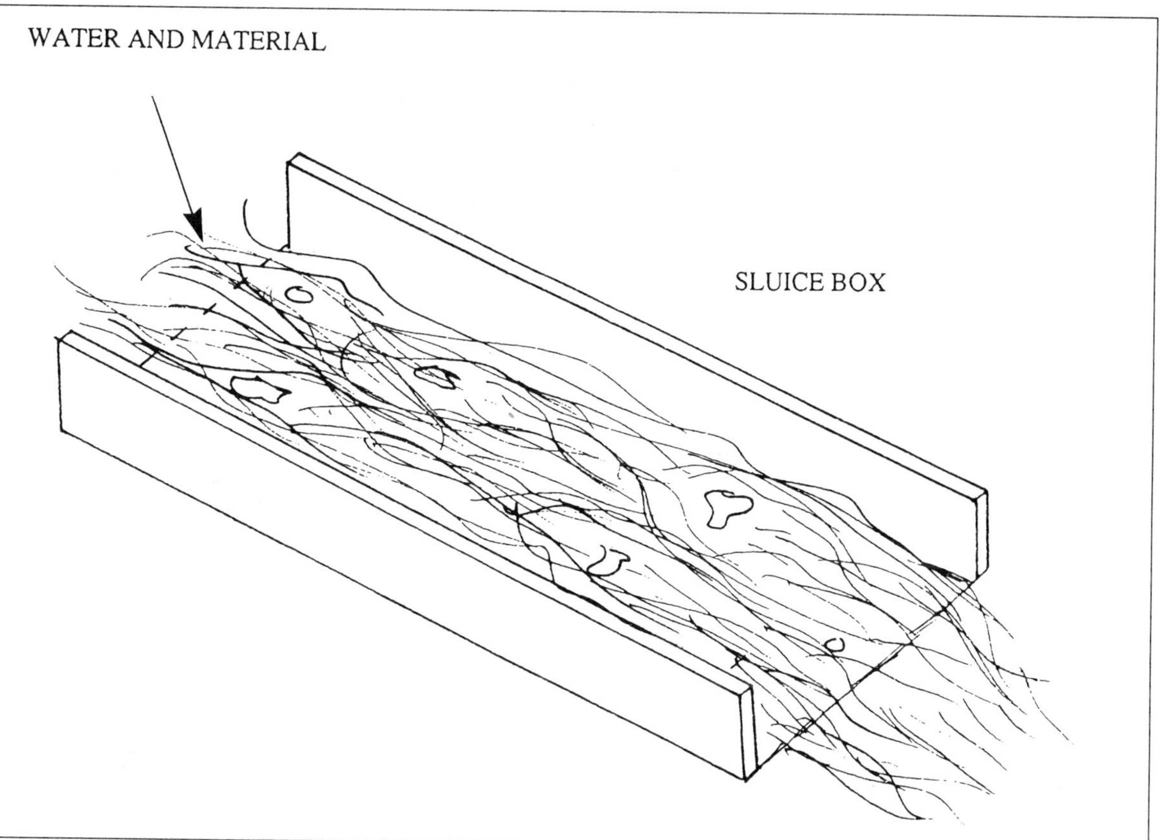

Fig. 5-2. Riffles catch the gold out of the material that is washed through the sluice box.

loaders and the like.

For a one or two-man sluicing operation, a medium sized sluice box can be built (covered later),

Fig. 5-3. Modern day, store-bought type of sluice box.

or one can be purchased from a mining equipment dealer. Figure 5-3 shows an example of a modern-day medium sized sluice box.

The store-bought sluice box, as shown in Figure 5-3, has a few significant advantages over homemade wooden-type boxes. It is made out of lightweight aluminum, so is easy to carry around and will sink when you place it in the water, instead of floating off like the wooden ones tend to do. It comes with a nifty set of quick-release latches which hold down the riffles, so cleanup is easier and faster. Many of these models also come with a flare positioned at the head of the sluice which is designed to channel more water through the box. This means they can be used in slower moving streams of water and still work properly. These flares also come in handy when a miner wants to place more than one sluice box in succession to add additional recovery when rich ground is encountered. The store-bought models also generally recover gold quite well, or can be made to do so with a few minor modifications.

SLUICE BOX HISTORY

Photo Courtesy of Siskiyou County Museum

Fig. 5-4.

Some of the first sluices were made by early California miners, who dug trenches down to the bedrock or sometimes cut a trough into the bedrock itself. They would then channel water through the ditch and over the natural bedrock traps. Streambed material would be shoveled into the stream of water, which would allow the gold, which is heavier, to drop into the bedrock irregularities *(riffles)*. Lighter worthless material was washed away by the flow of the water. This method, known as *''ground sluicing,''* was developed in an effort by the 49'ers to find a faster means of recovering gold out of the streambed material than with the use of a gold pan. The idea of ground sluicing was originally based on the principle that if the bedrock irregularities trapped gold once, then they would do so again and again.

Ground Sluicing is still being done in some developing countries. I witnessed a substantial ground sluicing operation last year on the Osa Peninsula in Costa Rica. The natives were using a three-inch dredge to clean up bedrock at the end of each production period.

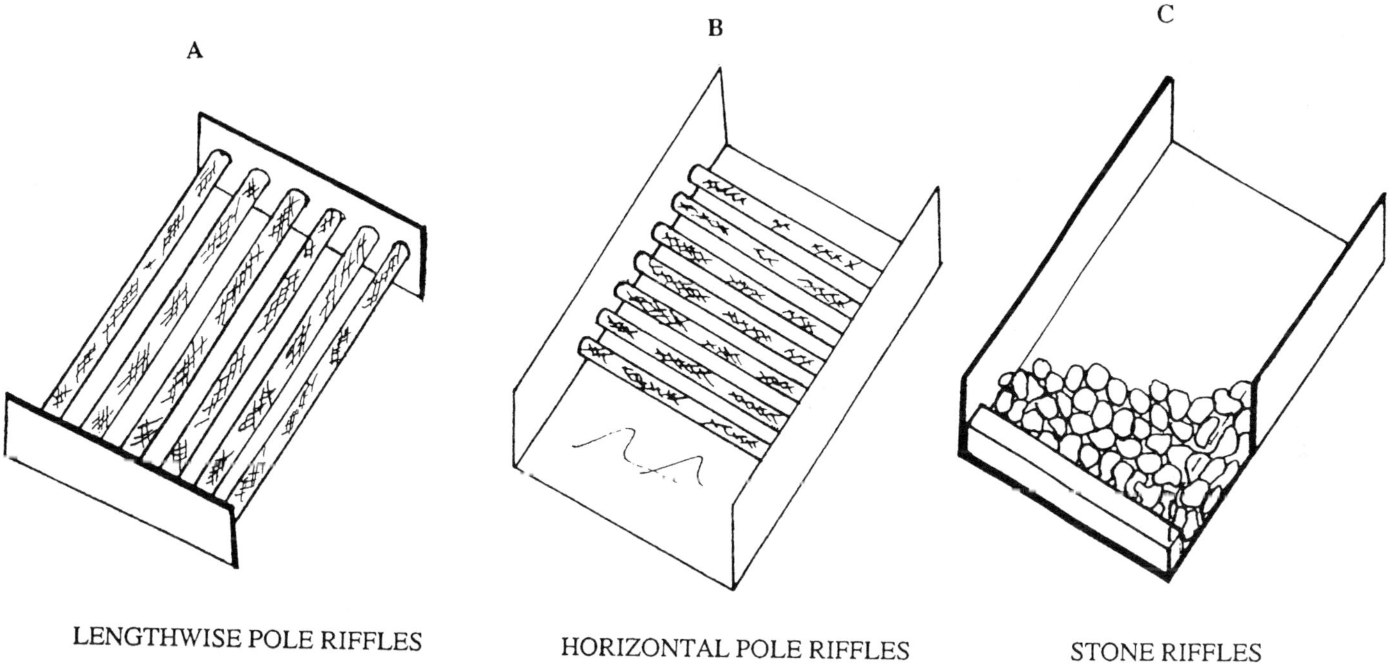

Fig. 5-5. Examples of early day riffles.

Over the years, many different kinds of riffle systems were experimented with, all which trapped gold to some extent; because the yellow metal is so heavy that it doesn't take much to trap its larger sized pieces. Figure 5-5 shows examples of some of the more primitive types of riffle systems that were commonly used by the old-timers on the early frontier.

Parts A and B in the Figure 5-5 examples show riffles which were usually made of thin, straight, round logs--called *''pole riffles.''* These were used to good effect by placing them in both crosswise and lengthwise directions in the sluice box, as shown in the above examples. Part C above shows *"stone riffles,"* which were often used where wooden riffles would not be strong enough and no metal was available. Most sluice boxes of the old days were much longer than today's modern sluices are, and they often employed the use of more than one kind of riffle in a single box. The idea was that if one particular set of riffles didn't catch the gold, another type might.

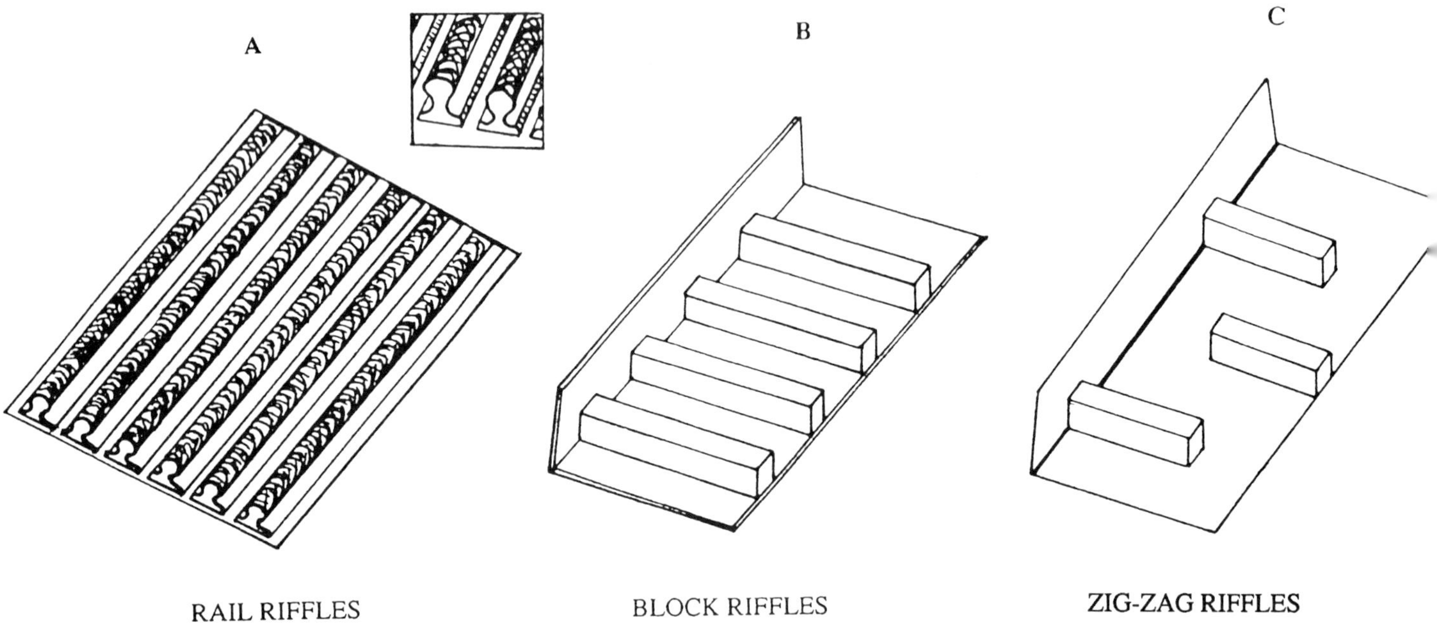

Fig. 5-6. Successful riffle systems used by the earlier miners.

Civilization followed just behind the gold rush, and with it came more tools and supplies. And so recovery systems were improved over time, with particular attention being focused on trapping the tremendous volume of fine particles of gold which were much more difficult to recover. Figure 5-6 shows examples of some of the more developed riffle systems used in the early days.

"Rail riffles" (Part A) were used both in lengthwise and crosswise directions in sluice boxes. They were also used in both right-side-up and upside-down positions--all to some effect. Rail riffles are still being used today in heavy equipment surface bench operations up in the Yukon Territory.

"Block riffles" (Part B) were popular amongst many of the old-timers as a good recovery system, and are also still used in the homemade sluice boxes seen out in the field today. Another version of the block riffle were the *"zigzag riffles"* (Part C). These riffles extended about 1/2 to 2/3 of the way across the box (depending on the miner), which caused the water and material to flow around the riffles instead of over them. Many of the old-timers agreed that zigzag riffles were one of the best fine-gold recovery systems ever developed, and swore by it.

MODERN RIFFLES

Today, there are three main kinds of riffles being used--all which work very well.

The first type, and probably the most often used of the three, is called the *''Hungarian riffle''* (see Figure 5-7).

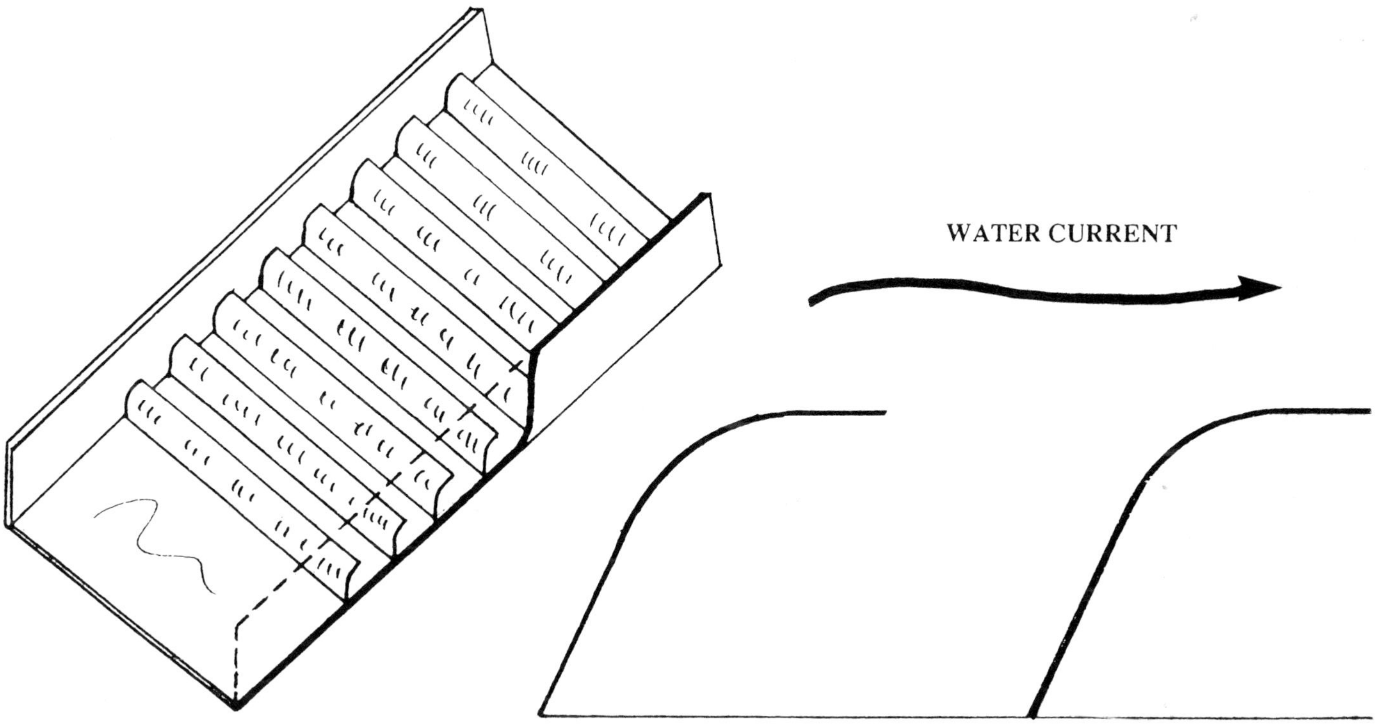

Fig. 5-7. Hungarian type riffle.

Hungarian riffles (sometimes called *''Lazy L Riffles''*) are usually made of a steel alloy. These are designed to recover the gold out of large volumes of material being washed over the riffles. Hungarians are found in the sluice boxes on most dredges for this reason. Hungarian riffles do more than trap gold; they concentrate the heaviest materials passed over them. The riffle shape is designed to cause a *''back eddy''* just beyond the upper edge of the riffle, as shown in Figure 5-8. This back eddy just behind the riffle sucks material down into it and causes a vibrating and continuous sorting action, keeping the heaviest materials and allowing the lighter materials to wash out as the new materials are sucked in. Lighter materials are then washed down and out of the box by the main force of water. It is this same concentrating action which causes the Hungarian riffle to recover so well.

The second kind of riffle, which is seen on some of today's larger-sized dredges and sluice boxes, is called the *''Right-Angle Riffle''* (see Figure 5-9).

Right-angle riffles, usually made out of 90-degree angle iron, are a favorite among larger-sized operations because of their superior strength. Angle iron comes in different sizes, but it is the

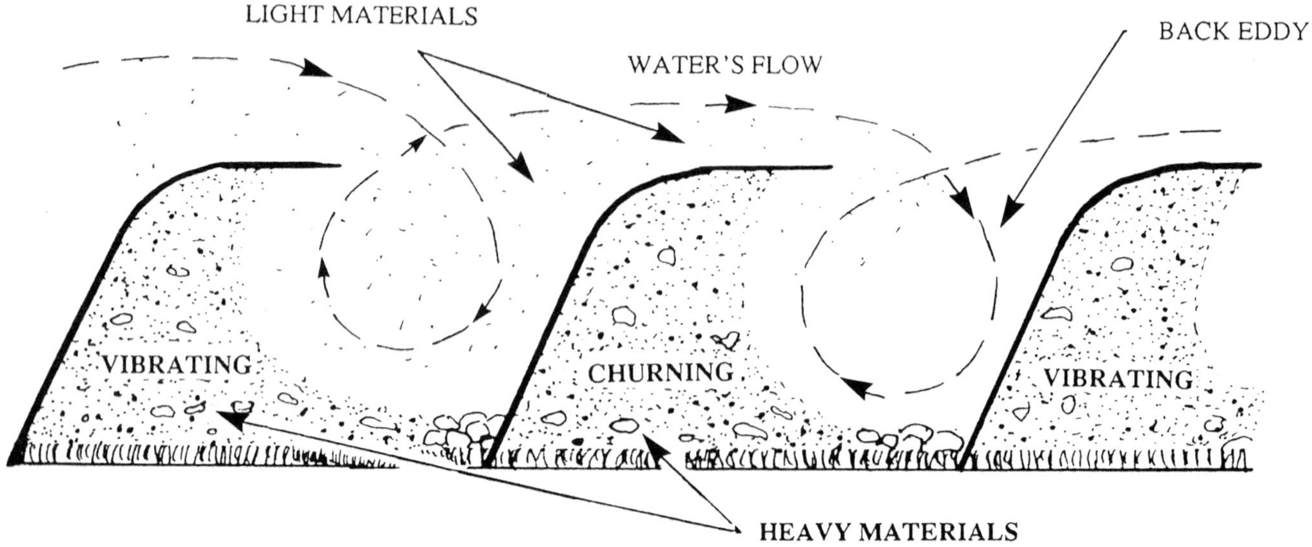

Fig. 5-8. Hungarian riffle action vibrates the heavy materials in and light materials out of the riffle.

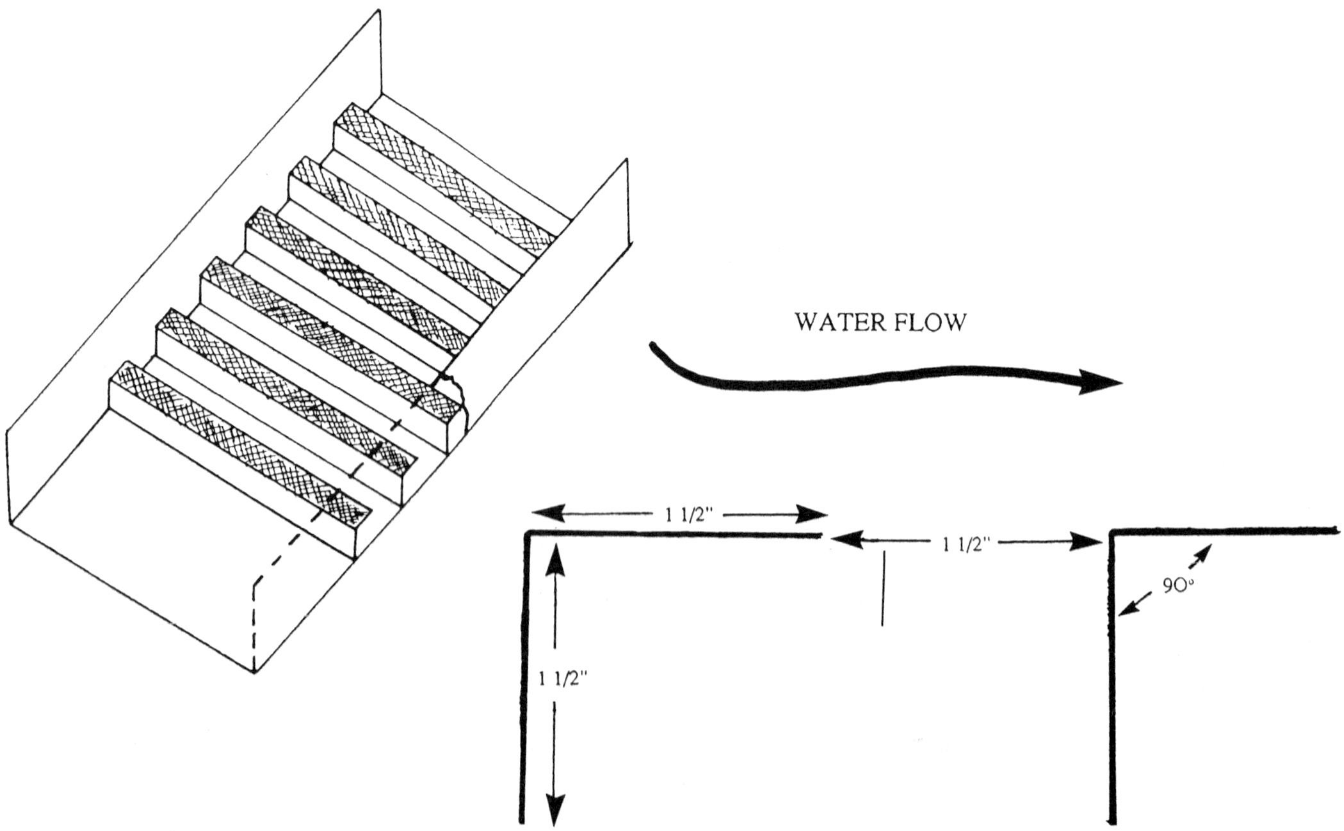

Fig. 5-9. Right angle-type riffle.

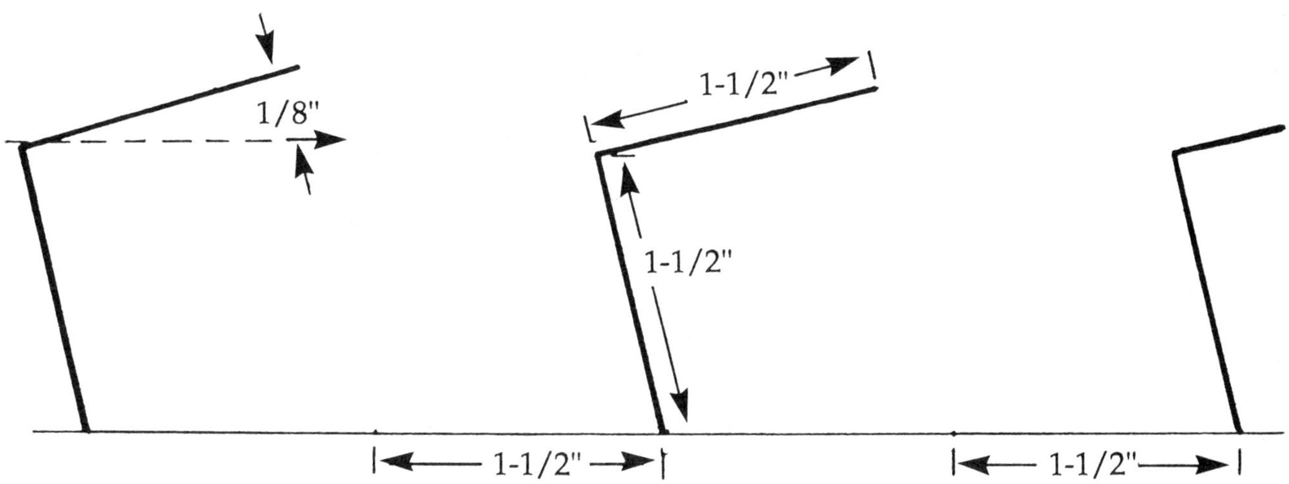

Fig 5-10. Right angle riffles work best when tilted forward slightly.

1-1/2 X 1-1/2 inch angle which is most often used. Larger or smaller sizes can be used when processing larger or smaller volumes of material with different sized sluice boxes.

The right-angle riffle is a concentrating riffle, too. Results are being obtained by placing the riffles an equal distance apart--that distance being the same or more as the width of the angle itself, as shown in the above diagram. These riffles concentrate best when they are tilted forward slightly. The proper tilt is about 1/8th of an inch in the case of a 1-1/2 X 1-1/2 inch riffle (see Figure 5-10). Better results are also being obtained in some situations by placing the tilted riffles about twice that distance apart. This is becoming more popular among some of today's miners.

The third type of riffle often seen in today's sluice boxes is expanded metal, the turned-up kind, as shown in Figure 5-11.

Fig. 5-11. Turned up kind of expanded metal can make an excellent recovering riffle system.

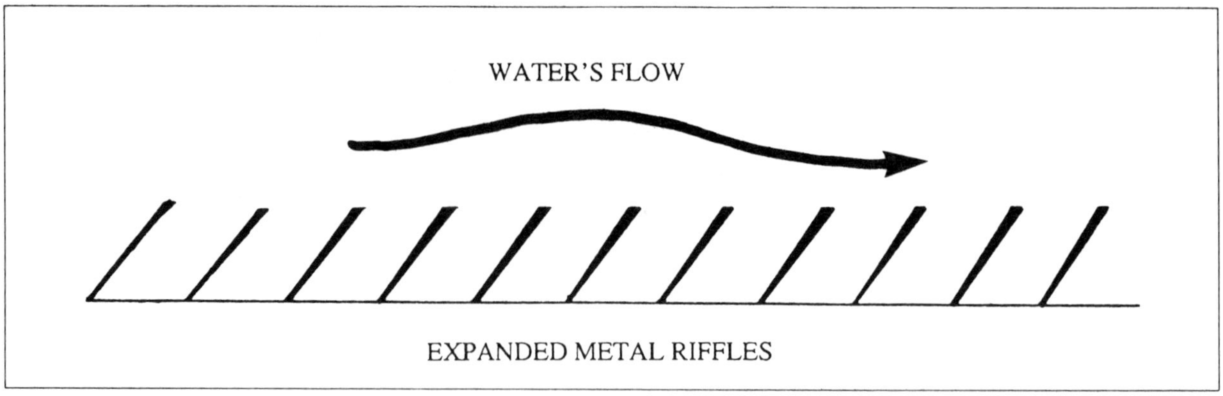

Fig. 5-12. Turned up points of expanded metal-type riffles should point with the flow of current which moves through the box.

Expanded metal is used with the turned-up side pointing WITH the direction of the current as shown in Figure 5-12. In this way, a similar back eddy is created just behind the turned-up points, which causes the expanded metal to concentrate and recover gold very well (see Figure 5-12).

Expanded metal, when used under the proper conditions, makes for an excellent recovery system--especially in the case of fine gold. Best results are obtained in recovery systems which have a mild flow of water passing over the box.

The reason for this is that it only takes a mild water velocity to keep expanded metal concentrating properly. Too much water velocity has a tendency to sweep expanded metal clean of a portion of its heavier concentrated material--including gold values. The reason for this is because the expanded metal riffle is so short in height. The best place to use expanded metal as a riffle sys-

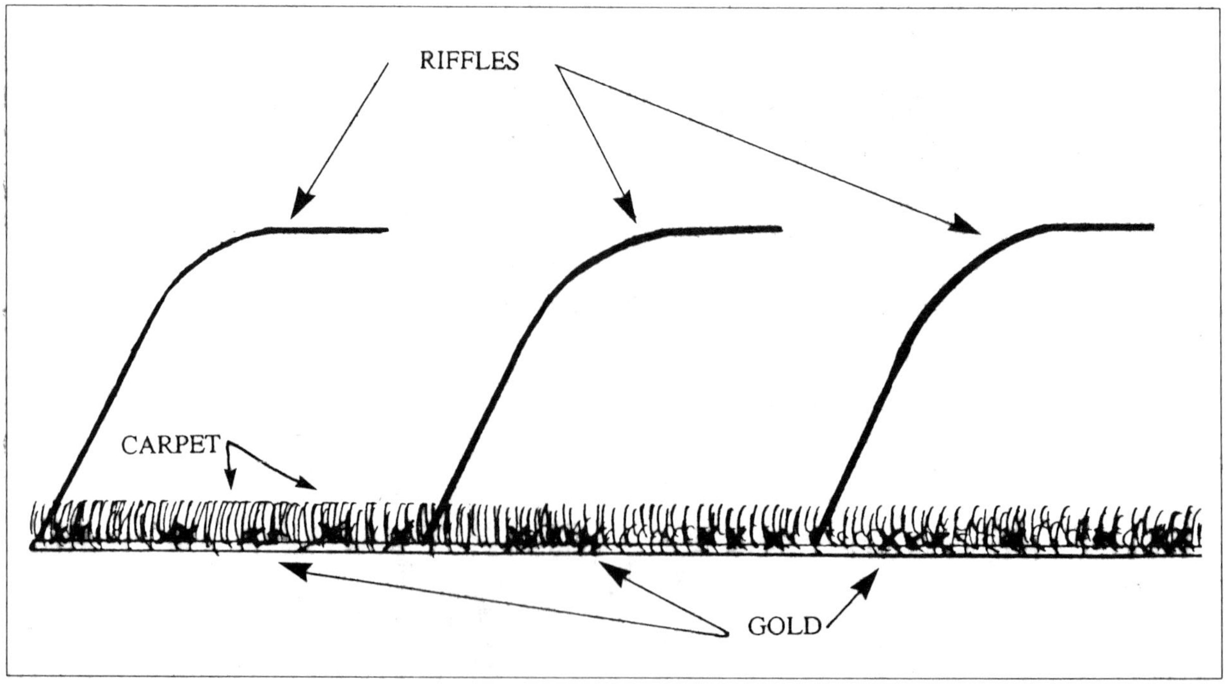

Fig. 5-13. Underlying carpet or matting will trap gold permanently in the sluice box.

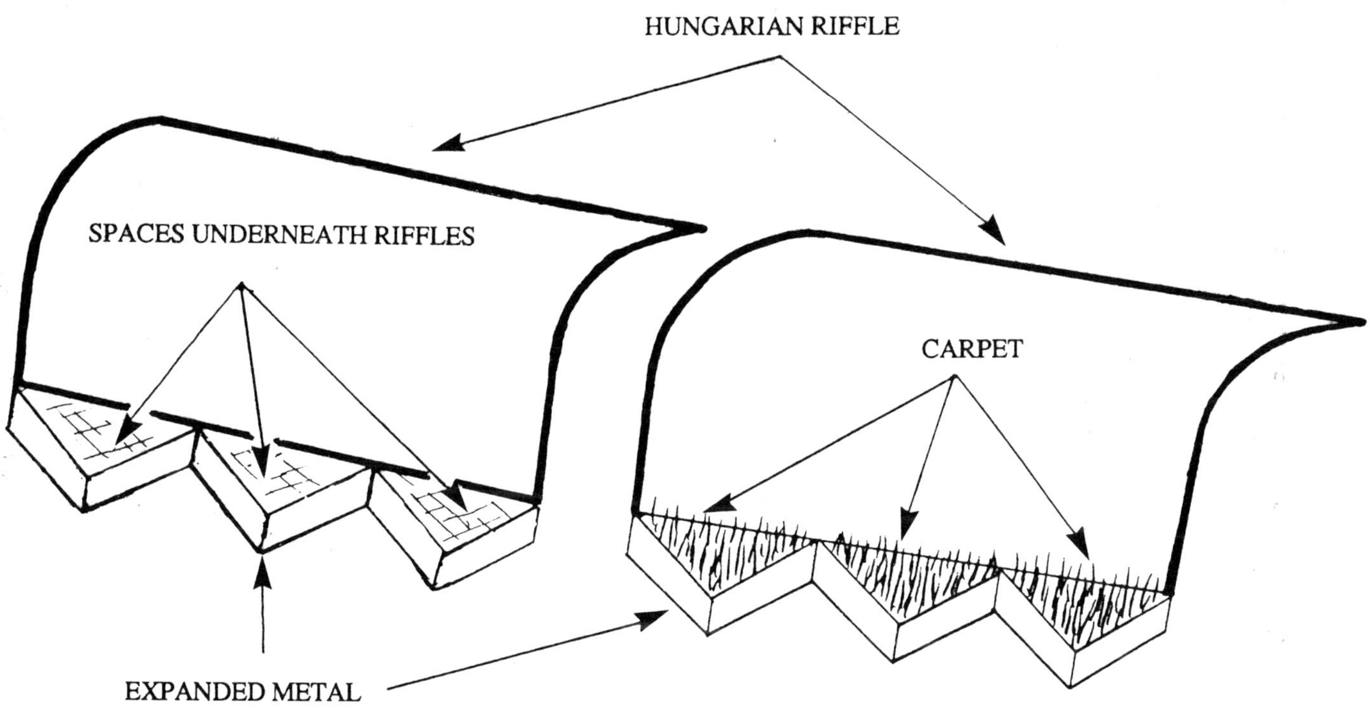

Fig. 5-14. Recovery system on the right is likely to recover gold better, because spaces underneath the overlying riffles are filled by carpet.

tem is in small sluice boxes or in the larger sluices where all of the larger sized material has already been screened out, so that a mild water velocity is all that is needed in the box to move material over the riffles. In these cases you will find that expanded metal recovers gold extremely well.

Most of today's sluice boxes also contain a form of rough and porous matting or carpet underneath the riffle system. Its purpose is to give the gold something along the bottom edge to work its way down into and become permanently trapped in the sluice box (see Figure 5-13). Various types of indoor/outdoor carpets are being used for this. The grassy types work well. Some miners prefer the use of several layers of burlap, or *''miners cloth,''* which is a heavier form of burlap.

One thing to remember about any set of riffles, especially the three kinds that are described above, is that the entire surface of the bottomside of each riffle should be set firmly against the bottom matting of the sluice box during operation. If a space is left between the carpet and the riffles, a portion of the back eddy effect which normally occurs behind the riffles will be lost. This causes some of the concentrating action to cease. Gold recovery is likely to suffer as a result.

Sometimes expanded metal is used in conjunction with other types of riffles. In the case where expanded metal is laid down underneath a top set of riffles, like a set of Hungarians, it is a good idea to make sure the underlying carpet is thick enough that it will stick up through and around the expanded metal and fill the space between the expanded metal and the bottom edge of the overlying riffles, as shown in Figure 5-14. Otherwise, some of the concentrating action in the upper set of riffles may be forfeited, and a loss of fine gold recovery may result.

SETTING THE PROPER WATER VELOCITY THROUGH A SLUICE BOX

As a general rule, the optimum slope setting of a sluice box is around one inch of drop per linear foot of box. However, this can change depending on several factors--like the volume of water being used, or the average shape, size, volume or weight (specific gravity) of material you are attempting to process.

There is no exact formula for setting the proper water velocity through a sluice box which will work optimally under all conditions for all the different types of riffles being used today. Therefore, rather than give you a formula, I will attempt to give you an understanding of what effects the proper amount of water velocity will cause in a box and also what the effects are of too much or too little water velocity. In this way you will be able to act from direct observation to ensure that your, or anyone else's, sluicing device will be recovering gold to the fullest extent possible.

In setting up a sluice, if feasible, it is desirable to have enough water flow to move the material through the box as fast as you can shovel it in at production speed.

Most of the riffles being used today are designed so that a concentrating action takes place behind the riffles. By increasing or decreasing the amount of water velocity over a set of riffles, the amount of water action behind each riffle is also increased or decreased--which has an effect on

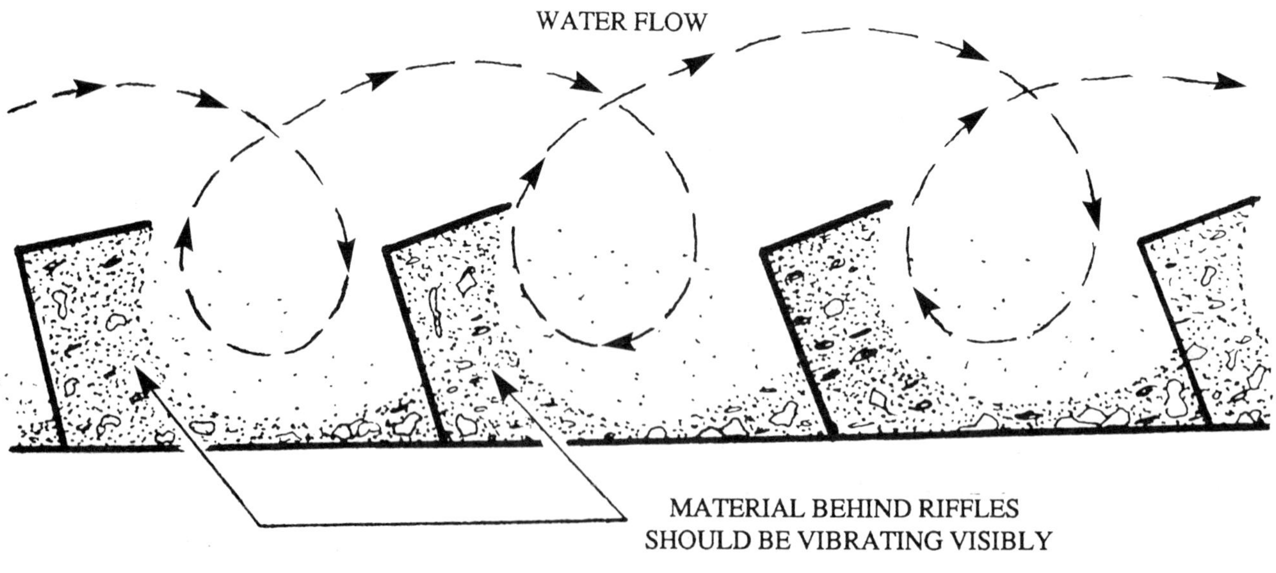

Fig. 5-15. The correct amount of water flow will run riffles about 1/2 full of material.

the amount of concentrating action taking place. Water velocity can be increased by either putting more water through the sluice box or by moving the same amount through faster. Optimally, the

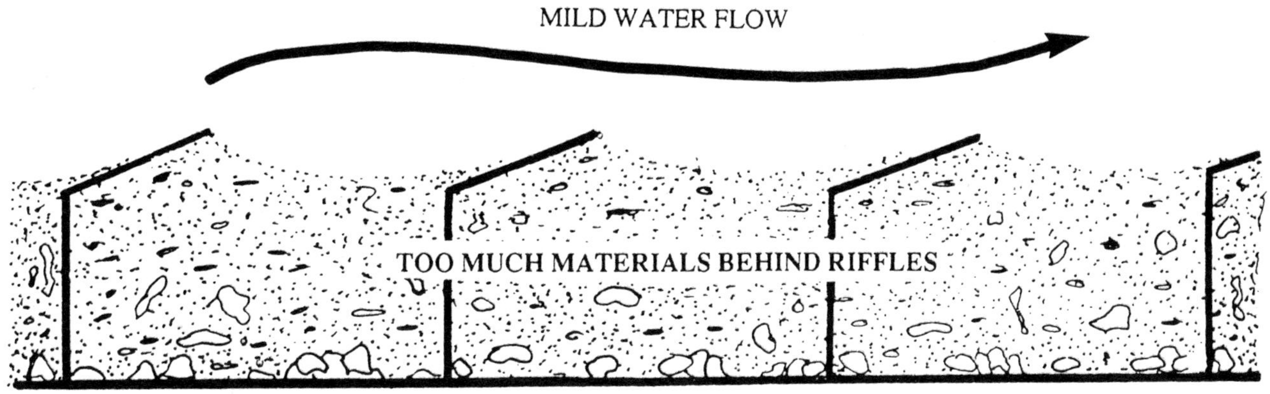

Fig 5-16. Overloaded riffles means not enough water velocity over box.

water flow is just enough to keep the concentrating action going behind each riffle, yet not so much that the riffles are being swept clean of their concentrated material.

How much water velocity is directed over the box determines how much material will stay

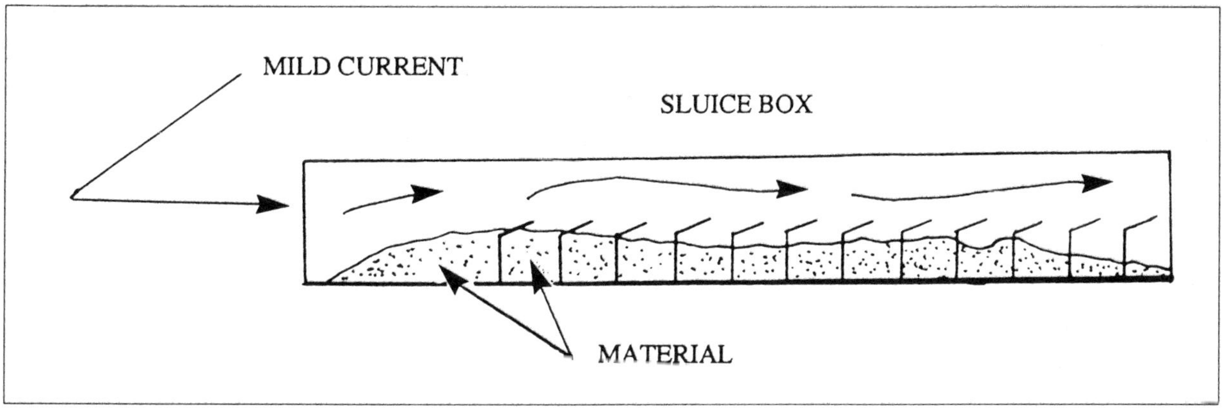

Fig. 5-17. Too little flow can allow the entire sluice box to load up with material.

behind the riffles. When the correct amount of water force is being put through the sluice, its riffles will run about half full of material, and the material can be seen dancing and vibrating behind the riffles (concentrating) when the water is flowing (see Figure 5-15).

If too little flow of water is directed through a sluice box, not enough water force can get into the riffles and they will fill up with material. In this case, little or no concentrating action will take place and gold recovery will be poor (see Figure 5-16). When this happens, little or no visible vibrating action behind the riffles will be seen and material will not be moving through the box fast enough to allow you to shovel at production speed without loading up the entire box, as shown in Figure 5-17.

Too much force of water through a sluice box will put too much turbulence behind the riffles.

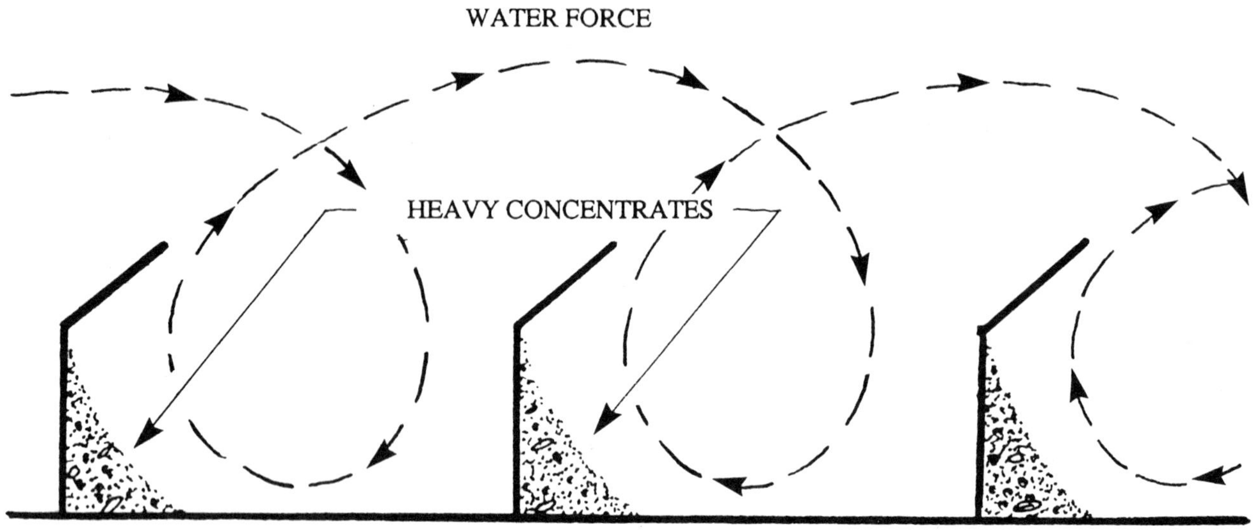

Fig. 5-18. Too much water velocity over a sluice box will sweep the riffles clean of some of the heavier concentrates, and a loss of gold may result.

This will cause some of the heavier concentrated material to be swept out of the box (see Figure 5-18).

When this happens, gold recovery will also suffer, because the spaces behind the riffles are not calm enough to allow a percentage of the finer pieces of gold to settle. You'll notice in this case that the dancing action is occurring behind each riffle, but less material will collect behind the riffles because of the increased amount of turbulence there. When you have too much water velocity, as material is shoveled into the box, you will notice that it passes through very quickly and has little time to make contact with the riffles.

All the above points remain true when adjusting to get the proper amount of water flowing over an expanded metal riffle system. However, when using such a system it is necessary to remember that the riffles are very short; it doesn't take very much water velocity to make them concentrate properly. The correct amount of flow is usually found to be just enough to move the material over the box to keep up with your shoveling.

In the final analysis, once you have your sluice box set up the way you think it should be, run a sizable portion of gold-bearing material through the box and then pan out some samples of the tailings. If you don't turn any gold up from the tailings, you are set up properly. If you are finding gold in the tailings, some changes are in order. Another test is to mix some pieces of lead in with some material, run it through the sluice, and see where the lead stops.

Sluice boxes process material best when receiving it from a steady feed. Too much material dumped at once into a sluice box will overload the riffles and choke off the concentrating action. This will cause gold to wash right through the sluice box as if there were no riffles at all. On the other hand, it is not good practice to run volume amounts of water flow over a sluice box without some material being constantly or regularly fed through. The reason for this is that the scouring action from the water flows will continue to further concentrate materials trapped behind the riffles, causing heavier materials to be washed out of the box. A sluice box operated for extended periods with no new material being fed to it does stand an increased chance of losing some of its fine gold values. How much gold loss will depend on a multitude of factors, such as the type of riffle design, how much water flow, the type and weight of concentrates and the size and purity (specific gravity) of the gold.

When a sluice box is set properly, an occasional larger stone or rock will become lodged in the riffles. These should be picked or flipped out of the riffles with minimum disturbance to the remaining portion of the sluice box.

On most suction dredges, the volume of water being moved through the sluice continues at the same steady flow during production speed. So changing the water velocity is done by changing the slope of the sluice box itself--which will speed up or slow down the flow of water over the box.

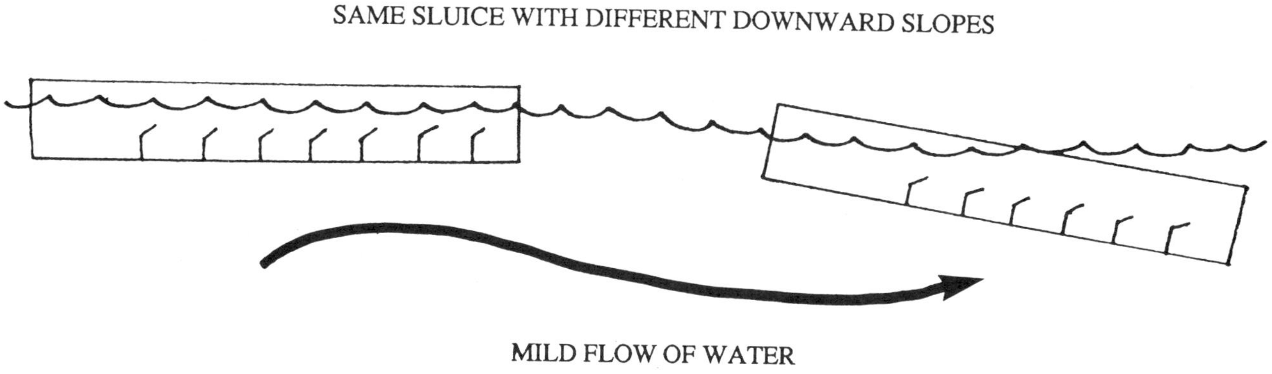

Fig. 5-19. In a mild flow of water, just changing the slope of the sluice usually has little or no effect on increasing the amount of water velocity through the box.

It's a different matter when dealing with a single sluice box being placed in a stream or creek for its water flow. In this case, the water velocity can be adjusted by either changing the slope of the box, by varying the volume of water being directed through the box, or by placing the sluice at different sites in the stream or creek where the water is moving under an optimum condition for your sluicing needs. Getting the right flow of water to pass through a sluice box out in the field is not difficult, but it is usually necessary to try different ideas until you find what works best in each situation. For example, in a location where the water is moving slowly, you might be able to direct more water through the sluice and gain the amount of water velocity that you need. In a stream where the flow is moving more swiftly, the water velocity through your box can be adjusted by changing the volume of water directed into it, and/or by varying its downward slope.

Usually, you will have little trouble arriving at the correct velocity through your sluice box when placing it in a fast stream of water. You can use river rocks to make a foundation in the stream so your box can sit level from side to side, and by allowing different amounts of water volume through the box, and by changing its downward slope, you will quickly get the required combination. It's good to have a length of nylon cord along with you for securing the sluice box to a rock or some other object upstream. This prevents the box from being moved off its foundation by the force of water. Sometimes it's necessary to pile a rock or two on top of the box to hold it in place. This is especially true when you are using a sluice made out of wood. Shovel gravel into the box while trying the different combinations to see what effects the changes have on the water velocity.

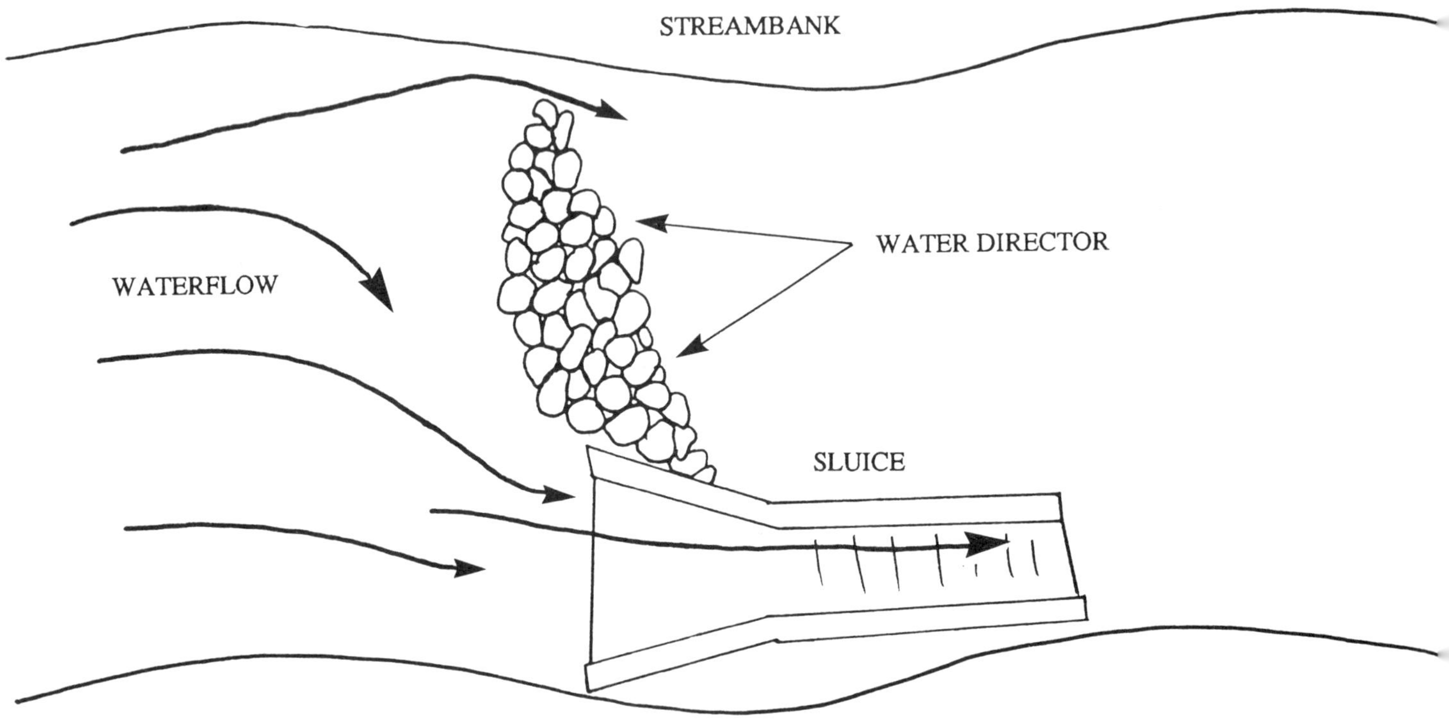

Fig. 5-20. Sometimes more water can be directed through a sluice box in order to get the necessary amount of water velocity.

In a situation where you have to set your sluice into slower water, you will find it is generally more difficult to get the flow you need, because you have to first create more water force than is presently there.

If the flow of the stream itself is not enough to move material through your box, you will sometimes find that changing the slope of the box within the stream has little or no effect on speeding up the flow through the sluice (see Figure 5-19).

In this situation, there are several things that might be done to channel enough flow through your box so you can run material through at production speed. Sometimes the flow of water in the stream itself is enough, so that by setting up a "*water director*" in the stream, you can move enough water through the box to give you the desired result, as shown in figure 5-20. A water deflector, or barrier, like this can sometimes be built by throwing river rocks out into the stream to make more

Fig. 5-21. Sometimes you can build a small dam to bring the water level up just a bit and get the needed water velocity through your box.

water flow into and through the sluice.

If the flow in the stream is not moving fast enough so a water deflector can be used, but some water flow is occurring, sometimes you can get the water velocity needed by building a small dam. By doing so, and by placing your sluice where the moving water spills over the top, sometimes you can get more than enough water flow through the box to meet your needs, as shown in Figure 5-21. It really doesn't take very much volume of water through a medium-sized sluice box to get the right amount of velocity, if the water is moved through the box at speed. In the case of a small dam, as shown above, the water level might only need to be raised up slightly to increase the downward slope of the box enough to create the needed water velocity. How high the dam needs to be depends mostly on how much water is flowing in the stream or creek.

A sheet or two of thin plastic, or a plastic tarp, comes in handy when you are building a dam or water director in a stream. Plastic helps prevent the water from pouring through the holes in your man-made barriers.

A water director or dam can most often be used with good result wherever the water in a stream or creek is moving and is not so deep that the barriers can't be built easily.

If the water at the work site is moving too slowly, or for some reason a water director or dam will not work in that location, it will be necessary to either set up your sluice in a different location where the water is moving faster, or use a pump to put water into your box. Or, in some situations

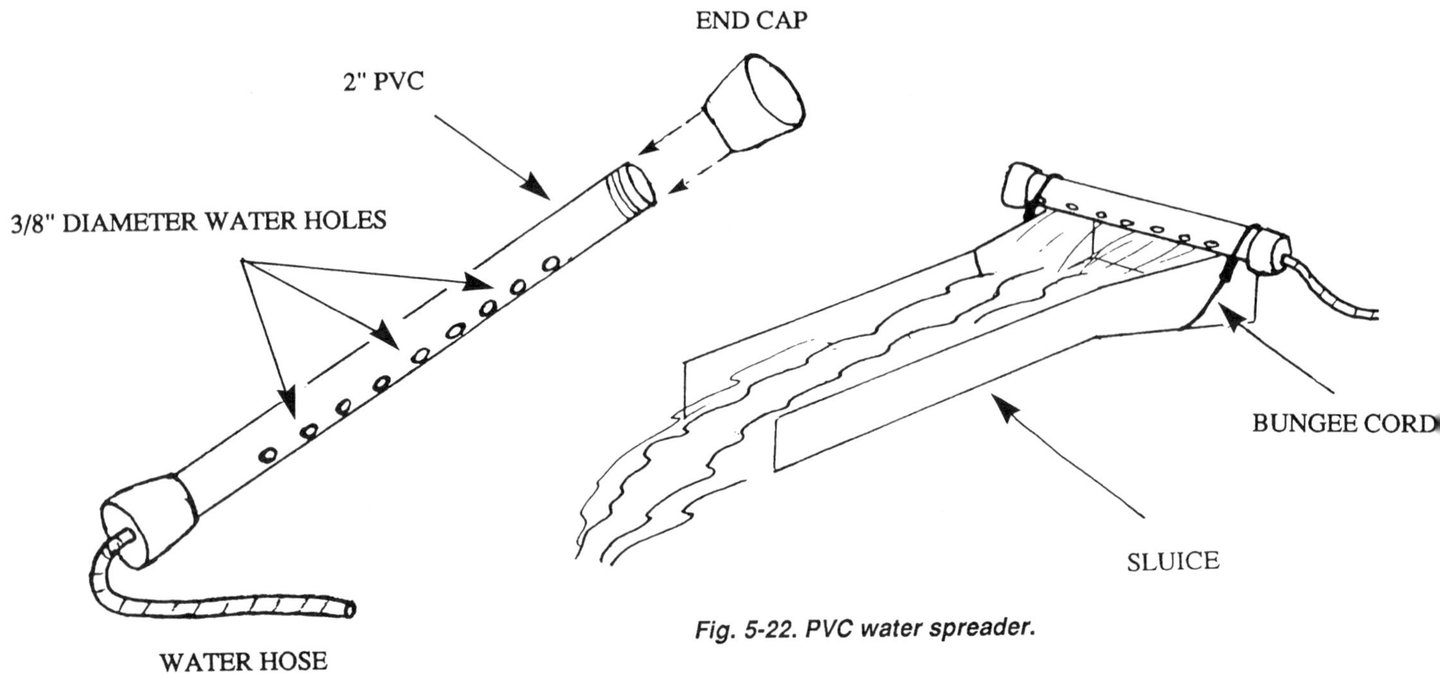

Fig. 5-22. PVC water spreader.

it is possible to siphon water to your box from a higher point upstream. Siphoning can be done effectively with the use of reinforced garden hose or PVC.

PUMPING WATER TO A SLUICE

Engine-pump units are available that will pump enough water to supply any size sluice box. A medium-sized sluice box can usually be fed an ample supply of water via a large diameter garden hose by a smaller and more economical pumping unit. Sluicing with a pump is very popular among more serious minded miners, because it allows the sluice to be placed in the optimum position for production purposes. Sometimes the paydirt you are interested in mining is not directly near moving water. In this case, pumping or siphoning the water to the sluice can be more efficient than hauling the paydirt over to the nearest running water. Pumping water to the sluice also allows the miner to move his box into better positions as progress is made into the streambed. This allows him to continue optimum production all the time, instead of having to carry the material as progress takes him further and further away from the stream of water.

What size engine-pump unit an operation will need depends on the size of the sluice box being used, how far the water will have to be pumped, and how far upward the water will be pumped. Elevation of the work site above sea level, and outside air temperatures, can also affect engine performance. Most dealers of mining equipment have the specifications on their various pumping units and will be happy to help you find the best equipment to fill your needs.

When pumping water to a sluice box through a hose, it's usually necessary to have some kind of a ''*water spreader*'' to effectively distribute the water at the head of the box. A water spreader can be easily put together with the use of a short length of two-inch diameter PVC pipe, as shown in Figure 5-22. In putting such a device together, an end cap should close off one end of the

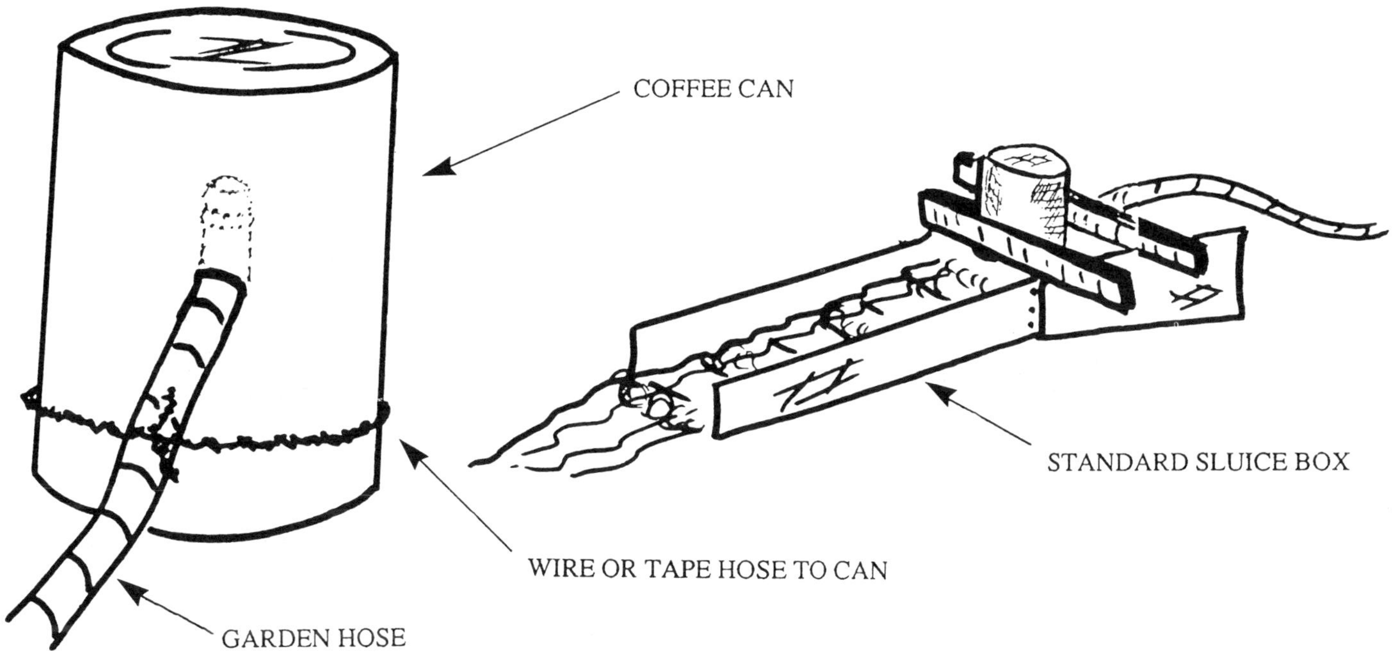

Fig. 5-23. Coffee can water spreader.

spreader, and the other end should be capped with a fitting which will connect to the hose that is being used to pump or siphon water to the box.

Holes should be drilled in a straight line along the length of the spreader. These should be an inch apart and about 3/8 inches in diameter, or larger, depending on how much water is needed. Generally, 3/8 inch holes are about right, but they can be enlarged if necessary. This water spreader can be clamped to the head of your box if it is to remain there permanently, or it can be attached with the use of a heavy rubber bungee cord if it's only a temporary affair. One reason why this kind of water spreader works so well is that it is relatively out of the way. Another reason is that the line of holes can be turned to adjust the direction in which the water hits the head of the box so that optimum performance can be obtained.

If you are out in the field and have a coffee can handy, a workable spreader can be made by punching a hole toward the upper edge of the can large enough to feed your hose through, as shown in Figure 5-23. The can is used upside down with the hose stuck through the hole and fed up into the can. The hose can then be wired or taped to the outside of the can to ensure that it stays in place. Small coffee cans work best for this when being used with medium sized sluice boxes. Nail the can to several pieces of wood long enough to extend across the head of the sluice box, as shown in Figure 5-23.

CLASSIFICATION OF MATERIAL

It takes more water velocity to move the larger material than it does to move the smaller material through a sluice box. Yet, to the degree that the water velocity over a sluice box is increased, there

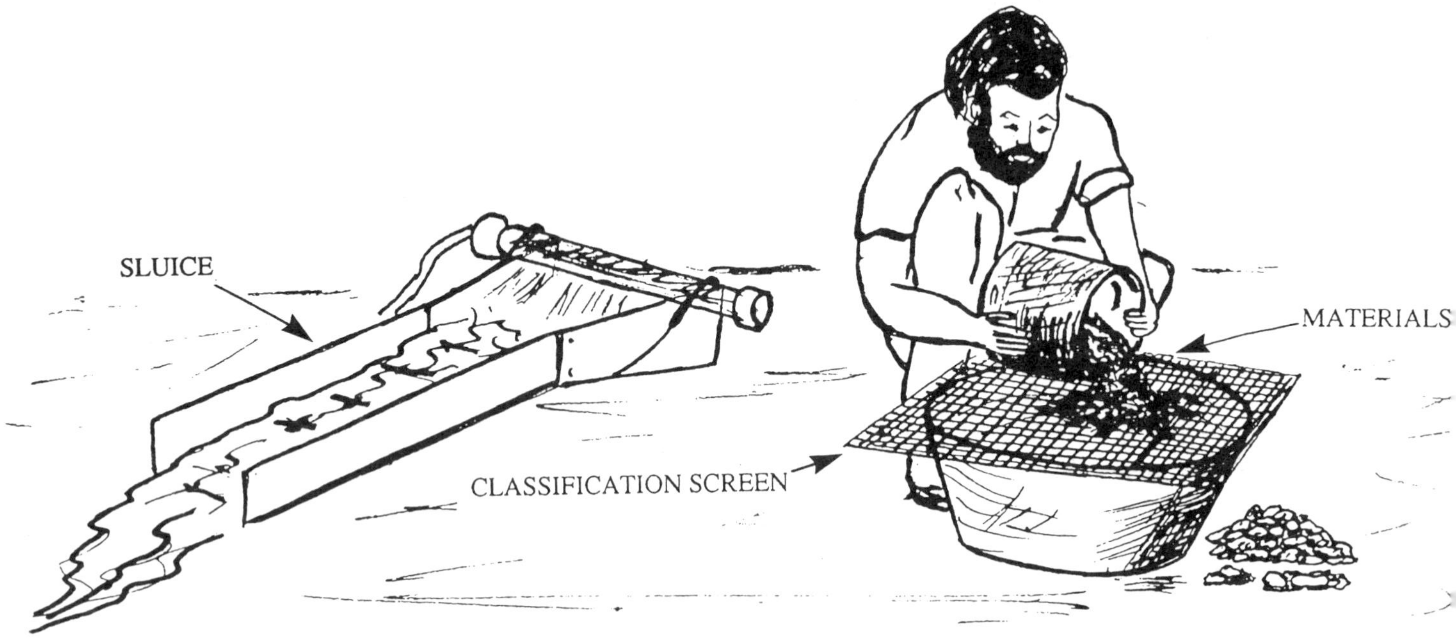

Fig. 5-24. Classifying material before running it through a sluice box will improve fine gold recovery.

will be a loss of a certain percentage of fine gold recovery. Or, to the degree the water velocity over a sluice box is slowed down, there will be an increase in fine gold recovery--as long as there is still enough flow over the box to keep the riffles concentrating.

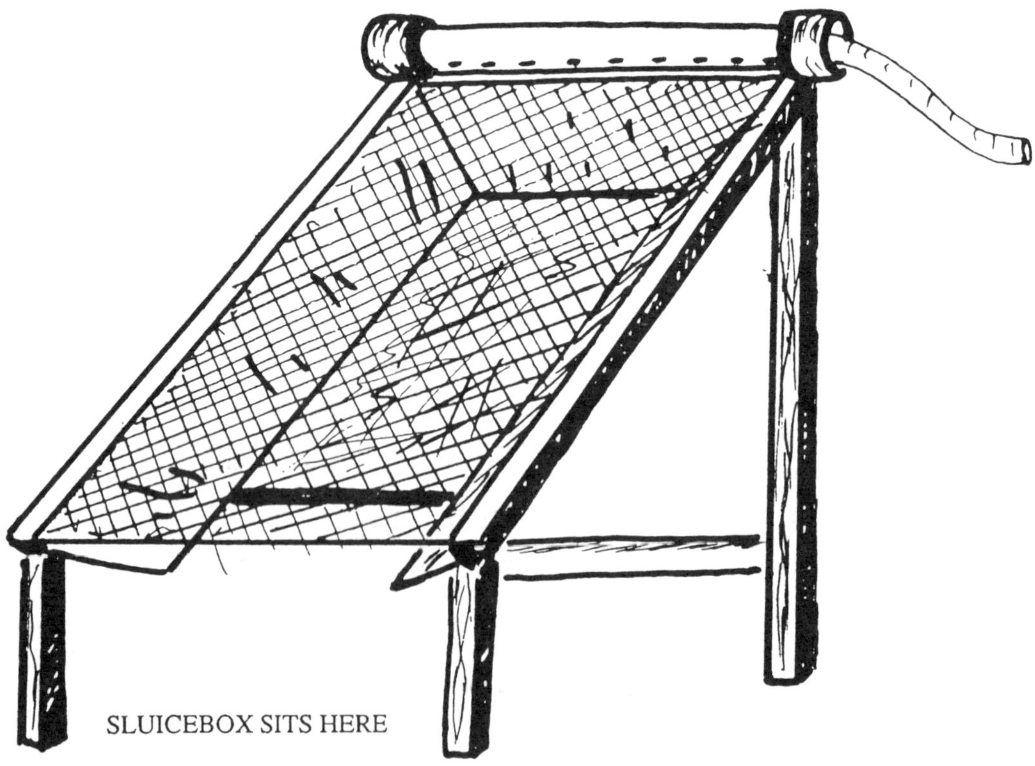

Fig. 5-25. Homemade screening device used to speed up classification and improve gold recovery.

When larger rocks are pushed through a sluice box by water force, they also create greater turbulence behind the riffles as they pass over, which causes an additional loss of fine gold.

In order to improve gold recovery, it is common practice to run material through a screen to classify out the larger rocks before running the material through a sluice box. In this way, less water velocity is needed through the box, which allows for a more orderly flow, and an increase in fine gold recovery. The action of screening materials is called *"classification."* Materials which have been passed through a classification screen are called *"classified materials."*

Fig. 5-26.

Half-inch mesh screen is commonly used in small and medium-sized sluicing operations, because the screen is large enough so classification will take place quickly, yet no materials greater than half an inch in size will get into the sluice. In this way a much slower current of water is required through the box, and fine gold recovery will be improved.

Classification for a sluicing operation can be done in any number of ways, one of which is to place a piece of strong half-inch mesh screen over a bucket, and shovel or pour through the screen into the bucket while sweeping the larger material off to the side, as shown in Figure 5-24. Once the bucket is filled with classified material, it can be poured into the sluice box at a uniform rate. Don't just dump the whole bucket into the sluice all at once. Too much material is likely to overload the riffles and cause the loss of gold recovery.

In a situation where it is necessary to haul material for a short distance to the sluice box, sometimes the best method is to classify the material directly into a wheelbarrow and cart only the classified material to the box.

Perhaps one of the best screening methods is to build a classification device that you can shovel into, and which will stand directly over the top of, and drop the classified materials into, the head of your sluice box, as shown in Figure 5-25.

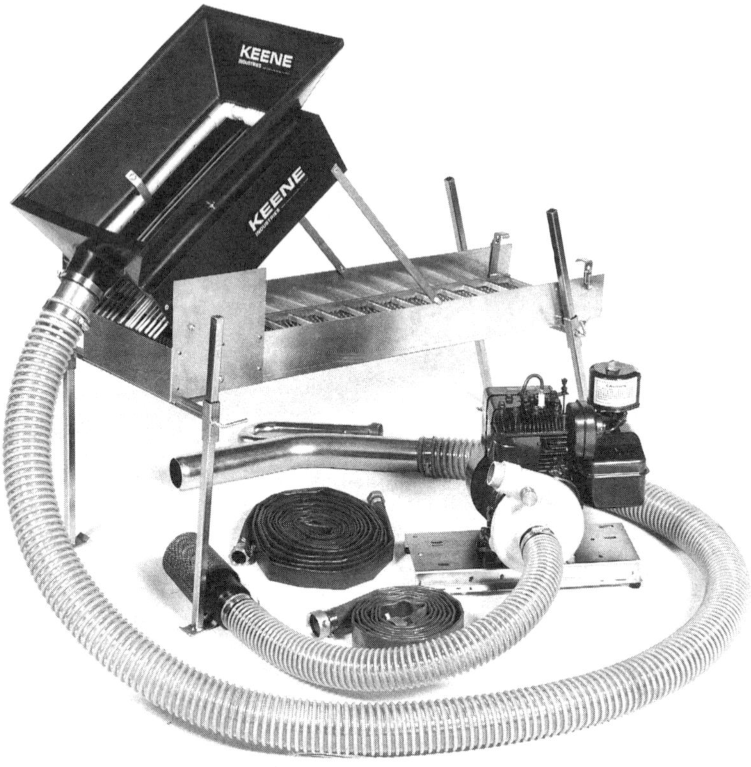

Fig. 5-27. Hydraulic Concentrator ("Highbanker").

The device should be built with the screen set at an angle. This way, the larger material is helped to roll off of the screen as the paydirt is shoveled onto it. Smaller materials should fall through the screen and be directed to fall into the head of your sluice box. This is actually a miniature model of the big classifiers used by many of the large-scale heavy equipment bench mining operations. A classification device such as this is rather easy and economical to build and will speed up a production sluicing operation when screening is being done to improve gold recovery (see figure 5-26).

HYDRAULIC CONCENTRATORS

A *''hydraulic concentrator''* (see Figure 5-27), also called a *''highbanker,''* is a highly portable sluicing device containing all of the best features mentioned thus far in this chapter. These units have classification screens which are designed to keep the larger materials out of the sluice. Yet they also classify fast enough to keep one or two people busy shoveling at production speed.

The manufacturers of the model shown in Figure 5-27 say that their unit will pump water to the concentrator as far as 200 feet above the source of water. This allows for a lot of operating mobility.

The concentrator is designed so the water flow over the sluice is adjustable, to get the best recovery possible. The hydraulic concentrator is very easy to set up compared to having to set up a sluice in a streambed.

A typical highbanker will handle up to about two cubic yards of material per hour, which is the equivalent of any medium sized sluice box when set up under optimum conditions. The hydraulic concentrator is not very heavy, about 50 to 75 pounds in all, depending on size including the engine-pump unit. It can be used successfully as a sampling tool when you want larger volume samples and the areas being tested are not too far off the beaten path. The hydraulic concentrator can also be used as a production machine with excellent results; and for this reason, and the other factors mentioned above, it is perhaps the best and most popular all around, medium sized, portable, surface sluicing equipment on the market today.

WHEN TO DO CLEANUP

Some miners like to *cleanup* their sluice boxes after every hour of operation. Some prefer to do cleanup at the end of the day. Others will go for days at a time before cleaning up. This is all a matter of preference and seldom has much to do with the actual needs of the sluice box. Some of the large-scale operations which ran during the early 1900's used to allow the lower two-thirds of their boxes to run for months at a stretch without cleaning them up, and without worry of losing gold. However, it is true that sluice boxes were longer in those days.

There is a method of determining when a sluice box needs to be cleaned up in order to keep it operating at its utmost efficiency. If the majority of gold is catching in the upper third section of the sluice box, then the recovery system is working well.

After a sluice box has been run for an extended period of time without being cleaned, the riffles will have concentrated a large amount of heavy materials behind them. Sometimes a lot of heavy concentrated material in a sluice box will affect the efficiency of the riffles' gold recovery. This is not always the case; it depends on the type of riffles being used and how they are set up in the box. The true test of when a set of riffles is losing its efficiency because of being loaded down with heavy concentrates is when the gold starts being trapped further down the length of the box than where it normally catches. When this occurs, it is definitely time to clean up your box. Otherwise, clean them when you like.

Expanded metal riffles, being short, will tend to load up with heavy black sands faster than the larger types of riffles. However, a large, visible amount of black sand being present is not necessarily a sign that you are losing gold. Gold is four times heavier than black sand; and in some cases, the black sand will have little or no effect on gold recovery. Again, it depends on how the system is set up, on the type of material being run, the purity (and therefore weight) of the gold, as well as other factors. The best way to evaluate your recovery system is by direct observation of where the gold is being trapped.

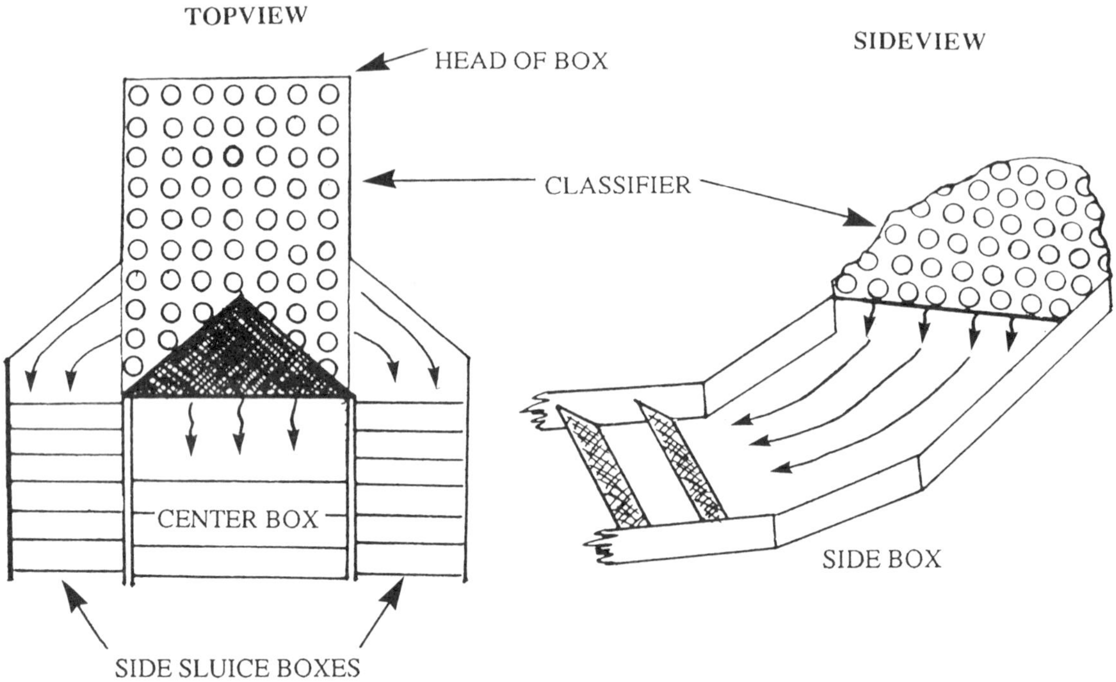

Fig. 5-28. On a triple sluice recovery system, most of the finer sized values will drop down through the classifer and be swept into the side boxes, where the flow of water is moving much slower.

Fig. 5-29. Triple sluice dredge in action.

TRIPLE SLUICES

The idea behind the triple-sluice recovery system is to have a classification screen at the head of the box. This allows the smaller material to be separated out and run through side sluice boxes where the water flow is controlled to move more slowly. This reduced flow allows the side boxes to work more efficiently and capture more of the gold. Most gold, platinum, and gem stones are of a size that will drop through the screen at the head of the box and end up in the slower currents of the side boxes (see Figure 5-28). As a result, the gold recovery is excellent in triple-sluice recovery systems.

The center box in a triple sluice has a sufficient water flow to move the larger materials on through the box, yet the flow is not so strong that it will wash the larger pieces of gold out of the box.

These boxes are well used in the case of suction dredges, where large amounts of streambed material are being processed and classification is necessary in order to recover a large percentage of the fine gold (see Figure 5-29).

On the other hand, other dredge recovery system designs are also being used in the field, with excellent gold recovery, utilizing a set of riffles or expanded metal overlaid with a protective classification screen. Fine gold drops through the screen and is trapped by the riffles or expanded metal in slower current underneath.

TOOLS

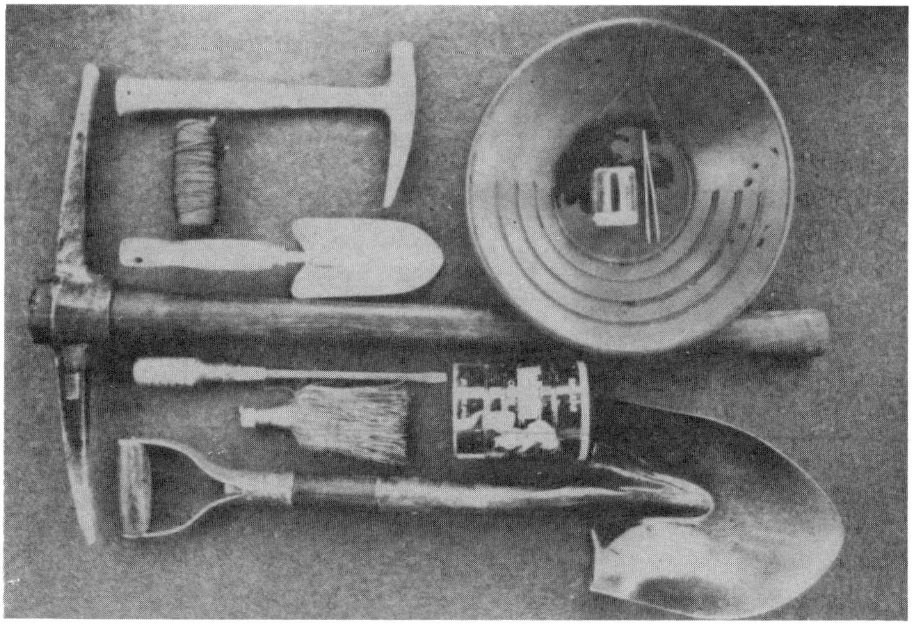

Fig. 5-30. Additional sluicing equipment.

The following tools are useful to a sluicing operation: shovel, pick (comes in very handy for breaking material out of the streambed), rock hammer (to help break up bedrock and clean out crevices), screw driver, garden trowel (for fine work in digging out bedrock irregularities), whisk broom or Mack-Vack (for cleaning bedrock), plastic sheets (for building water barriers), nylon cord (for anchoring sluice in running water), coffee can or bucket (for concentrates), gold pan (for clean-up), tweezers or snifter bottle, and glass or plastic vial ... for the gold.

BUILDING A SIMPLE SLUICE BOX

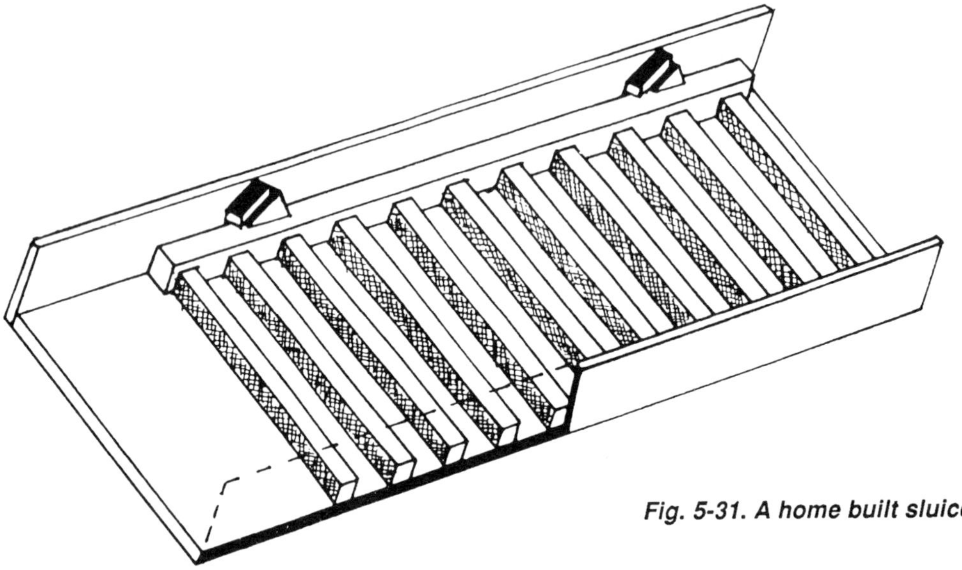

Fig. 5-31. A home built sluice box.

A very simple, but effective, sluice box can be built out of wood which can be used in a small one- or two-man operation with good results.

The sluice is put together as follows: Using marine or a good grade of outdoor plywood, cut the

Fig. 5-32.

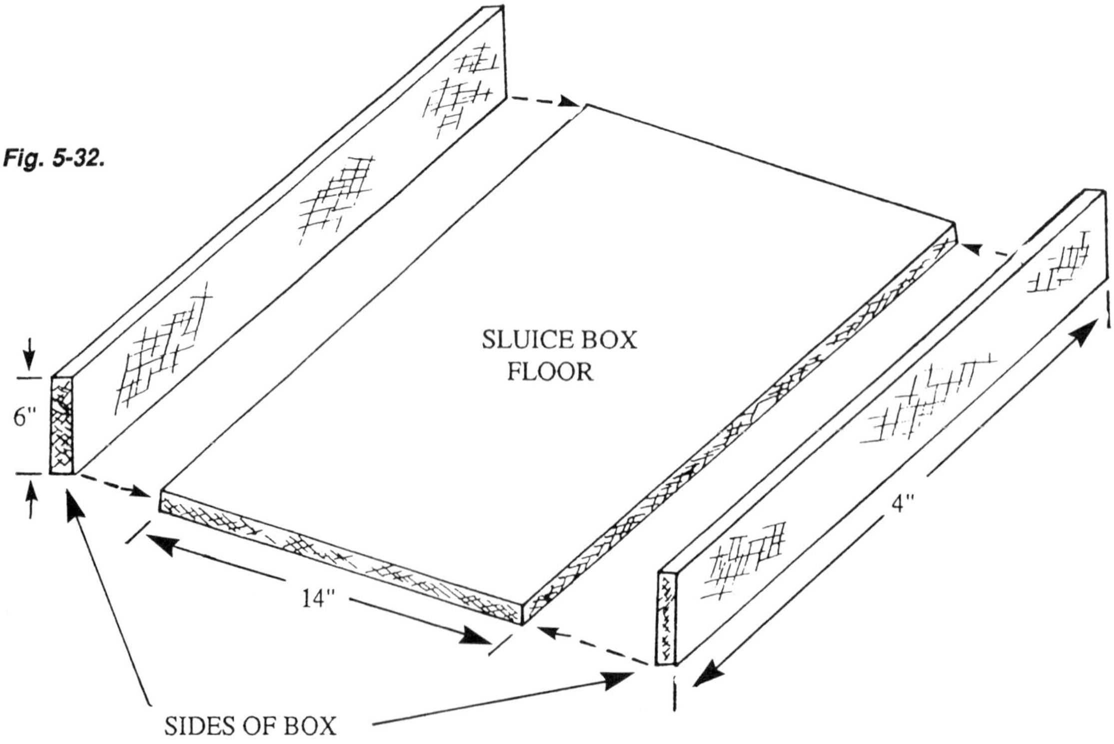

rectangular pieces for the basic floor and walls of the box to the dimensions as laid out in Figure 5-32.

The sideboards should be screwed tightly to the bottom board, making sure that the sideboards are attached to the side edges of the bottom board, as shown in Figure 5-33. Once the sideboards are securely fastened to the bottom side, caulk the seams with silicone rubber. This is to insure that no gold is lost through the cracks. Use just enough silicone to fill the cracks--no more.

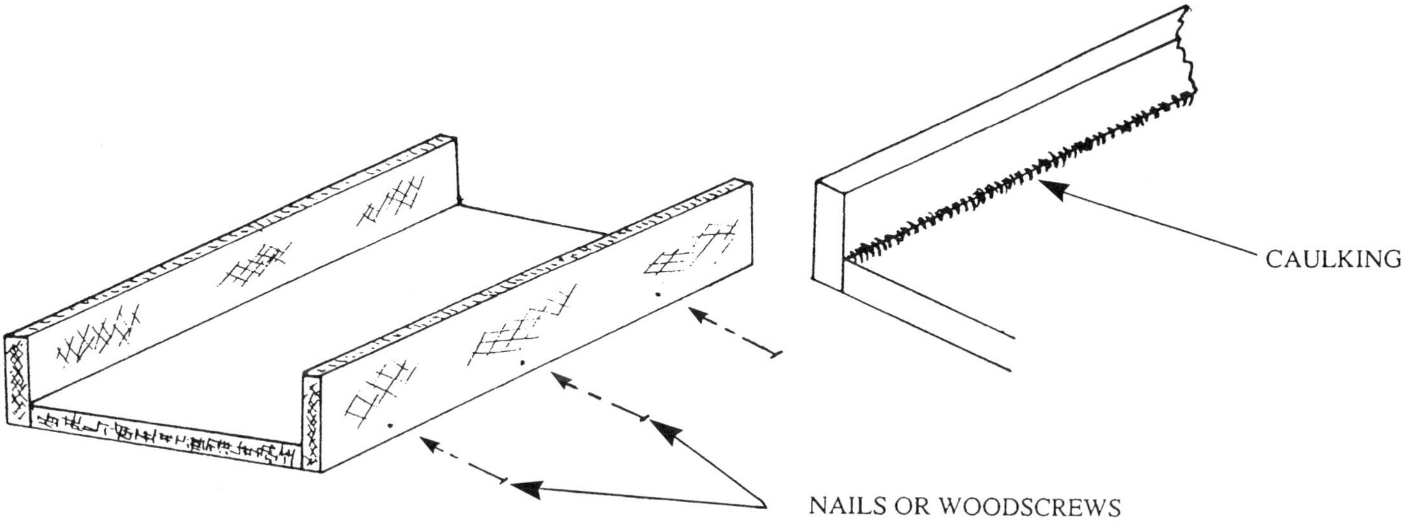

Fig. 5-33.

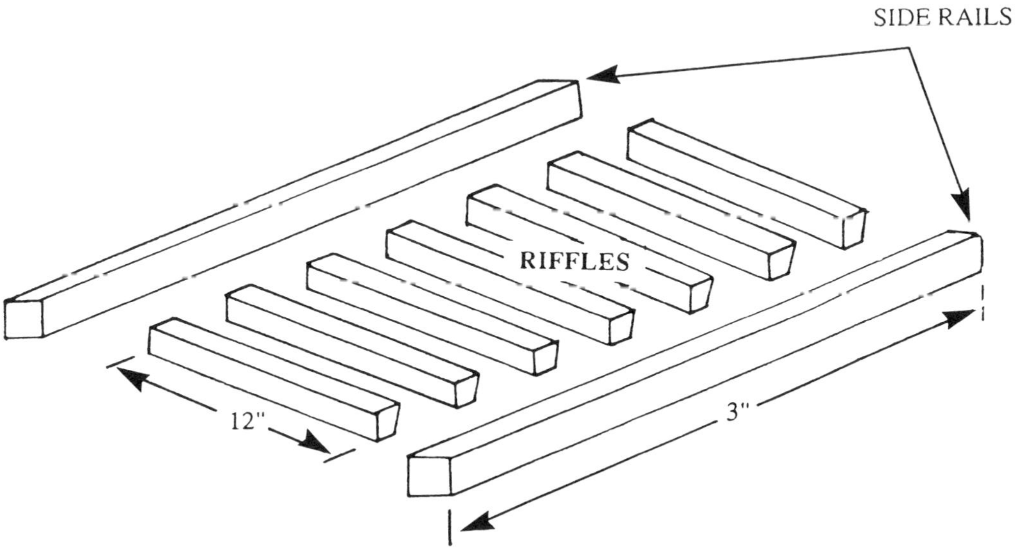

Fig. 5-34. All pieces made of 1 x 1inch oak.

Next comes the riffle system. It will be made of one inch by one inch pieces of oak. The riffles themselves should be cut 12 inches long, and the side rails should be three feet long, as shown in Figure 5-34.

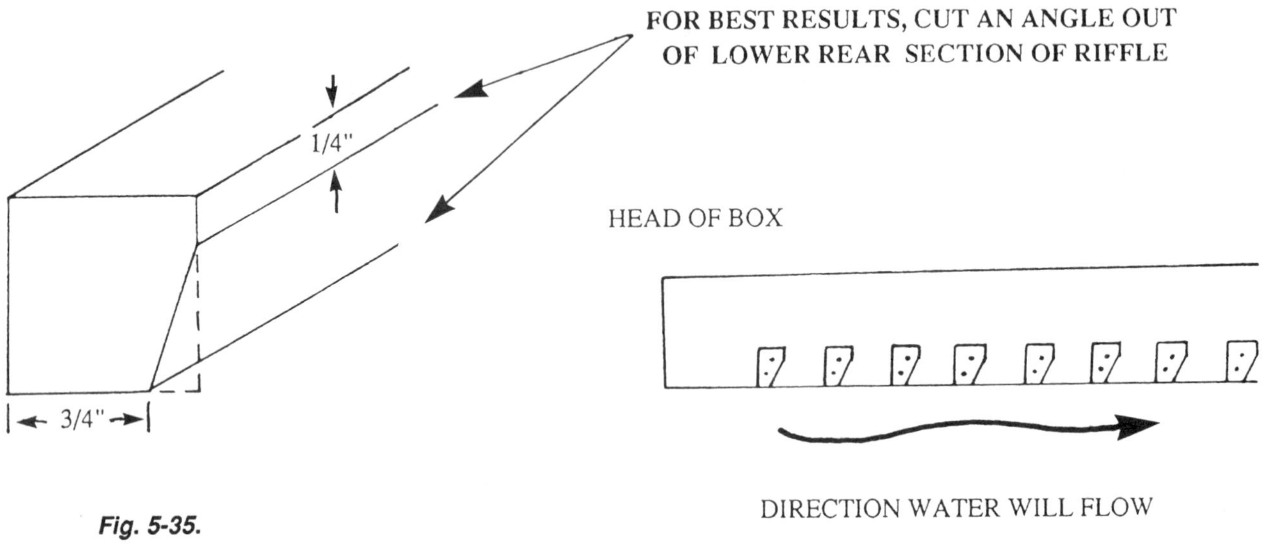

Fig. 5-35.

These riffles will be placed one inch or more apart down the length of the side rails. There will be 18 of them in all. Best results will be obtained from these riffles by cutting a small angle off the lower rear part of each riffle, as shown in Figure 5-35. The riffles will work without this cut, but better results are obtained with it, because it creates a better back eddy behind the riffles, causing more concentrating activity.

Screw each riffle to the side rails with the riffle angles all facing to the rear, along the bottom edge. The riffles should start at the forward edge of the side rails and be placed one inch or further apart, taking care that the bottom edges of all the riffles are even. Two screws on each side of each

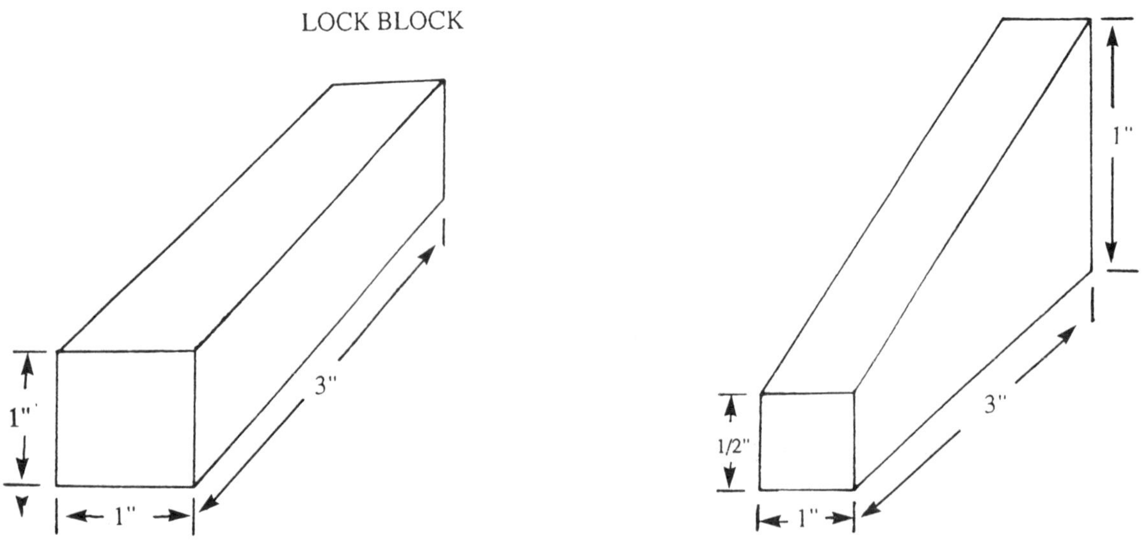

Fig. 5-36.

riffle should be used, as shown in Figure 5-35. Smaller holes should be drilled first to avoid splitting the wood.

Now cut a piece of indoor/outdoor carpet or several pieces of burlap the size of 3 feet by 14 inches. This will be placed in the lower section of the box, with the riffles laid on top.

Next, four riffle locks need to be made, and Figure 5-36 lays out the dimensions for them.

To assemble the riffle locks, lay the matting in place within the box with the riffles on top facing the right direction. Put the wedges on top of the side rails, pointing forward, about six inches from each corner of the riffle system, as shown in Figure 5-37. The lock blocks should be placed over the wedges and bolted to the sides of the sluice box. Be sure to attach the lock blocks up and away from the side rails as far as possible, but not so far that you cannot tap the wedges into place to hold

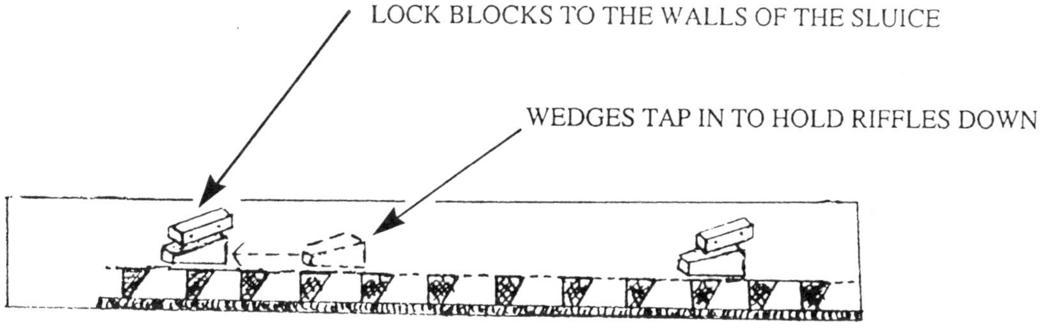

Fig. 5-37. Installing lock blocks.

the riffles securely down. The reason you want the lock blocks up away from the side rails is so there will be plenty of room to slide the riffles out when you want to remove them during cleanup.

Now the box is complete and the riffle system can be locked in tightly by tapping the wedges into place and removed by tapping the wedges out. Notice that the head of the box has no riffles. This is where the material is fed during sluicing.

PLASTIC SLUICES

There is something very positive to be said about the modern plastic sluice boxes coming out on today's market. Le' Trap is putting out an excellent plastic sluice. It is lightweight. It is durable. And, it recovers gold better than the conventional store-bought aluminum models, or the

Fig. 5-38. Le' Trap sluice.

homemade models using any of the riffle designs mentioned above--perhaps with the exception of expanded metal.

The limitation of plastic sluices is that they cannot take a severe pounding. They will also process only a limited volume of material.

One nice thing about a plastic sluice is ease of cleanup. There are no riffles or carpets to remove. You simply pour the concentrates into a tub while splashing a little water over the riffles.

Because the recovery of fine gold is so good in the Le' Trap plastic sluice, many miners using hydraulic concentrators are setting up a Le' Trap sluice to process tailings from the concentrator. They simply set up the plastic sluice directly in line to catch the water and material washing out of the hydraulic concentrator. This way, they are able to test their recovery systems and pick up a little more gold, gold which is finer than the conventional highbankers are able to recover.

Le' Trap sluices are also being broadly used by gold dredgers and more serious placer operations to concentrate the gold from the final cleanups of their recovery systems. The concentrates from an intermediate or large dredging operation, or that of a commercial placer operation, may be as much as 10 or 20 gallons or more in volume (two to five 5-gallon buckets). This amount of concentrate can be further concentrated in a Le' Trap plastic sluice with almost zero gold loss. The result is all or most of your gold, with only a quarter-gallon (a third of a gold pan) volume of remaining concentrates.

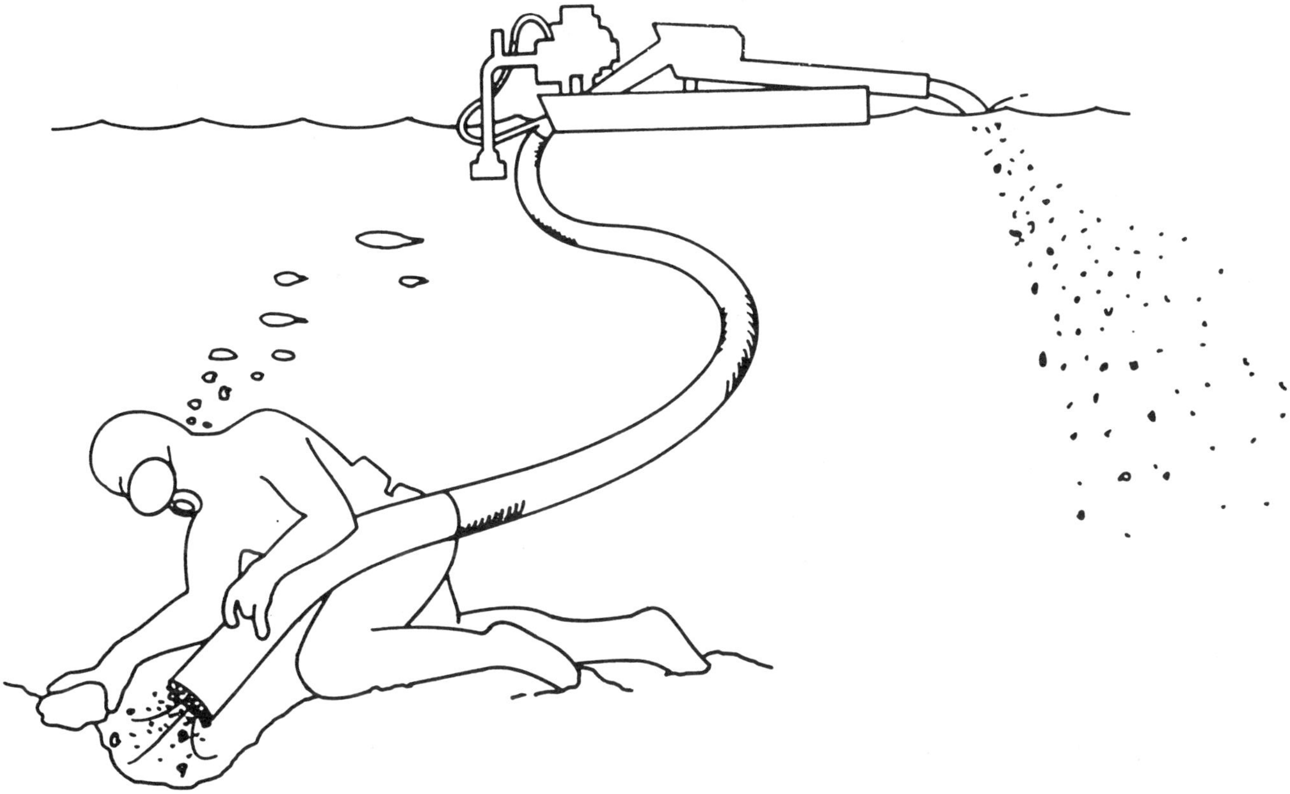

DREDGING FOR GOLD

CHAPTER VI

Suction dredging for gold is a relatively new subject in the field of gold mining, with a large amount of development in equipment and procedure within the last several years.

Because a prospector can process so much more material with the use of a gold dredge than with hand operated sluicing gear, and because of the relatively inexpensive dredging equipment available today, gold dredging is probably the best thing that has ever happened in the field of gold mining for the small-scale operator (one- or two-man operation).

With the use of a gold dredge, under the proper conditions, a single person can process up to 20 cubic yards or more per hour of gold-bearing streambed material, which could be as much or more than 100 times the material he or she could process using pick and shovel methods under similar conditions at the surface.

He or she? Yes, women are operating gold dredges too, and some of them are recovering gold in large quantities!

Gold dredging today has put the small-scale operator in the position of being able to process larger portions of streambed. This puts the person in a position of being able to find and recover large paying quantities of gold. The only type of gold mining activity that can out-produce a suction dredging operation is a professional high-bench operation employing heavy earthmoving equipment. Such operations can cost hundreds of thousands of dollars to get started.

A one or two-man dredging operation costs very little to get started in comparison to the amount of gold that can be recovered in a short period of time, once the operators have gained an understanding of how to locate gold deposits.

Volumes of gold have continued to be washed into the present streams, rivers and creeks. For the most part, underwater gold deposits are, and have been, rather inaccessible to surface mining operations. Therefore, the person with a suction dredge is in a good position to locate and recover substantial placer gold deposits.

It takes far less effort to suction dredge a much greater volume of streambed material than it does when using a pick and shovel. The reason for this is the dredge operator does not need to pound away and lift material out of the streambed. The suction dredge is an underwater vacuum cleaner. Processing the streambed material is just a matter of sucking it up into the hose. The dredge does all the rest. Those larger rocks and boulders that are too big to be sucked up are much easier to move underwater than they are at the surface. The dredge operator does not even have the weight of his own body to move around, because it also is in a state of suspension in the water, the diver wearing only enough additional weight to counteract the bouyancy of his wet suit and remain stable on the bottom.

During recent years, as the market value for gold has increased, so has the interest in gold mining, and many people have turned to suction dredging. As a result, a large competitive market in suction dredging equipment has sprung up. This has been the cause of many, many improvements in the field. It has also brought the cost of equipment way down.

Most dredging is done during the hot summer months, and is done under the cool, clear water of mountain streams. It is thrilling to have the capability in your hands to process so much more streambed material. This increased sampling and production capability gives the operator a fantastic opportunity to visibly uncover more and larger quantities of gold than was ever possible by earlier methods. Gold dredging is more like treasure hunting than anything else. As a result, the gold rush of the 90's is largely due to suction dredges.

I'm not kidding you; once you get good at it, it's like hunting treasure--which you know you are going to find. It's exciting and fun--even during the cold winter months for me.

Don't misunderstand me in thinking that gold dredging is not hard work. There's plenty of hard work involved. Generally speaking, any gold mining operation will produce only as well as the operator is willing to put his or her energy into it. It's just that so much more can be done with a suction dredge. The same amount of energy produces a lot more in a dredging operation. So the guy or gal who is really willing to put out in an effort to succeed well at gold dredging, usually does! And that's one of the things that makes suction dredging so popular.

WHAT A GOLD DREDGE IS

Suction gold dredges come in a wide range of size and variety, but all of them basically break down into five major components. These are the flotation system, engine/pump assembly, jet/hose system, recovery system, and the air breathing system.

FLOTATION

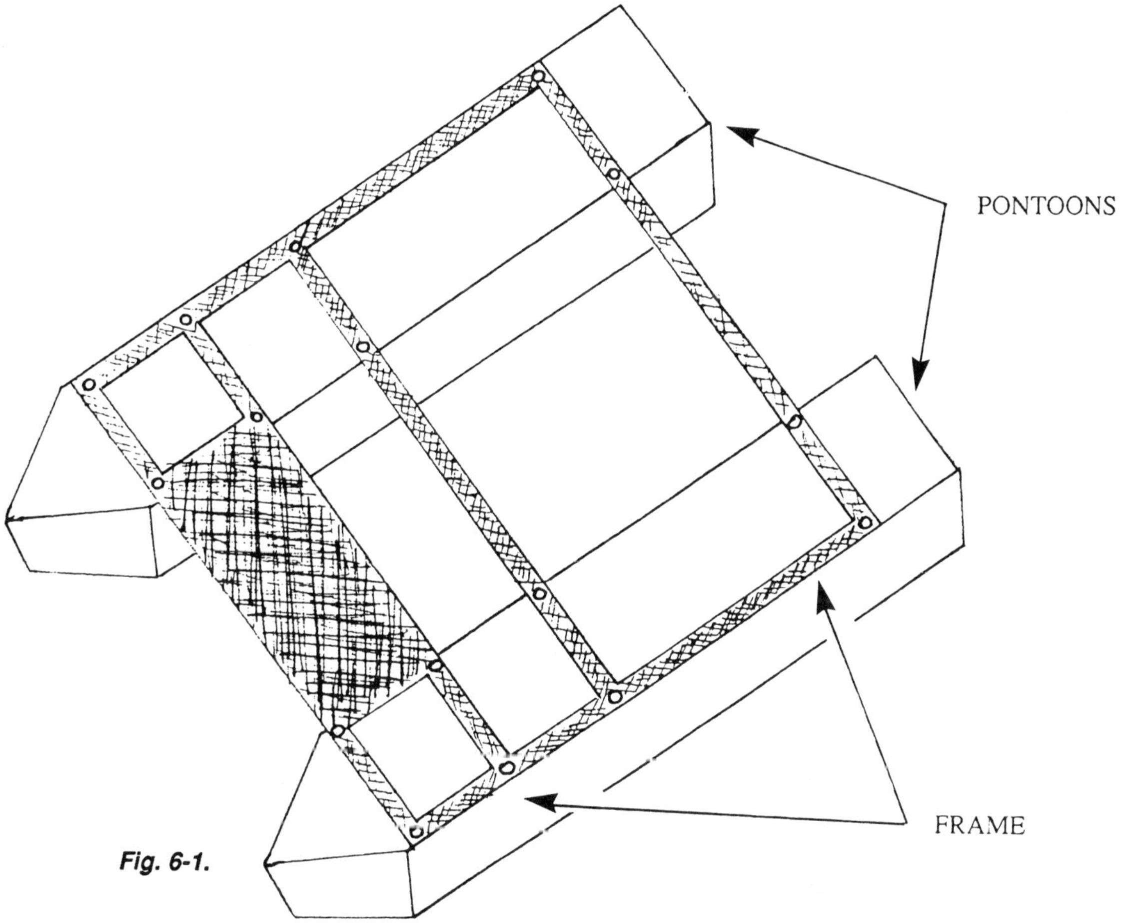

Fig. 6-1.

Today's dredges, meaning the entire units, are usually floated either on a set of inner tubes, or on a set of pontoons. The larger the dredge, the more flotation is needed. Inner tubes are used on some of the smaller and intermediate size dredges, and have the advantage of deflating. They can therefore be packed away and hauled around more easily than the larger pontoons can.

Pontoons today are usually made of lightweight Styrofoam, aluminum, durable plastics, or some combination of these. Some homemade rigs use empty steel or plastic drums.

Plastic pontoons have several advantages. They are durable; they won't deflate on you at the wrong time and allow your dredge to sink. Some pontoon designs float a dredge very well--even in swift water conditions. Innertubes generally don't handle fast water very well!

When shopping around for a dredge, look over the flotation system to ensure that the floats can be securely fastened to the dredge's frame, and that they are wide enough to give the dredge good stability on the water.

PUMP/ENGINE

Portable gasoline engines are being used to power most suction dredges found on the market today. These are the same kind of engines found on lawn mowers and industrial equipment. Portable gasoline engines are light and easy to pack around, and take very little maintenance to keep operating.

Some of the larger dredges use car engines to get the additional needed power. These are popular on large professional production dredges.

Some dredge manufacturers use more than one small engine to create the power needed for their larger dredges.

Most often, to supply the necessary water flow, a centrifugal pump is used, which bolts directly onto the engine (see Figure 6-2).

Fig. 6-2. Engine/pump assembly

Different sized dredges have different sized engine/pump assemblies. Larger amounts of water and pressure are needed to power a larger sized dredge.

Fig. 6-3. Pumps which are mounted above the water's surface suck water up to themselves through an intake tube or hose.

Centrifugal pumps are most often mounted on the engine in an above water location. An intake hose is used to get water to the pump, as shown in Figure 6-3.

Centrifugal pumps which have been mounted above water need to be primed before they are operated. This is because a centrifugal pump is made to pump liquid and so it must be filled with liquid in order to start pumping. Any air pocket in the pump or substantial vacuum leak on the intake side stops the pump from working properly. Once primed and started, the pump will suck water to itself and no further service will be needed to keep it running properly.

On most of today's gold dredges, priming is being done in three different ways. Some of the larger professional rigs have primers which use the vacuum off the engine's intake manifold. Priming the pump is a matter of starting the engine and turning a valve or two to vacuum water up the intake hose into the pump. Some intermediate-sized dredges have a combination intake tube and primer. These have a screw cap at the top. The cap is removed and water is poured into the primer from the surface in order to prime the pump before starting (see Figure 6-3). On these, you have to remember to put the cap on tightly before the engine is started. Otherwise the prime will be lost.

Many of the smaller gold dredges use the intake hose and foot valve as a primer. Such hoses are equipped with a screening device at the lower end to prevent pebbles and rocks from being sucked up into the pump. They also have a one-way flow valve *("foot valve")* so water can enter the intake hose but is prevented from running back out again. To prime such a hose, you take it in your

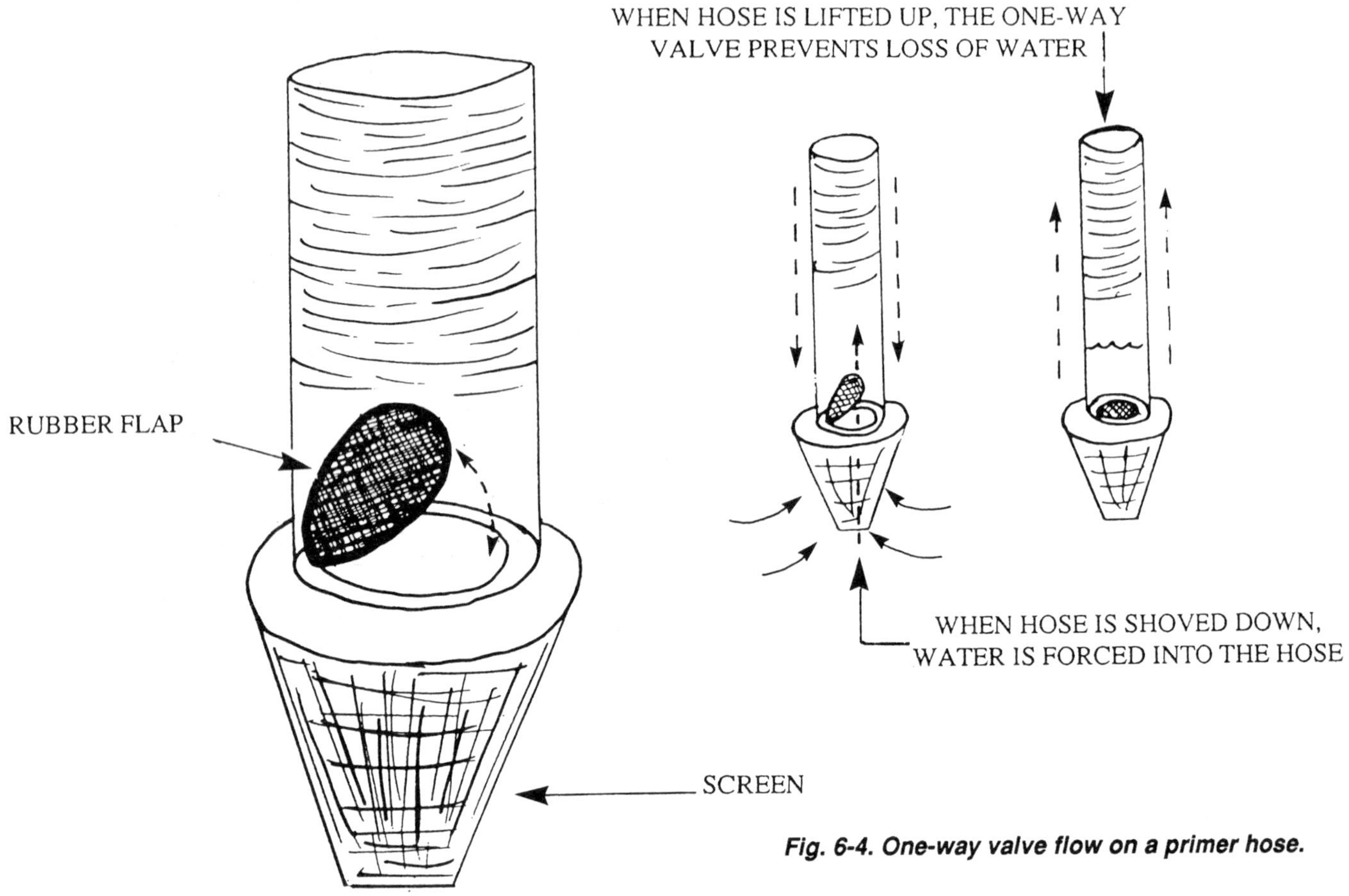

Fig. 6-4. One-way valve flow on a primer hose.

hand and quickly push it up and down under the water. This action forces water up into the intake hose. The one-way valve prevents the water from flowing back out again. After a bit of pumping in this fashion, you are able to get enough water up through the hose to prime the pump. This procedure is sometimes done while the pump is running at low speed or sometimes before starting the engine, depending on the model (see Figure 6-4).

Some of the smaller dredges on the market use vertical shaft engines. The pumps are mounted on the underside of the dredge beneath the water. No priming is necessary with this design.

Some home-built dredges use jet-boat pumps, which mount underwater too and need no priming.

Centrifugal pumps require little or no maintenance. One thing to remember is they should not be operated for an extended period of time or at high speed without water being run through them. The reason for this is that these pumps use a rubber seal which fits snugly on the engine shaft to keep water from leaking out around where the shaft enters the pump. This seal depends on the water passing through the pump to keep it lubricated and cool. If the pump is operated for a while without water passing through, the seal can become too hot and thereafter lose its efficiency in preventing water from leaking around the shaft. The way to avoid this is to never run the pump for more than about a minute or so at a time without water. You can tell if water is being pumped by looking to see if it is running into the dredge's sluice box.

If it does happen that the pump's seal on your dredge becomes defective, it is not a difficult job to replace it. An extra seal is good to have on hand in case this should occur out in the field. It is done by unbolting and removing the outer pump housing and then the impeller. Very few impellers on today's centrifugal pumps are pressed on. Most of them are screwed onto the shaft. Usually it's necessary to get a firm grip on the engine's flywheel and hold it steady, so that the impeller can be unscrewed off the shaft. Once the impeller is removed, the pump's inner housing can be unbolted and removed from the engine. The seal can then be replaced on the inner housing. The pump goes back together in the opposite way it came apart.

The engines being used on most of today's dredges require very little maintenance. Usually they do not come with oil in the crankcase, so they will need to be filled to the proper level first, before being started. The oil level should be checked periodically, and a log of operating hours should be kept so the oil, spark plug, and breaker points can be changed according to the manufacturer's instructions. Air and gas filters should also be checked and cleaned every once in a while.

When the pump is primed and the engine is started, water will be pumped through a pressure hose to the jet.

JET SYSTEM

The earliest suction dredges pumped water, mud, sand, clay, rocks and gravel--all through the pump itself. Needless to say, that must have been extremely hard on pumps! I have also seen this method widely used in South American dredging operations.

Most of today's suction dredges use a *''venturi system,''* in which a volume of water under pressure is pumped into a steel tube at an angle. As shown in Figure 6-5, this causes a vacuum effect (venturi effect) just behind where the water enters the main tube (jet), which pushes/pulls the water, sand and gravel up through the suction hose and main jet. In this way, no streambed material needs to be run through the pump.

The jet system is probably the most important component of a suction dredge. The idea is to get the highest pressure and greatest volume that is attainable out of the pump, in order to get the most suction power at the nozzle. This is done by reducing the size of the jet eductor(s) just before it enters the main tube of the jet. The highest volume--highest pressure mixture is difficult to get right without trial and error on the part of the dredge builder. This is the main reason most dredgers buy their equipment instead of building their own. Different pumps have different pressure and volume capacities, which require different sized power jet eductor reductions.

These changes also affect how wide the sluice will need to be in order to accommodate the amount of water being pumped through the box--which calls for additional trial and error on the part of the dredge builder. A wider box, because of more water, may require a stronger frame and more flotation.

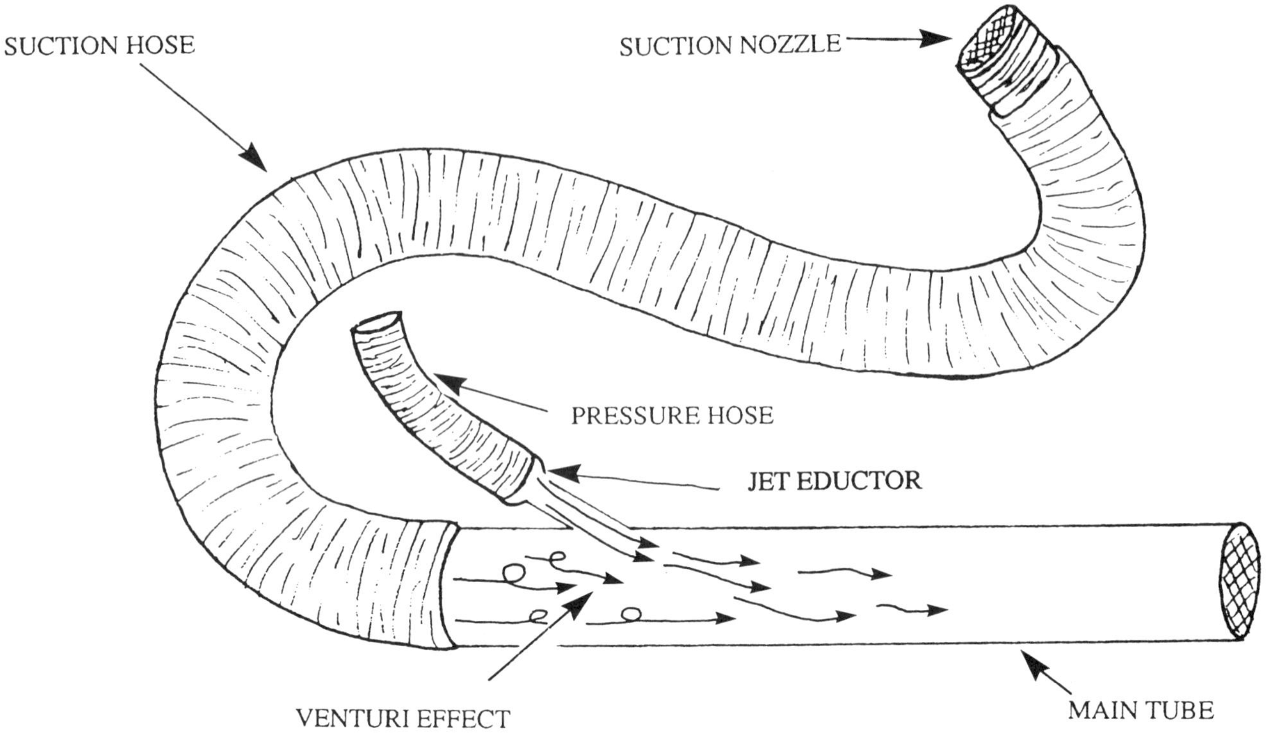

Fig. 6-5. Venturi effect creates a suction through the jet and suction hose.

It is possible to build your own dredge, and it's really not too difficult, if you are naturally inclined in that area. However, I assure you that it would be hard to build a better dredge than some of the improved models which are being offered on today's market, and almost impossible to do it at less cost than it would take to buy one. Many miners have built their own dredges, and some of them are excellent machines, mine included. Also, many of them are complete failures, my first ones included. If you are naturally inclined toward building your own dredge and choose to do so from scratch, speaking from experience, I recommend that you put aside some extra money and prepare to spend at least a portion of a dredging season to work out the bugs to get the entire dredge operating to YOUR satisfaction.

There are several different kinds of jets being used on the dredges found out in the field today which you should be familiar with.

Nozzle Jet: The nozzle jet was the first type of jet to be used on suction gold dredges (see Figure 6-6).

In this case, water is pumped through a pressure hose down to the suction nozzle itself where it is directed into the suction hose to push the water and material through. The nozzle jet was a great improvement over pumping material directly through the pump. This kind of jet system is most often seen on smaller-sized suction dredging equipment, which is designed to operate in shallow water. One of the main advantages to this kind of system is the suction nozzle can be raised out of the water and put back under again without the worry of filling the entire suction hose up with air. You

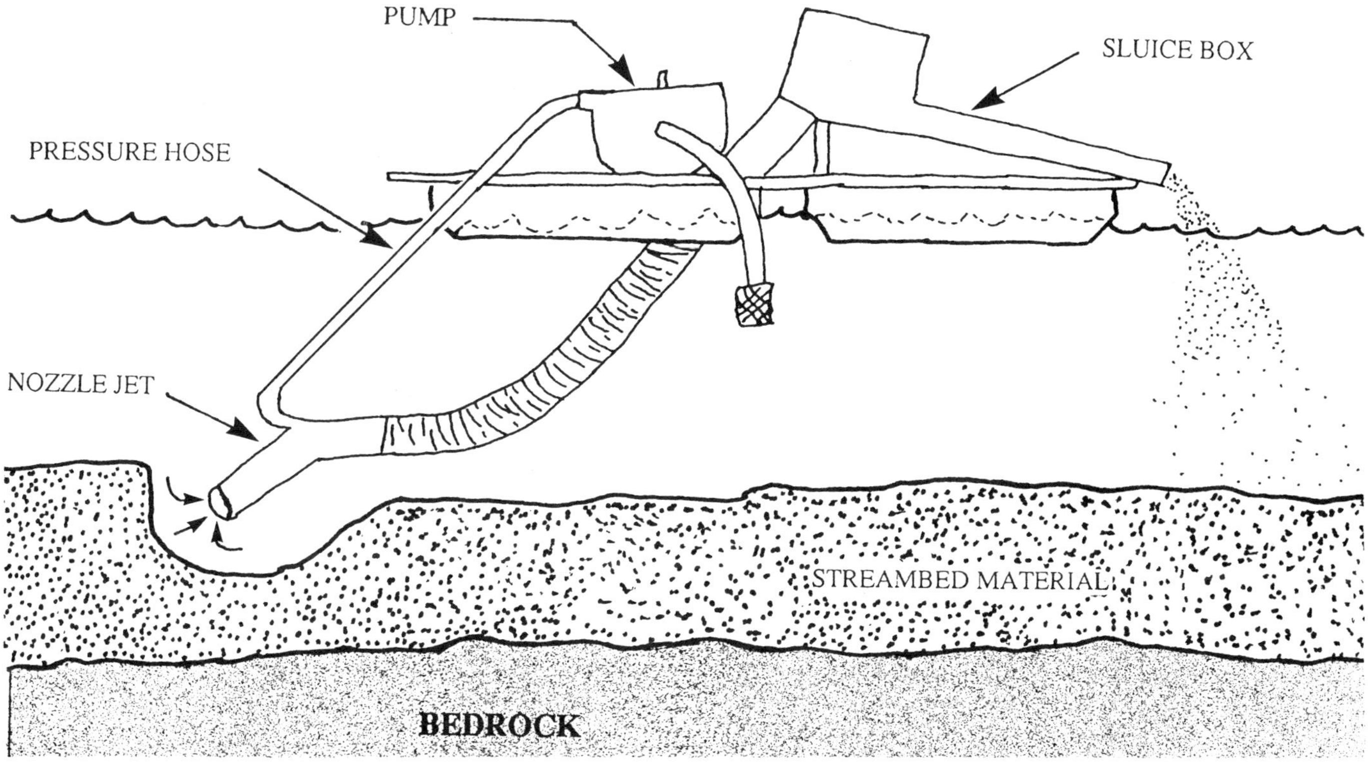

Fig. 6-6. Nozzle jet type gold dredge.

will not find this system on many intermediate or large-sized dredges because of the inconvenience and bulkiness of manipulating two hoses around while underwater.

Power Jet: The power jet, as shown earlier, utilizes a volume of water being directed into a main jet tube under high pressure to cause a suction up through the suction hose (see Figure 6-7).

This kind of jet system is most often used on the intermediate and large sized production dredges.

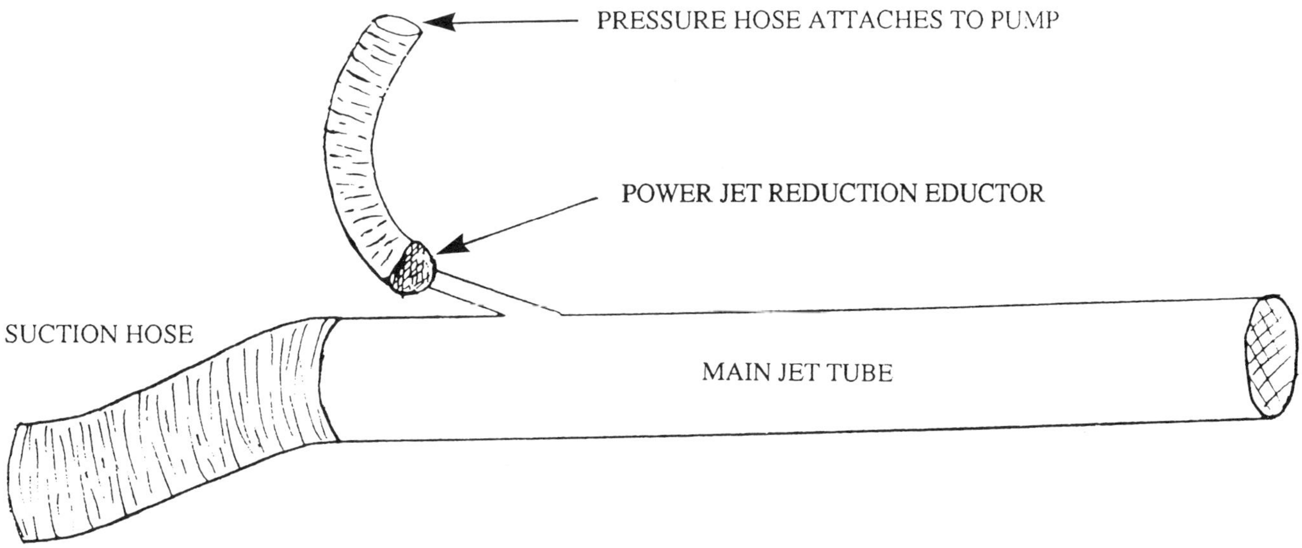

Fig. 6-7. Power jet.

One advantage to this kind of system is that the operator only has one hose to manipulate around while working underwater, as opposed to the two hoses which are involved with a nozzle jet system.

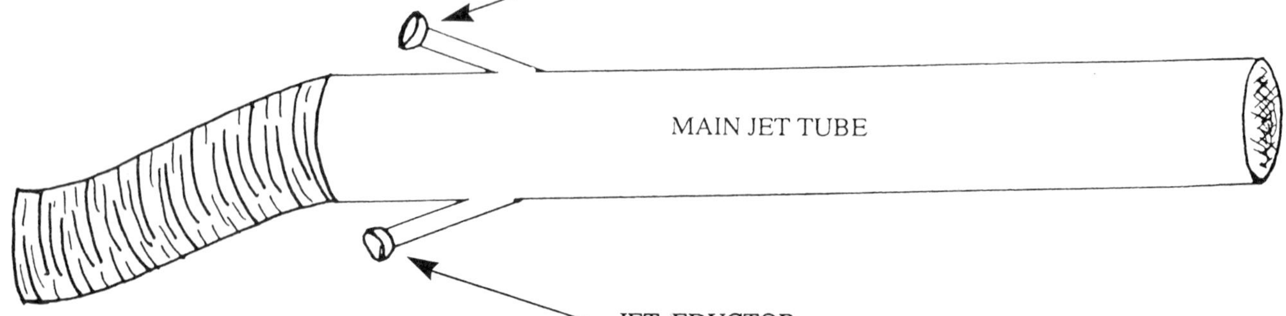

Fig. 6-8. Some power jet systems utilize more than one jet eductor.

There are power jets which use more than one jet eductor, some using as many as four or five of them to cause a more orderly and effective mixture in the main jet tube. This is especially true of dredges which employ more than one engine/pump assembly (see Figure 6-8).

Couple Jet: In theory, the best type of jet would be one in which the eductor was positioned directly in the center of the main jet tube. This would allow optimum mixing, with the minimum amount of friction loss. However, this type of jet cannot be used on a dredge system, because it would obstruct the flow of rocks and material through the main jet.

There is another type of jet being used by some dredgers today called the *''couple jet.''* In this design, a volume of water is forced into the main jet tube from the outside surface of the tube (see Figure 6-9).

This kind of jet is usually constructed to allow the jet tube to slide further in or out of the outer housing, allowing adjustment to the optimum pressure/volume setting for the pump and engine assembly being used.

Sometimes, after a couple jet has been used for a while, a few rocks or pebbles can become jammed in the space between the jet tube and the outer housing. These, when accumulated, can cause a power loss to the suction nozzle. To remove them, it is necessary to loosen up the jet tube and slide it back just a bit, and run water through the jet. Then the tube can be slid back to its normal

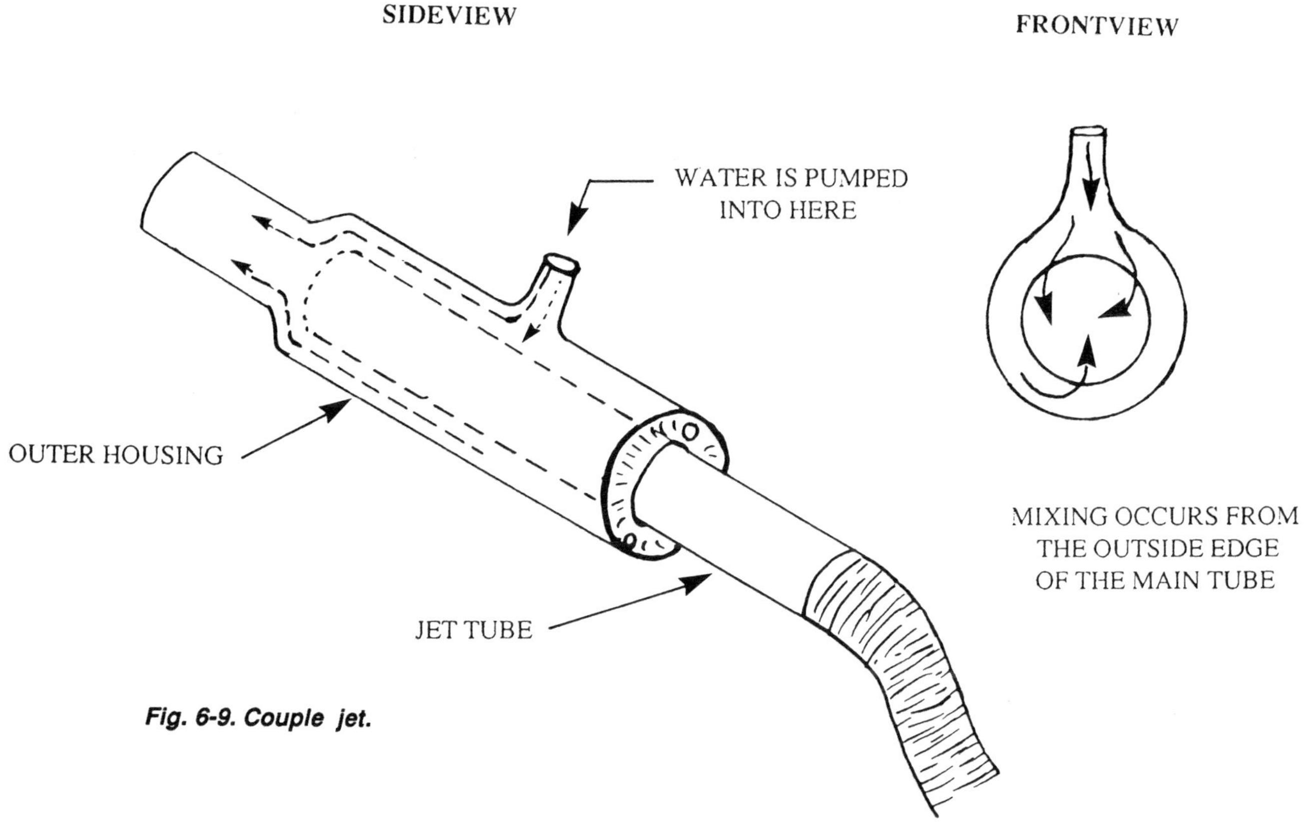

Fig. 6-9. Couple jet.

position and the pebbles should be gone.

Some jets are designed so the suction hose slides onto the jet and clamps on tightly. A nozzle clamps onto the other end of the suction hose. This is the nozzle which the operator moves around to suck material up into the suction hose. A more recent innovation is the *oversized jet.* This is where a power jet is made from steel pipe with an inside diameter the same size or slightly larger than the suction hose. This sharply reduces the occurrence of rock plug-ups inside the jet during dredge operation. Oversized jets commonly use a hinged exterior quick connect hose clamp to attach the suction hose--which is more convenient than the conventional slide on and clamp method outlined above.

Suction hose comes in different sizes. For example, six-inch suction hose has an inside diameter of six inches. Four-inch suction hose will have an inside diameter of four inches, and so on. Suction nozzles almost always have an inside diameter which is slightly reduced in size. This prevents rocks of the same size as the inside diameter of the suction hose from being sucked up and jamming inside the hose. The idea is to slightly reduce the size of the material that passes in to allow a little leeway inside the suction hose. In this way, there are far less plug-ups, and more production.

Review: The engine powers a pump, which forces a volume of water under pressure into the jet. This creates a venturi effect, causing the water and material to be sucked up by the nozzle, through the suction hose and jet, to be poured into a sluice box.

RECOVERY SYSTEMS

The majority of dredges today use a sluice box to recover gold out of the material. Most of the basics concerning sluice boxes were covered in the last chapter, all of which apply to the sluice boxes found on dredges. Still, there are a few additional points that should be mentioned.

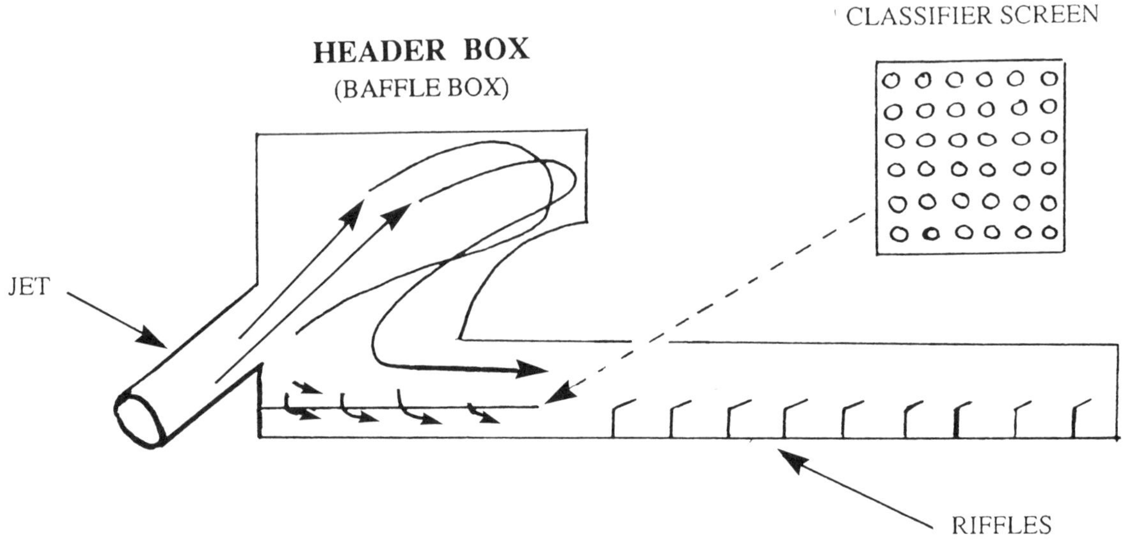

Fig. 6-10. Many dredges have a baffle box to slow down material and spread it evenly over the box.

First, some dredge designs have a *''baffle box''* (also called *''header box''*) at the head of the sluice, which is designed to break up the material further as it enters the box, and also to slow the water and material down and spread it out so it can flow evenly down the length of the sluice box

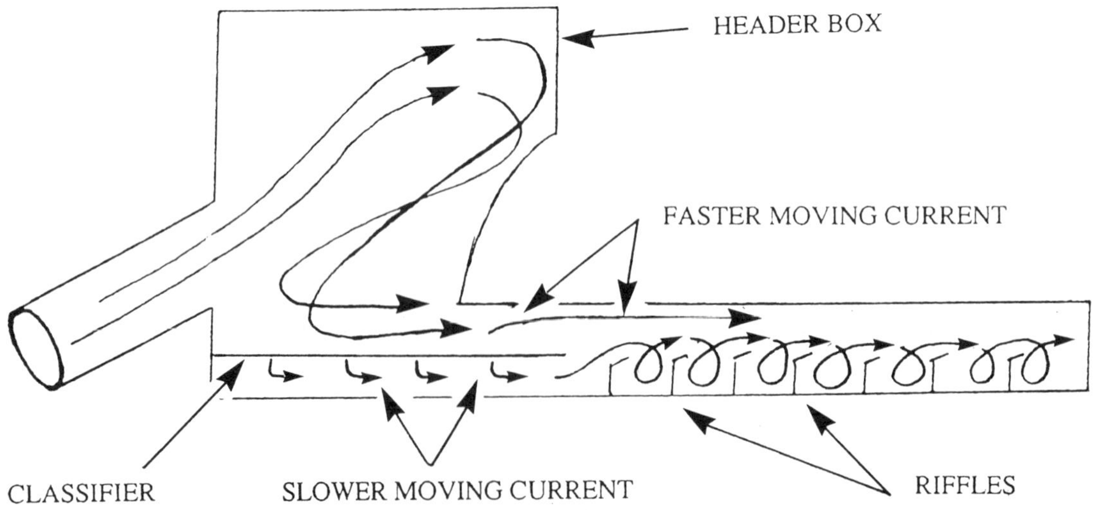

Fig. 6-11. Most of the gold should drop down through the classifier into the slower moving current, to become trapped behind the first few riffles.

(see Figure 6-10).

Sluice boxes on some gold dredges also have a classifier screen located at the head of the baffle box, as shown in Figure 6-10. This screen is usually made out of steel punch plate, and is designed to allow the smaller pieces of gold to drop down through the holes and become trapped under the classifier, or be pushed along by the slower current flowing underneath the punch plate. This makes it easier for the gold to become trapped behind the first few riffles, as shown in Figure 6-11.

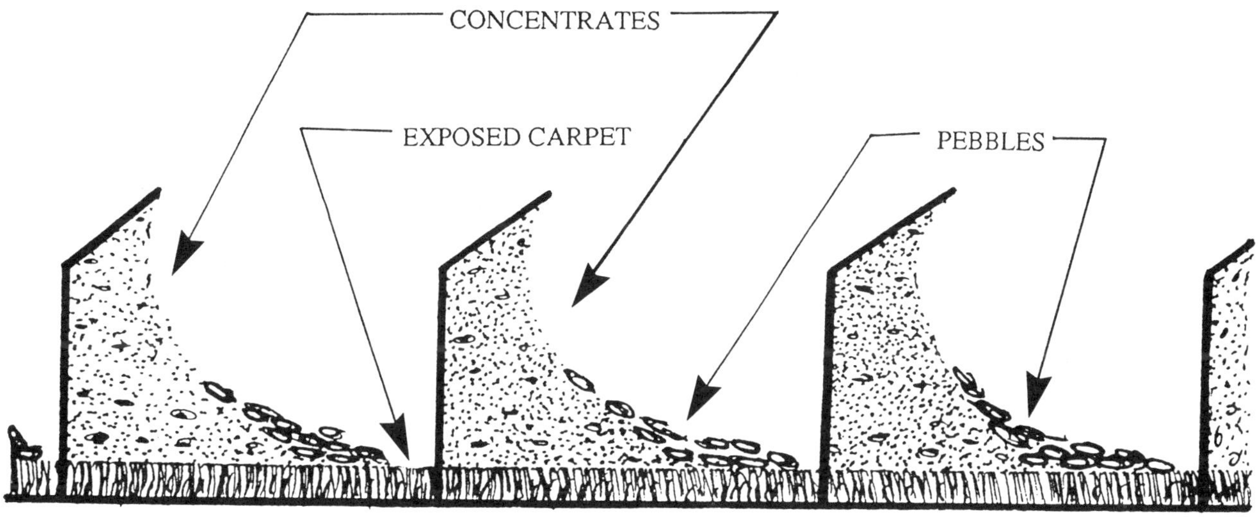

Fig. 6-12. The correct velocity over a dredge's sluice box is obtained when a small amount of carpet is exposed in front of each riffle after running.

At operating speed, a gold dredge has a steady flow of water moving through its sluice box, so water velocity adjustment is normally done by changing the slope of the sluice box. Downward slope determines how fast the water will travel over the box, which affects how much concentrating action will be created behind the riffles.

Hungarian-type riffles are almost always used in the store-bought gold dredges. With a suction dredge which is using either Hungarian or right-angle riffles, the proper slope of the box is obtained when a small amount of carpet is showing just in front of each riffle in the box (see Figure 6-12). To test this, run a good amount of material through the dredge at operating speed. Then allow the dredge to run at the same speed with no obstruction over the suction nozzle for about 20 seconds or so without pumping any material. Then shut down. If no carpet is showing in front of each riffle in the sluice box, it will be necessary to increase its slope slightly. If more than just a little bit of carpet is showing in front of each riffle, then you should lessen the slope of the box slightly. Always set the proper slope of your sluice box to catch properly when the dredge is running at production speed. However, the water velocity must be enough to keep rocks and material from loading up in the sluice box.

When a sluice box on a gold dredge is recovering properly, the majority of gold will be found behind the first three or four riffles in the box; and in the final analysis, this is the way to tell if the correct adjustments have been made. And, of course, it's always a good idea to test your tailings with a gold pan to make sure. You can also use pieces of lead to test your recovery.

Once the sluice box is properly set, it does everything for you automatically. It's just a matter of going down to the bottom and sucking material up into the nozzle. It doesn't really matter from how deep; except that when you are underwater you will need air breathing apparatus.

AIR SYSTEMS

The earliest gold dredge operators used scuba equipment to remain underwater for extended periods of time. In fact, it was the evolution of scuba equipment which brought on the development of suction dredges.

Today, scuba equipment is not used very often in dredging activities. The *''hooka air system''* on a dredge consists of a low pressure air compressor, which is powered by the same engine that powers the dredge. While the engine is operating, low pressure air is pumped through an extended air line to a regulator, which the diver breathes from while dredging.

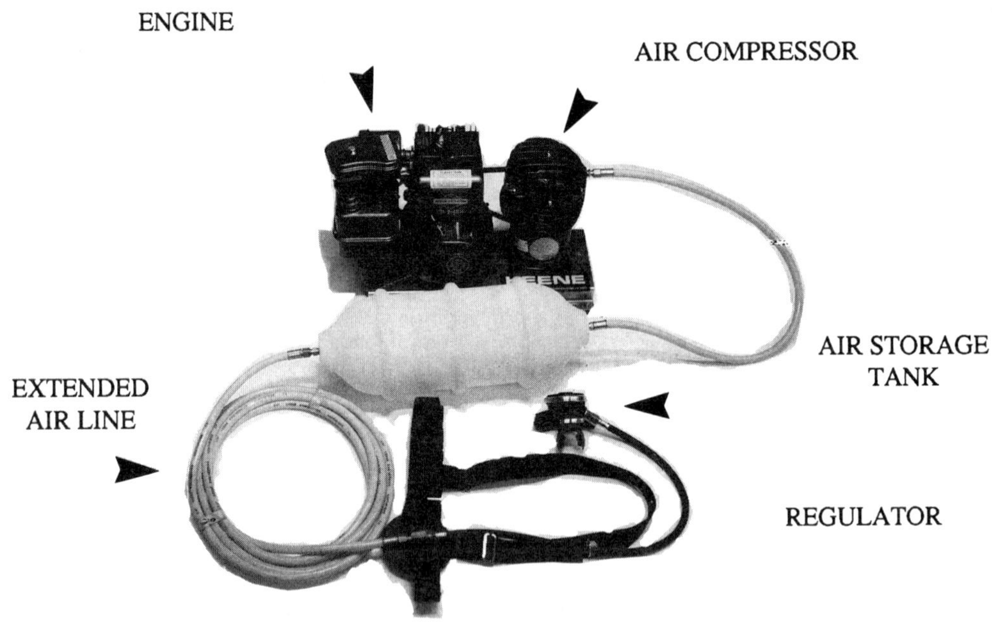

Fig. 6-13. Air breathing hooka system for a suction dredge.

There are various sized compressors available for the different sized dredges offered on the market. Larger compressors are able to pump more volume of air and more air pressure when it is needed, in case more than one diver is operating the dredge, or when work is being done at greater depths. Some of the larger compressors will pump enough air to supply up to three or four divers at depths to 30 feet.

Most of these compressors are constructed with sealed-type bearings, so that no oil is needed for lubrication. In this way, the larger scrubbing air filters used on the high pressure scuba compressors are not required to filter out oil vapors.

These hooka air systems also usually contain a small air storage tank, which is designed to store a volume of reserve air for the diver. So if the dredge should run out of gas and quit, or stop running for some other reason while the diver is underwater, he or she will have a small supply of air to breathe enroute to the surface (see Figure 6-13).

As a safety feature, on most hooka air systems, a one-way flow *"check valve"* will be attached to the airline just before it connects to the regulator. The check valve basically consists of a ball and spring seal, which forces a ball bearing against a seating inside of the valve, as shown in Figure 6-14. The check valve is positioned on the air line to prevent air from being sucked back up the line

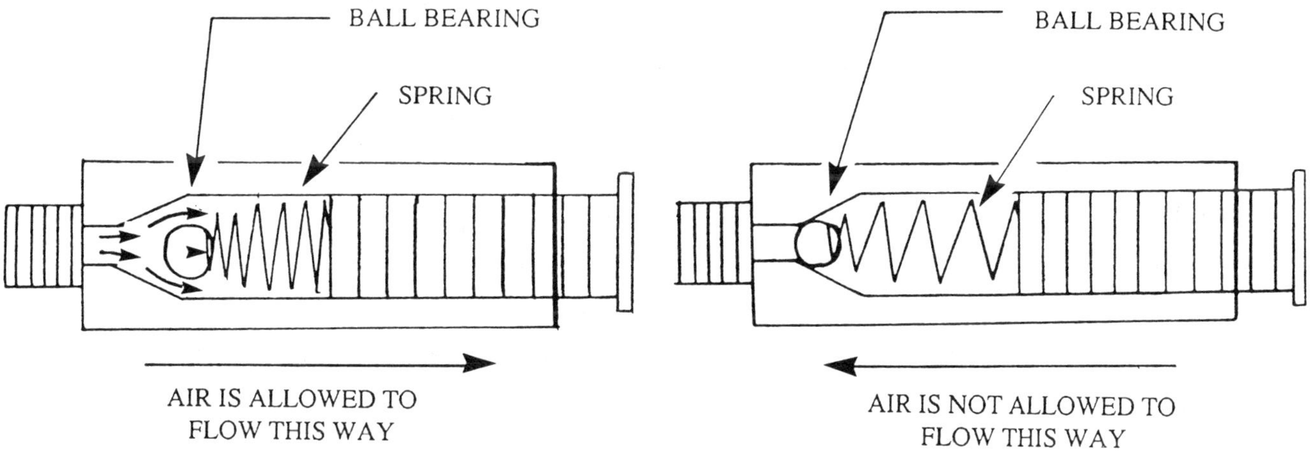

Fig. 6-14. One way flow "check valve."

from the diver. In this way, if the air line happens to sever or burst at the surface, the pressure differential between the depth at which the diver is dredging, and the surface, is not permitted to suck air back up through the hose from the diver. It is a good safety feature and will be found on most dredges on the market.

The regulators used on these hooka systems are the same components which are used as the final stages in scuba apparatus. However, they are modified to breathe easily on low pressure air. Regulators normally used for scuba purposes will generally not work very well on a hooka air system until modified for low pressure air. The regulator fits into your mouth and you breathe in air just as you would at the surface. You also exhale into the regulator, which has its own one-way exhaust ports to get rid of the air you exhale into it.

The regulator can be taken out of your mouth while underwater and then replaced, but will have a small amount of water inside of it. This can be blown out through the exhaust ports by exhaling into the mouthpiece, or by pushing the *"purge button"* located on the front face of the regulator, as shown in Figure 6-15. When the purge button on the regulator is pushed, it will cause an automatic flow of air into the regulator, which will force any water there out of the exhaust ports.

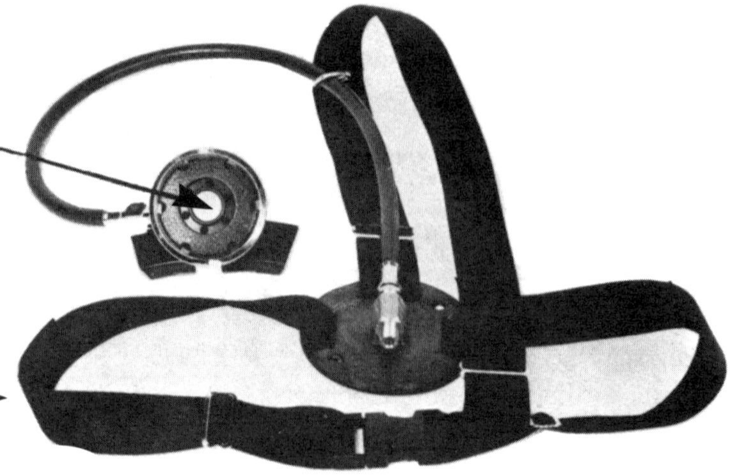

Fig. 6-15. Purge button (in this case is white) is located on the center of the regulator's front face plate.

The hooka system on a gold dredge requires very little maintenance. The small air filters should be replaced or cleaned in soap and water every once in a while, depending on what kind they are. Care should be taken not to spill gasoline onto the compressor's air filter(s) when the dredge is being refueled. Sometimes the air filter is located just below the gas tank, and it's necessary to place a bucket or something over the top of the filter to prevent spillage from getting onto it. Needless to say, the resulting gasoline fumes, being pumped down to the diver through a hooka air system, can make for a very uncomfortable dive. If gasoline is spilled on an air filter by mistake, it should be removed and thoroughly cleaned before the next dive is begun.

SETTING UP A GOLD DREDGE

Carry your dredge to the location where you will enter the water, near the location that you intend to dredge or sample, if possible, and assemble it. Be sure that all of the clamps are securely fastened, that the engine has the proper amount of oil, and the gas tank is filled. Float the dredge to the site you want to work and tie it off according to the water currents in play at that location.

If there is more than just a mild current flowing down river, it will probably be necessary to tie the dredge off from a position upstream to keep it from floating down river with the current. If there is little or no water flowing in that location, you will also probably need to tie the dredge off from the rear to keep it from motor-boating forward on you from water flowing out of the sluice box.

Once the dredge is tied off, ensure that the suction hose is full of water and that the suction nozzle is under the water.

Prime the pump and start the engine. Allow the engine to run at low speed and warm up while you put your diving gear on--weights, hood, mask, etc.

Once you are ready for the dive, accelerate the engine to operating speed and start vacuuming up the riverbed materials, gold and all.

DREDGE SIZES

The size of a suction dredge is labeled by manufacturers and just about all miners according to the inside diameter size of its suction hose. Dredges come in a wide range of sizes. Some government agencies, such as the California Department of Fish and Game, for the purpose of regulation, label the size of a suction dredge according to the inside diameter of the suction nozzle, which is always smaller than the suction hose.

A larger suction hose requires more power to make it suck properly than does a smaller suction hose. This means a larger pump/engine assembly is needed, and that more water will be pumped. So a larger dredge will also require a larger sluice box to handle the additional flow of water and material, and an additional amount of flotation will also be needed to float the entire works and keep it stable above water.

A larger dredge will pump more streambed material and also will move larger rocks--which means that less rocks will need to be moved out of the way hand, and so much more production can occur. However, the larger a suction hose diameter is, the thicker and heavier it is, and so it becomes more stiff and unwieldy, and takes more energy to move around.

Fig. 6-16. 2-inch Backpack dredge.

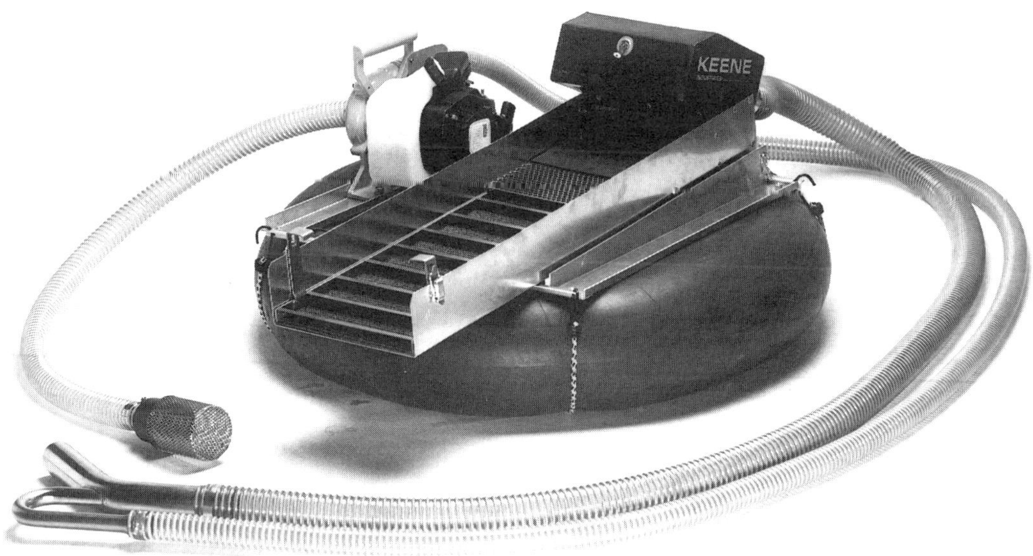

Additionally, the larger a dredge is, the more energy it takes to move it from place to place. For this reason, larger gold dredges are mainly used for production dredging. Intermediate sized dredges are mostly used for sampling and recreation, because they can move a healthy volume of material but are not too difficult to pack around. And small dredges are usually best suited for the smaller streams, and the remote locations where accessibility is a problem.

One of the smallest dredges found on today's market is the "2-inch Backpack dredge" (see Figure 6-16). This model is powered by a two-cycle two horsepower engine and pump, with the

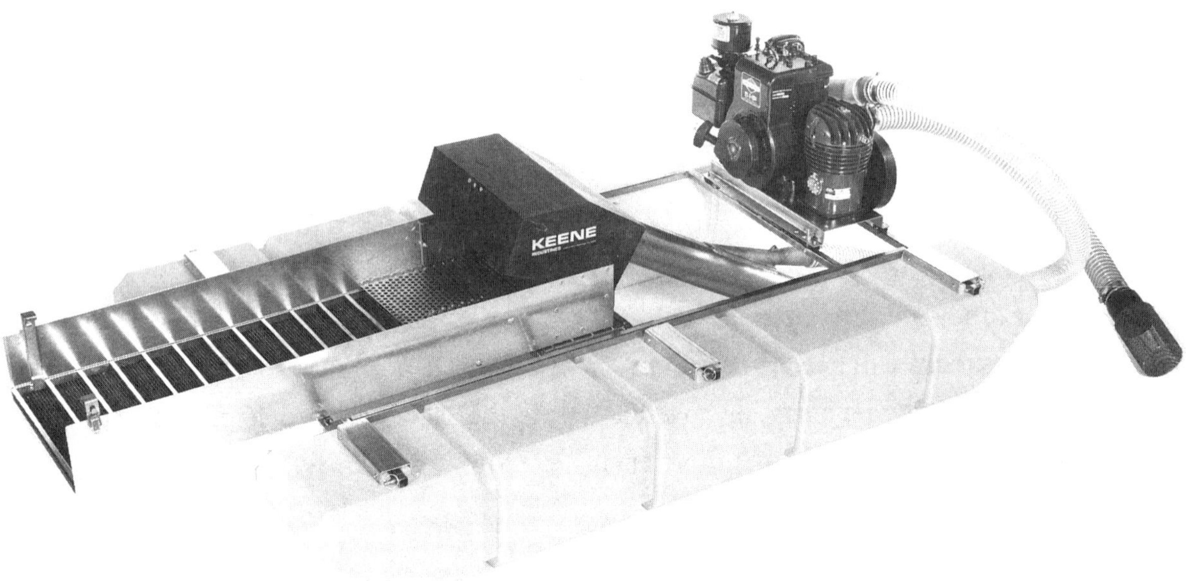

Fig. 6-17. Four-inch intermediate gold dredge.

entire dredge weighing about 40 pounds. It is very economical on gasoline and can be rigged easily into a backpacking unit, which makes this dredge an excellent machine for sampling the small streams and creeks in high country and other remote locations. The 2-inch Backpack dredge can process up to two cubic yards of gravel per hour* under ideal conditions. This is about the equivalent of what a hand sluicing operation can do under similar good conditions.

Intermediate-sized suction dredges are available in a wide variety. Figure 6-17 shows a 4-inch dredge, an intermediate, weighing around 160 pounds. This dredge breaks down into individual pieces--each which can be carried by a single person. It can also process up to 12 cubic yards of streambed material per hour under the most ideal conditions*. This makes it a very effective sampling machine for the more serious individual who is attempting to find the more substantial underwater placer deposits. This dredge is also excellent recreational gear for the hobbyist who wants a piece of equipment that will create a significant effect on a streambed, to locate paying quantities of gold. I have known professionals, myself included, who have used this same intermediate as a production machine and made a living at it.

The 5-inch sized dredge is perhaps the best all around one-man dredge that can be found on the market today (see Figure 6-18). This dredge is just slightly heavier than the 4-inch covered above, yet it can process a third more volume of streambed material because of the larger size. The extra

**** See section "DREDGE CAPACITIES" explaining production capability limiting factors on page 136.***

Fig. 6-18. Five-inch triple sluice intermediate sized production gold dredge.

1-inch sized diameter of its nozzle intake allows larger rocks to be sucked up--which means just that many less rocks needing to be tossed by hand. Because of this, a 5-incher is able to process double the volume or more* over what a 4-incher can produce in experienced hands. This makes it a more efficient machine for its size and weight. Because of all these features, the 5-inch dredge is not only the best all-around recreational dredge on the market, but also, it's one of the best all-around one-

Fig. 6-19. Six-inch production gold dredge.

** See section "DREDGE CAPACITIES" explaining production capability limiting factors on page 136.*

Fig. 6-20. Eight-inch production dredge.

man professional rigs, too. It is light enough that a single person can carry all of its individual parts around alone if necessary. Plus, with excellent recovery, it is able to process enough material to make a worthwhile piece of production equipment under the proper conditions.

The 6-inch dredge is a machine which has been designed to give excellent gold recovery, and can move up to about 25 cubic yards of streambed material per hour* under ideal conditions (see Figure 6-19). A 6-inch dredge can normally be operated by one strong person, but some dredgers prefer to team up when using a dredge of this size. The 6-incher can be packed around easily by one or two people when broken down into separate component parts. Therefore it can also be used for sampling purposes. Yet, six-inch dredges are utilized well as production machines, because of the large amounts of streambed material they can process with excellent gold recovery.

An 8-inch dredge is definitely considered a professional production machine by most dredgers (see Figure 6-20). However, it is considered recreational-scale mining equipment by many governing agencies--in comparison to heavy-equipment, large-scale mining operations. Some 8-inch dredges are powered by two or more lightweight portable gasoline engines. Others are powered by automobile engines. Well-powered 8-inch dredges are capable of processing up to about 40 cubic yards of streambed material per hour* under ideal conditions--with good gold recovery. On the twin engine power systems, the entire dredge, when broken down into individual components, can be packed around by a one- or two-man team. Again, a single strong man can operate one of these on his own, but it really takes two if it is to be operated efficiently for extended periods of time day after day, like a professional dredge ought to be. This is the kind of dredge that you bring on once you have found a sizeable deposit. And it makes little difference whether the gold

See section "DREDGE CAPACITIES" explaining production capability limiting factors on page 136.

is of coarse or fine size, because when set up properly, this dredge will recover well.

The dredges which have been shown above are just a few of the many different kinds and sizes available on the market. Those above have been chosen in an effort to show some of the various

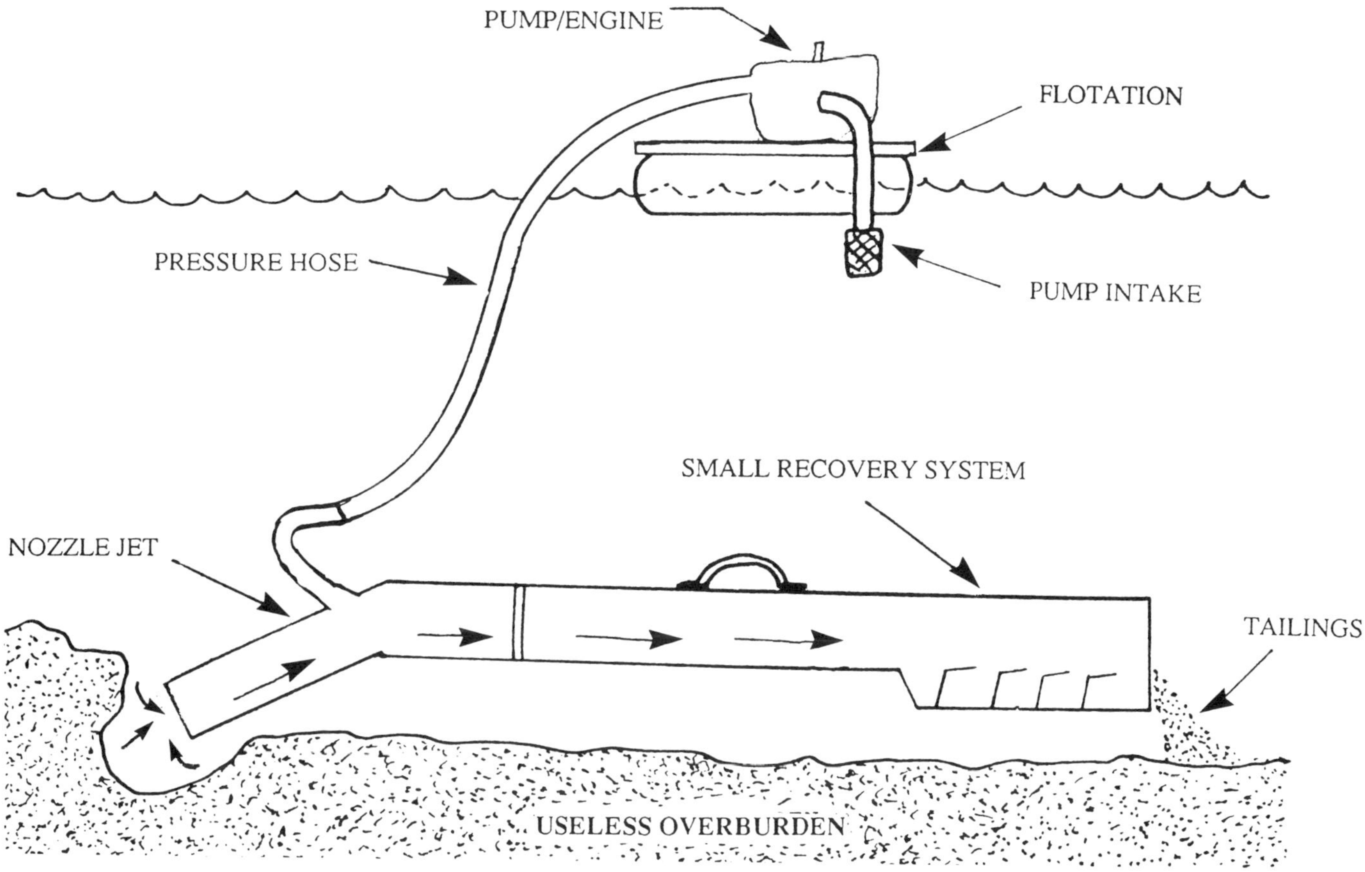

Fig. 6-21. Underwater gold dredge.

models which are available, and to illustrate the wide range of dredge sizes--from the mini-dredge to the intermediate and up to the large production machines, to give the reader some idea of what they are and how they are best utilized according to their sizes.

UNDERWATER DREDGES

Underwater-type suction dredges have the advantage of being able to pump a greater volume of material with less power than surface gold dredges. The reason for this is that the surface-type dredge, as covered earlier, pumps its water and material from the bottom of the stream or river up to the sluice box above the water's surface. It takes a lot of power to lift up a volume of streambed material any distance--especially when it is being lifted 8 or 10 inches above the water. The underwater dredge pumps streambed material to a sluice box located just behind the suction nozzle, as shown in Figure 6-21.

The underwater dredge is an example of a nozzle jet being put to very efficient use. With this

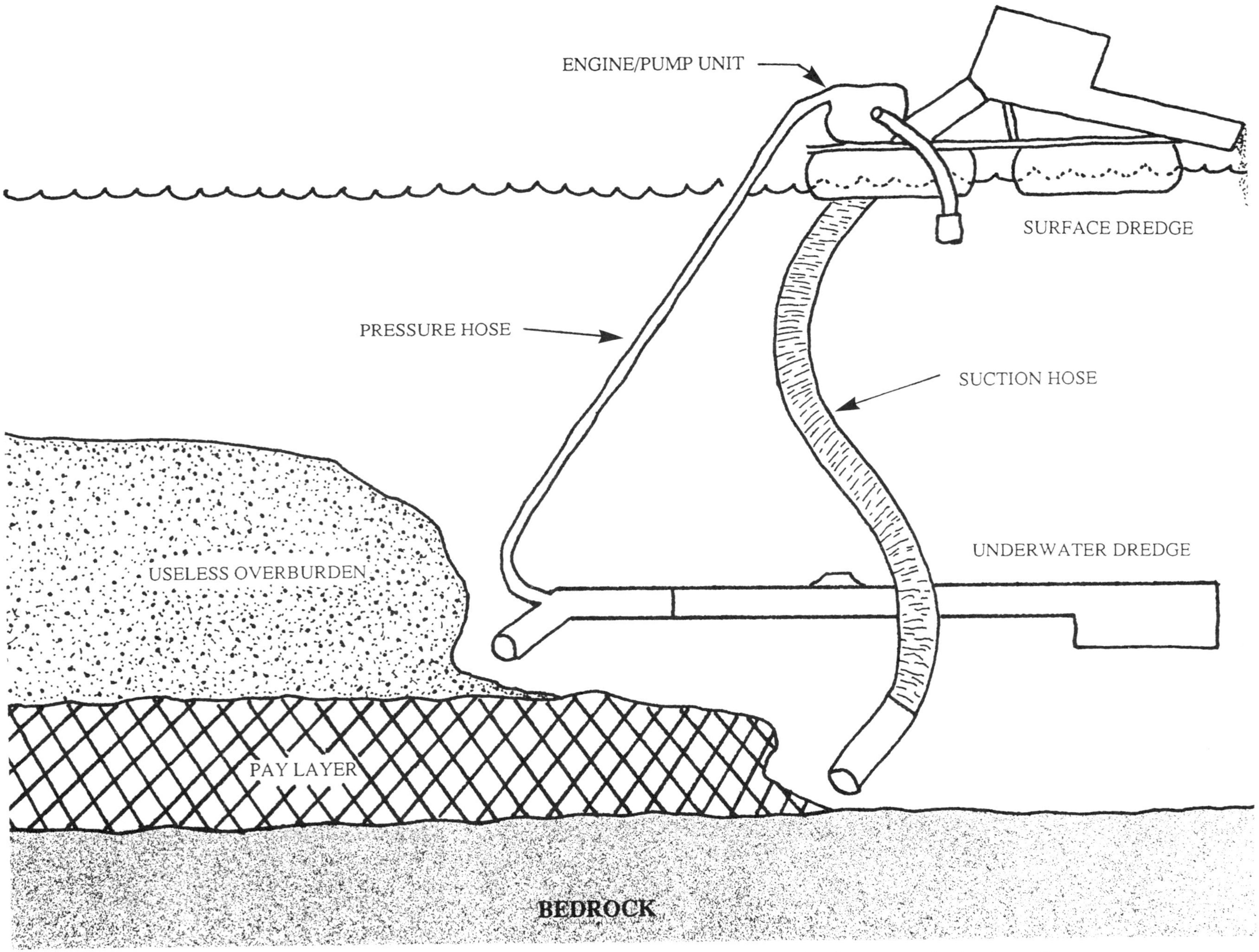

Fig. 6-22 Using a surface dredge and an underwater dredge in combination, to get the best results.

kind of system, an engine/pump unit at the water's surface is able to pump water down through a pressure hose and power an underwater dredge capable of moving about double the volume of streambed material that the same pumping unit could process when powering a surface dredge. For example, the power plant on the 5-inch dredge covered earlier is able to power a 6-inch underwater dredge.

The primary drawback of underwater dredges is that their recovery systems are very limited as to size. So they are not able to recover gold nearly as well as the surface type recovery systems.

The underwater dredge is more effectively used in conjunction with a good recovering surface dredge. It is not uncommon to locate a nice paystreak that is lying underneath five feet or more of flood layer *("overburden")*, which sampling tests have shown does not have any significant amount of gold. In this case, the idea is to move the overburden off the paystreak as quickly as possible so the gold can be recovered. An underwater dredge comes in handy here. The engine/pump unit on a surface dredge will be able to power an underwater dredge of greater size, which will be able to move the useless material out of the way approximately twice as fast. The underwater dredge is equipped with a small recovery system that will catch the medium and coarse gold--should some be present in the waste material and be overlooked by the dredger. Most often, if sampling has shown that a top layer of overburden does not have paying quantities of gold, then very little gold will be found in that top layer when it's being moved. So there is little need to worry about loss of gold when moving such a layer of overburden with an underwater dredge. Usually, it's a case of having found a nice paystreak underneath a large amount of useless overburden--too much for your surface dredge to move and still make it worthwhile to dredge the paystreak which lies underneath (not an uncommon occurrence). In this case, an underwater dredge can be attached to the power unit of a surface dredge and be used to move the waste material out of the way, resulting in the paystreak being recovered effectively and at a profit (see Figure 6-22).

SUBSURFACE DREDGES

Subsurface gold dredges are similar to underwater type dredges, in that water and material are pumped to a sluice box located below the water's surface. The recovery system on this type of unit is suspended just below the flotation system which supports the engine/pump unit (see Figure 6-23). Subsurface recovery systems are larger than those found on the underwater type dredges of the same hose diameter--about twice as large. So a subsurface dredge offers better gold recovery than the underwater dredge does, but still not as good as the surface dredge. But, because the material and water are not pumped above the water's surface, a subsurface dredge can also pump about twice as much volume as a surface dredge being powered by the same engine/pump unit. So the subsurface dredge is the "go-between" of the surface dredge and the underwater dredge, it being able to pump more material than a surface rig and recover better than the underwater dredge.

Subsurface dredges are best used when a substantial amount of overburden needs to be moved

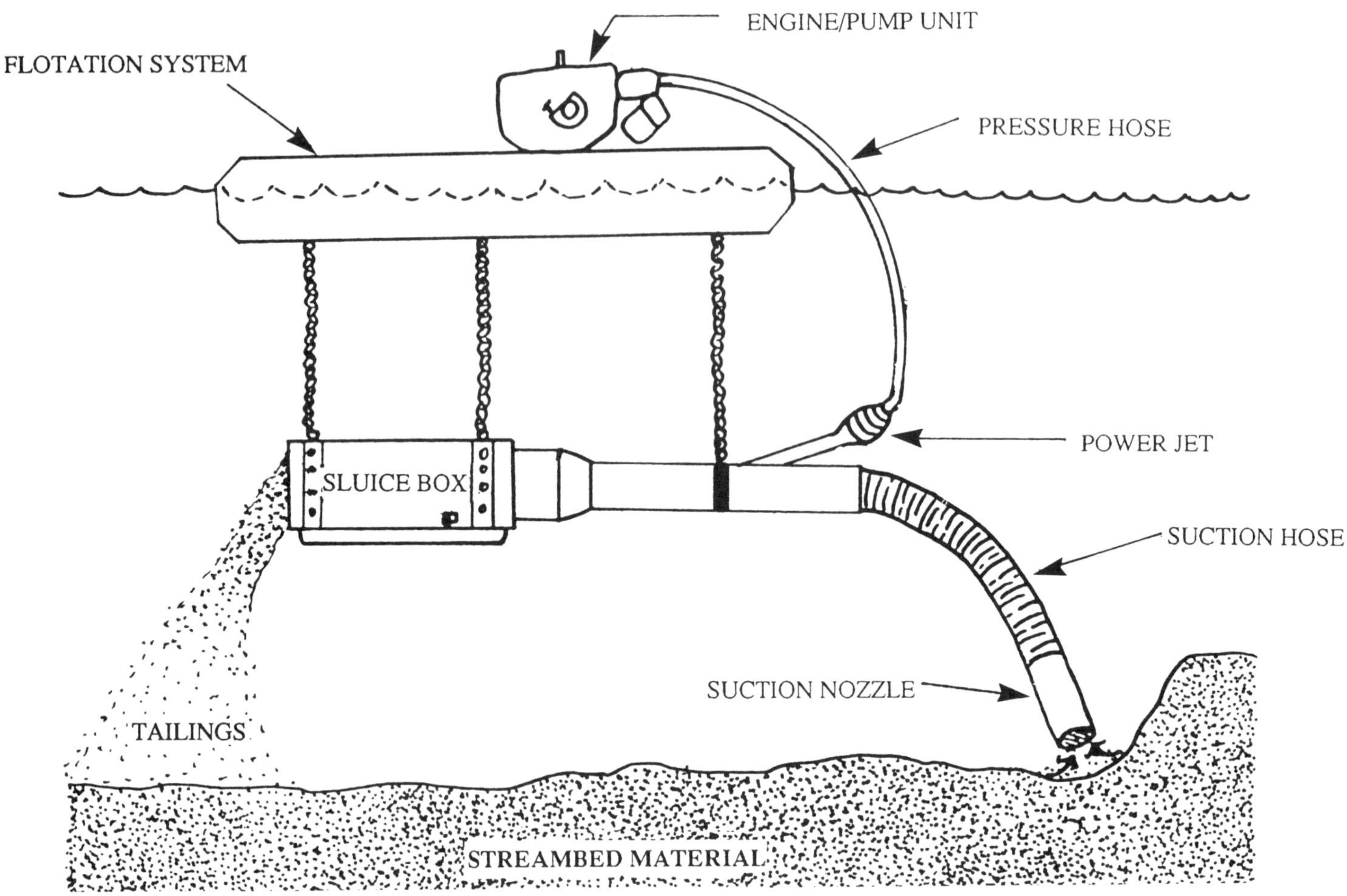

Fig. 6-23. Sub-surface type gold dredge.

off an underlying paystreak, yet the overburden has proven to pay to some degree. Also, subsurface dredges work best where the water current is flowing mildly, or not at all--otherwise, fast current can cause a pulling effect on the sluice box which hangs just below the water's surface.

DREDGE CAPACITIES

The production capability of any dredge varies with differing conditions. Hotter weather can cause a power loss to a dredge, as can dredging at higher altitudes. Also, the deeper you dredge, the more power is required because the material will need to be lifted a greater distance against the pull of gravity.

The manufacturer's specifications as to what a dredge is capable of producing are not going to be what you can actually produce out in the field. The manufacturer's specifications are given for the dredge when operating under ideal conditions. Classified material is used during these tests. Very seldom does one run into a large deposit of gold out in the field where loose, classified material is all that needs to be pumped. Usually the streambeds where one finds gold are composed of a wide range of different sized rocks and material which are packed tightly together. Streambeds like this need to be systematically taken apart. Larger rocks need to be tossed aside, and the boulders must be rolled out of the way. Only the smaller material is sucked up into the dredge as it is exposed. All this takes time and energy. So you are seldom, if ever, able to process as great a volume of material out in the field as the manufacturers can under totally controlled conditions. I wanted you to be aware of this so you won't get the false impression that you can go out into the field and move 12 cubic yards of material per hour just because the manufacturer says that his dredge is capable of moving that amount. How much material you will be able to move out in the field depends greatly on the consistency of the streambed material that you will be pumping and your skill at operating the suction nozzle and organizing your hole. The main idea is to get a dredge which has enough suction power to move material through your suction hose as fast as you are capable of feeding it into the nozzle. Most dredges available today are able to supply you with enough power to do this.

If you are planning to dredge gold under more severe conditions like at higher altitudes or in deep water you should get a dredge which has a larger engine/pump unit to have the additional power needed to operate under those conditions. Dealers of dredging equipment are usually quite knowledgeable as to what is needed by way of equipment to handle abnormal conditions. They can help you choose the proper equipment for your needs if you are uncertain what is required.

Small gold dredges normally come with 10-foot lengths of suction hose. Intermediate sized dredges usually come with 15 feet of suction hose, which is enough length to do the job well under the majority of circumstances. Most dredging is done in less than 10 feet of water. Extended pieces of suction hose are available. Short lengths of suction hose with couplings are also available if you should want to extend the length of your hose.

The suction hoses on larger sized dredges--six inchers and up in particular--are commonly stiffer and much easier to manipulate around if you have 20 feet of suction hose instead of 15 feet. Eight-inch dredges are better off with 25-30 feet of suction hose. This is something to keep in mind when buying a gold dredge of this size, as they can be bought with longer lengths of suction hose if you want them that way. It costs a little more, but you'll get more production done, too.

PROTECTIVE SUITS

If you are planning to do a lot of dredging in cold water, it will probably be necessary for you to wear a protective suit to keep your body comfortably warm.

There are two basic kinds of protective body suits being used among gold dredgers--''*wetsuits*'' and ''*drysuits.*''

Wetsuits are more economical to buy, and generally work well during the summer months when the water has warmed up. These are designed to allow water to get inside the suit. Your body heat then warms that water up and it works as an insulator against the colder water which remains outside of the wetsuit. Wetsuits are available in a wide variety of sizes and design, and your local dealers will have information on these.

A drysuit (arctic diving suit) is similar to a wetsuit in that it is often made of neoprene rubber, only it is made with waterproof zippers and watertight seals at each extremity to prevent water from entering the suit. Some models available are constructed from hard rubber or nylon material, but these tend to be more restrictive of body movement while underwater. A drysuit will keep a person warm when diving in extremely cold water for extended periods of time, and it is almost a necessity (unless you are using hot water) for a dredger to own one if he plans to work during the colder months, when the river water gets ice cold. Thermal underwear can be worn inside a drysuit to give extra warmth when it's needed. Dry suits also come in a wide variety of size and design, which you should look into before spending the money. I highly recommend that you buy one with a relief zipper as part of the suit. This way, you don't need to take the whole thing off every time you have to relieve yourself. Drysuits generally cost two to three times as much as a wetsuit of similar quality, but they are definitely worth the extra money if you are planning to spend extended periods of time in cold water.

A more recent development in dredging is to create a heat exchange unit on the dredge's exhaust system. This heats water which is then pumped down into a standard or modified wetsuit. These *''hot water systems''* are becoming very popular among winter dredgers and those operating in cold water streams and creeks. Some manufacturers are now building hot water heat exchanger systems to mount on the various engines commonly being used on today's dredges. Splitter systems are being built with a series of small hoses and fittings, to direct hot water to hands, booties and hood to keep the entire body comfortably warm. This is covered in more detail in my *''Advanced Dredging Techniques''* volumes.

When wearing a protective rubber suit, you will also need to wear extra weight to compensate for the additional buoyancy caused by the suit. Lead weights on nylon belts are most commonly used for this purpose, and can usually be found wherever the diving suits are sold. If you want to save money, a mold is generally available and you can buy the lead from a scrap metal yard and make your own. Dealers who sell protective rubber suits usually have a rough idea of how much weight you will need. This varies with the different kinds of suits. Between 40 and 60 pounds of lead would be a fair estimate. You will want to have enough to sink you to the bottom and make you heavy enough that you can move around on your knees and feet. The more water current you are working in, up to a point, the more lead weight you will need in order to hold your position on the bottom. So it's a good idea to have some extra weights on hand for changing dredging conditions.

The proper nylon belts for the job always contain quick release buckles--which are very important to have in case the dredger needs to get to the surface in a hurry. It's good to know how to release this buckle, and to know where to find it on your body when dredging--just in case.

Some dredgers also prefer to use ankle weights to help keep their feet on the river's bottom. This is not necessary, just a preference among some dredgers.

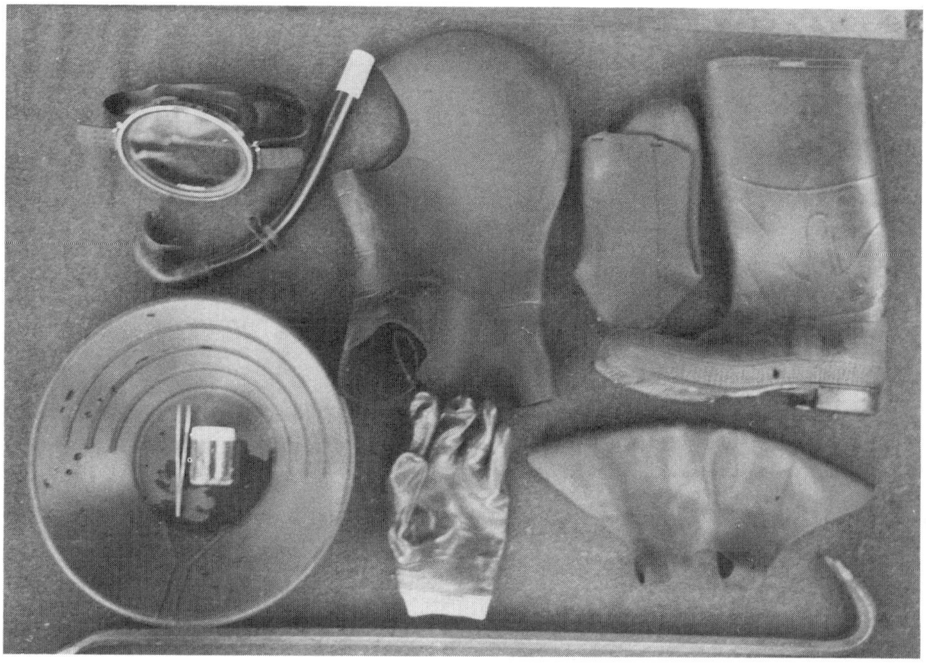

Fig. 6-24. Additional dredging equipment.

OTHER DREDGING EQUIPMENT

Some of the other pieces of equipment commonly needed in a dredging operation, as shown in Figure 6-24, are as follows: face mask, snorkel (if you are using a smaller-sized dredge having no hooka system), hood (to keep your head warm if one is not already attached to your protective suit), booties (to keep your feet warm if they are not already attached to your protective suit), boots (to fit over and protect your booties), knee pads (to protect your suit and your knees), rubber gloves (to protect your hands), crowbar or pry bar tool (to help pick apart the streambed materials and break up bedrock), rope (to tie off the dredge), spare gas can (for extra gas), classification screens and wash tub (to help with cleanup), gold pan (for cleanup), snifter bottle, tweezers and sample bottle (for the gold!).

DREDGING SAFETY*

Because a person is working in the water when dredging, most of the rules of swimming safety apply. And because one is sometimes underwater, many safety rules for scuba diving also apply.

The first rule of safety is to not do it alone. Have someone else around to keep an eye on you in case you should get into some kind of trouble and need assistance.

The activity of gold dredging often puts you in the position of needing to move boulders. There is the possibility of one of these rolling on top of you in some way and pinning you to the bottom.

For more information on dredging and diving safety, read "Safety In Gold Dredging" --a book by Dave McCracken.

Or a rock could fall out of the streambed material, bounce off your head, and stun you. These things are far less likely to happen if you pay attention to what you are doing, but the possibility is always there -- in which case it is good to have someone else around.

DIVING SAFETY

As you go further beneath the water's surface, there is an increase in the amount of water on top of you. This creates more pressure on your body. For dredging purposes, this increased pressure has little or no effect on the liquid and solid parts of the body, because they are not compressible. However, air space, with an increase of pressure upon it, will compress; that is, unless an additional amount of air is pumped into that air space to keep it the same size.

As an air space is taken further down below the water's surface, the outside pressure will continue to increase, so the airspace will be further compressed (see Figure 6-25).

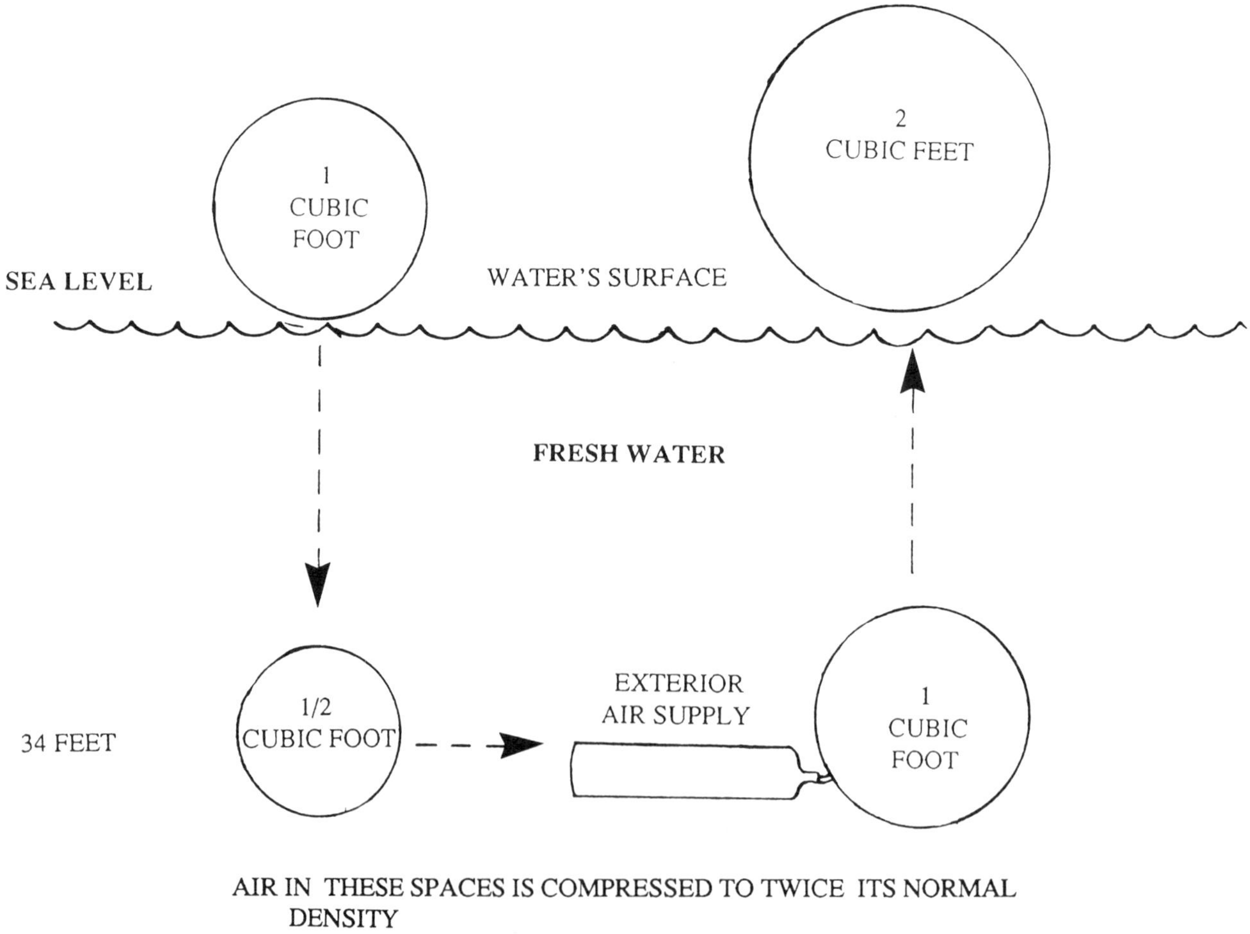

Fig. 6-25. The pressure increase at 34 feet in depth will compress an air space to half its normal size.

At a depth of 34 feet in fresh water, there will be twice as much outside pressure on an air space as there would be at the water's surface (at sea level). At this depth, an air space would be compressed to half its normal size. If there is an exterior air supply available (like a hooka system), and additional air can be pumped or breathed into that air space to keep it the same size as it descends in depth, it would need to have twice as much air at a depth of 34 feet. Then, if that same air space is brought up to the surface without allowing the additional air to escape as it expands, the space will expand (or burst) to twice the size that it was at 34 feet in depth (see Figure 6-25).

The main air spaces in your body which will normally be affected by increased pressure during dredging are the ears, sinuses and lungs.

Ears: In talking about the ears first, see Figure 6-26 and take note that the middle ear and external ear are separated by the ear drum. The middle ear has an air space which is connected to the breathing passage by the *eustachian tube,* as shown in the above mentioned figure. At the water's surface, the middle ear space contains air at surface pressure. As descent is made further beneath the water, the outside water pressure increases, and the middle ear space will be compressed inward, unless more air is allowed into that space through the eustachian tube. Compression on the middle ear space puts pressure on the ear drum, making it uncomfortable or painful if one continues to go deeper without equalizing the ear pressure. Many people have experienced this when swimming to the bottom of a swimming pool.

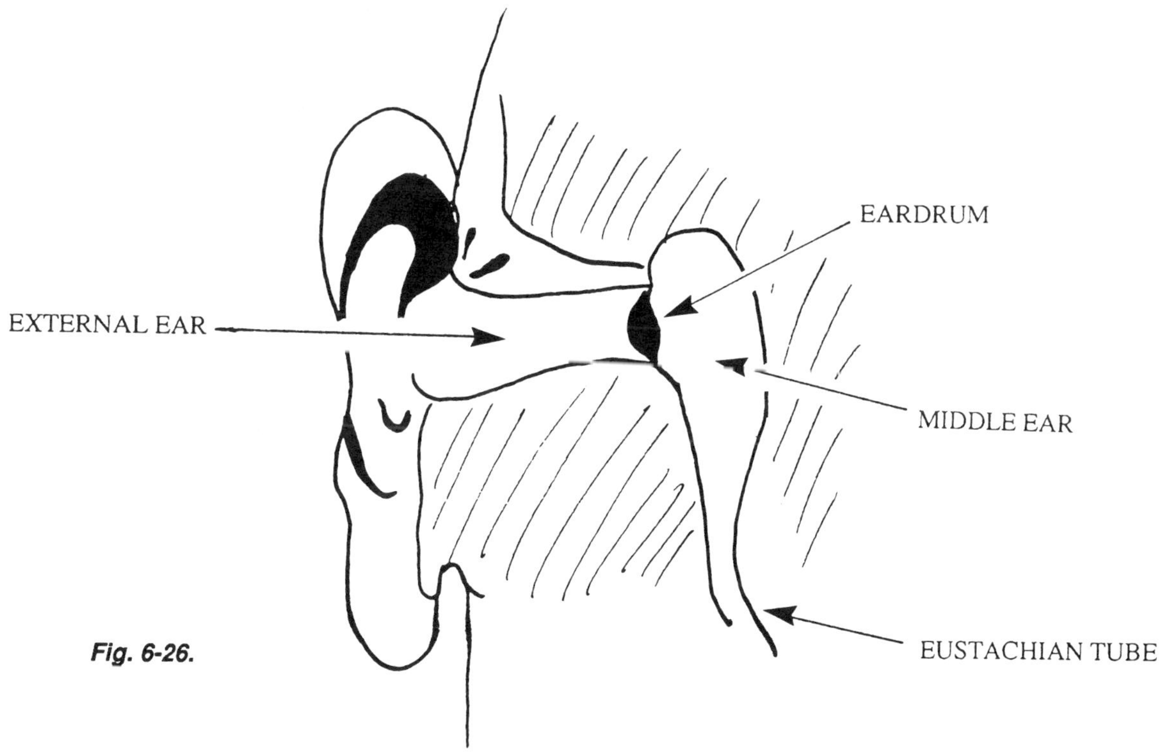

Fig. 6-26.

The eustachian tube does not always allow air to pass freely into the middle ear. Few people have easy control over this passage. Sometimes it's necessary to do some jaw moving, swallowing, or some yawning to open up this air canal and equalize the middle ear pressure when going deeper under the water. Sometimes it's necessary to hold your nose and blow air up into the middle ear area in order to equalize the pressure. This almost always works, unless a person has a lot of sinus congestion. The idea is to blow more air up into the middle ears as the increased pressure of water depth makes them uncomfortable. No matter the depth, you ears will be comfortable and feel normal when the inside and outside pressures are properly equalized.

If you are descending and the pressure increases on your ears, and you are not able to equalize one or both of them, go back upwards a few feet and equalize them there. Then continue downward, continuing to equalize as you go. Sometimes it's necessary to go to the surface and get rid of some mucus congestion in order to free up the passages to the ears.

When middle ear pressure is not equalized, the ear will be uncomfortable. To descend further without equalizing can cause pain. To descend further than that without equalizing the middle ear pressure is to invite the ear drum to become ruptured. This in itself may not be a serious injury. But it will have you out of the water for a matter of weeks while it heals.

If too much pressure is forced on your eardrums, and one or both of them are forced to rupture, water may enter the middle ear. If the water is cold, it can offset a person's equilibrium and cause nausea, until the water is warmed up inside the middle ear. Sometimes, the equilibrium can be affected so severely that the diver cannot tell which direction is up. Again, should this occur, the condition should only last for a few moments--until the water in the middle ear is allowed to warm up. Regulators are especially built with large exhaust parts to accommodate nausea, should it occur to a diver while underwater. However, I'm sure it is not a very pleasant experience!

Really, there is no reason for a diver to ever burst an eardrum; because, as outlined above, the warning pain happens long before the point where this will occur. Equalization should be done before the pain starts.

As one ascends towards the surface, the additional air within the middle ear will expand and automatically release itself through the eustachian tube. So it takes no effort or attention on the part of the diver to re-equalize ear pressures on the way to the surface.

Tight fitting hoods: Sometimes a diver will acquire a tight fitting hood, which fits so tightly over the ears that it does not allow water to enter the external ear space. Instead, it causes a trapped air space in the external ear as shown in Figure 6-27.

In this case, the air space within the external ear will compress as the diver goes deeper underwater, and there will be no way to equalize that pressure. This will cause the ear drum to be drawn outwards, as shown in the above figure. When this occurs, the unequalized pressure on the eardrum will cause an uncomfortable sensation just like before, and pain, and a possible puncture if descent is continued. An *''external ear squeeze''* can be prevented by avoiding very tight fitting hoods. Or, if you should happen to already have one, make a few small holes in the hood where the ears are located. This way, water will be allowed into the external ears while diving. One other thing that can be done when you enter the water is to lift that part of the hood off your face long enough to ensure that water is allowed to pass into and around both ears.

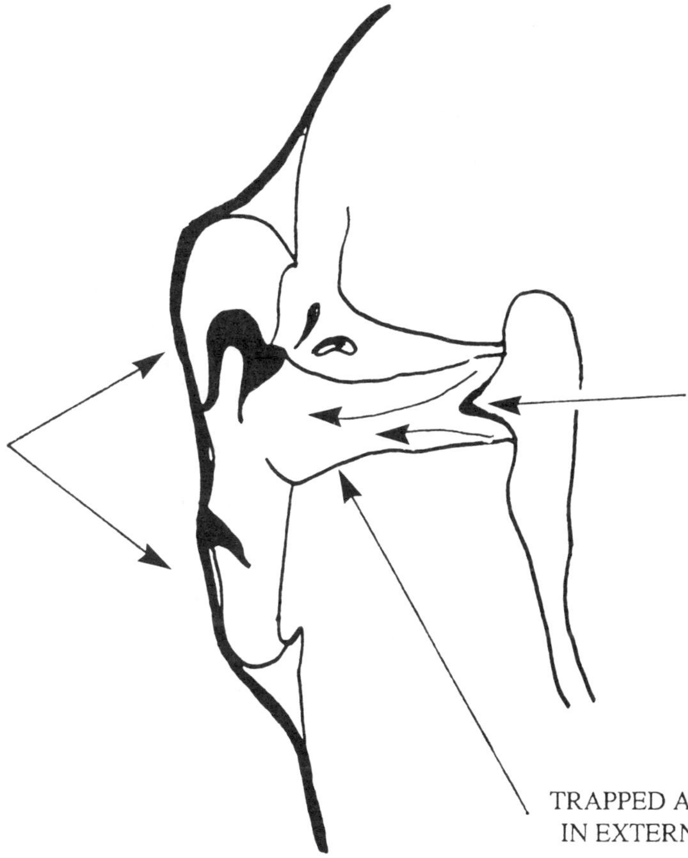

Fig. 6-27. A tight fitting hood can set up a trapped air space inside the outer ear--which cannot be equalized.

Earplugs: Never place earplugs in your ears when you plan to dive. An earplug will set up a trapped air space inside the external ear which cannot be equalized, in the same way as a tight fitting hood. Earplugs will prevent you from going more than just a few feet down without causing an external ear squeeze, and possible damage to the eardrums.

Sinuses: The sinus cavities are also affected by the increased pressure of depths underwater, but these automatically equalize themselves because they are connected directly to the nasal passages. The only time a typical diver will have trouble with sinus cavities is when he or she is having trouble with a lot of mucus congestion. When a diver descends underwater and his sinus cavities are too congested to equalize the pressure, he will feel the increased compression just inside the forehead, above the nose. When your sinuses won't clear (equalize) by themselves, sometimes you can go to the surface and get rid of some mucus and then go down successfully. Sometimes nasal sprays are a help. If this is not the case, it may be necessary to postpone the dive until your sinuses have cleared up.

Lung Cavities: The other main air spaces affected by pressure changes when dredging below the water's surface are the lungs. These automatically equalize pressure themselves, as long as the dredger continues to breathe normally as he or she goes down to and returns from a depth underwater. The primary danger with lungs in diving is holding your breath when coming to the surface. The reason for this is that when a person goes deeper underwater, the pressure on his lungs increases, causing them to compress. As he breathes normally from the hooka air system, additional air passes into his lungs to compensate for the increased pressure--which prevents his lungs from being compressed smaller in size. This is all automatic and the dredger will never feel any difference in pressures as long as he is breathing normally. At a depth of 34 feet in fresh water, when breathing from a hooka system, a diver will have twice as much air in his lungs as he would have at the water's surface (at sea level). His lungs will be normal in size, but the air inside is compressed to double its normal surface density and pressure. Now if that diver starts moving up, the outside pressure on his body will decrease, which will allow the air in his lungs to expand. If the diver breathes normally as he moves towards the surface, the excess air will be expelled and never noticed. If the diver holds his breath while ascending from 34 feet, there will be enough excess air in his lungs when he gets to the surface to expand them to about twice their normal size. Unfortunately, lungs will not withstand this kind of treatment. They will not expand to a larger size, but instead will burst to release the excess pressure. This is called an *''embolism,''* and can be a very serious injury. Diving physicists say that it only takes an average of about two pounds of excess pressure on a lung to cause an embolism, and that two pounds of excess pressure will occur from rising about four feet in depth while holding one's breath. However, for this to happen, the diver, or dredger, would already need to have his lungs filled to capacity.

Embolisms are more of a scuba accident; they seldom occur in gold dredging. In fact, I have yet to hear of a single case of it. Regardless, the possibility is there when working at depths of water greater than five or six feet while breathing from a hooka system. So you should be very aware of these principles and make sure to breathe normally when surfacing. Most embolism cases occur in scuba diving accidents when a diver panics and heads for the surface at full speed and forgets to expel the excess air on the way up. It's a natural tendency when in trouble while underwater for a person to not want to release that life sustaining air out of his lungs. Nevertheless, when breathing from a hooka air system and returning to the surface, it is necessary to expel the excess air from your lungs--regardless of the natural tendencies!

Bends: Bends--or *''decompression sickness,''* occurs when a diver breathes compressed air for extended periods of time at depths greater than 35 feet (at sea level), and then rises to the surface too quickly to allow time for the expanding nitrogen in the air to be flushed out of the body tissues. There is little chance of a gold dredger having difficulty with decompression sickness, because very little dredging is done at depths greater than 25 feet of water. If it happens that you decide to dredge for extended periods of time at depths greater than 25-30 feet, I highly recommend that you obtain a copy of the U.S. Navy decompression tables (available in any diving equipment shop), and calculate your dives accordingly. Also take into consideration that those decompression tables are figured for dives which are made in salt water at sea level. This means they will need to be adjusted to fit the fresh water and altitude at which you are diving. In my opinion he best defense against decompression sickness, if you are dredging deep, is to wear a decompression meter. This device, available from most diving supply shops, has a built-in computer which constantly monitors your exposure to depth and tells you how and when to surface to avoid the bends.

If you are planning to dredge at depths less than 25 feet, and most standard store-bought dredges are not very effective beyond that depth, then none of this really applies to you. However, the necessity of having to start figuring for decompression dives needs to be mentioned for those few who will venture to greater depths.

Summary of diving safety: Gold dredging is not a dangerous activity as long as you are aware of, and pay attention to, the above mentioned points on diving safety. Yes, to some the consequences of not paying attention to the safety rules may be a bit too gruesome to confront. But I would like to point out that the consequences of not driving a car properly can be even more gruesome, and there are many more safety rules to pay attention to when driving a car than there are when dredging for gold. It's primarily a matter of getting into the habit of breathing normally while underwater, and not forcing any more pressure on your ears past the uncomfortable stage before equalizing the pressure. You can practice equalizing ear pressure at the surface by holding your nose and gently blowing air up into the middle ear space. It's easy once you get the hang of it. Usually, swallowing will cause the excess air in the middle ear to release itself.

OTHER TIPS ON DREDGING SAFETY

One of the most important safety factors in gold dredging is to ensure that the exhaust fumes given off by the dredge's engine are swept downwind from the air intake of the hooka air compressor. Carbon monoxide has a cumulative effect, making one groggy and causing a certain amount of loss of judgment--which you really can't afford when working underwater. The necessity to avoid breathing in these gasses is increased as a dredger works at greater depth, because increased pressure will cause a greater density of air to be inhaled into the diver's system. Greater depths can make the adverse effects just that much stronger.

When setting up your dredge in a stream, pay attention to the wind direction. Sometimes the wind will change while you are dredging and cause some of the engine's exhaust fumes to be pumped down to you through the air breathing system. When this happens, you can usually taste the difference. Carbon monoxide by itself is odorless. But when emitted from an engine's exhaust system, it is usually in association with other byproducts which do have an odor. Exhaust fumes are rather foul tasting. If you are dredging along and suddenly taste exhaust fumes in your air supply, sometimes it's just a freak change in the wind only for a moment. If the fumes continue in your air supply, you should go to the surface and change the position of your dredge slightly, so that the

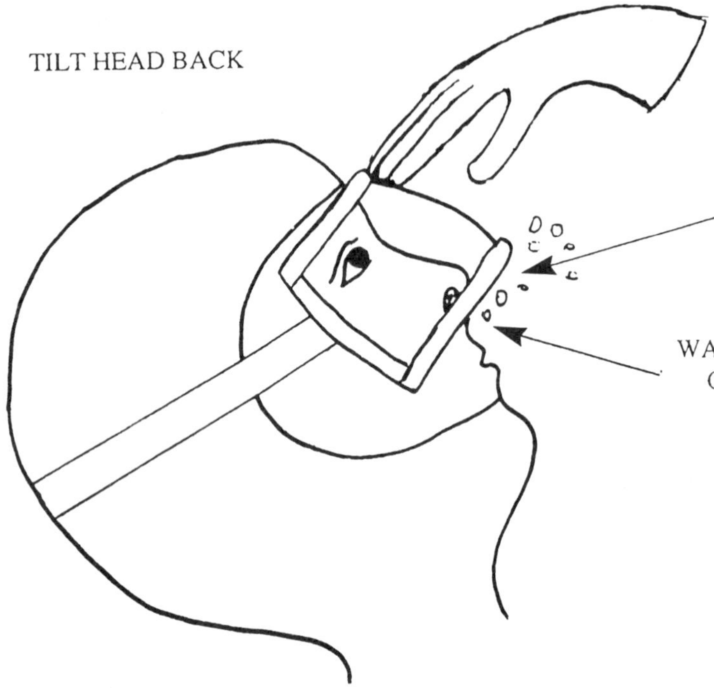

Fig. 6-28. Clearing water out of mask--tilt head backwards slightly, press top of mask firmly against forehead, and blow into mask through your nose.

compressor air intake is in more of an upwind direction. Sometimes you can run into a similar problem when positioning two dredges side by side. Continue to try new positions until you finally have all the divers breathing fresh air.

Clearing water from face masks while underwater: There are two things any gold dredger should be able to do well and should be prepared to do at any time. These are: (1) to be able to take a face mask off, put it back on, and clear all of the water out of it--all while underwater; and (2) to be able to do the same with a regulator. There are too many ways for your mask to get knocked off your face, or your regulator to become bumped out of your mouth, or both of these things to happen at once, while dredging for gold. You need to know how to handle it without it upsetting you.

In the case of the regulator, as mentioned earlier, it's just a matter of putting it back into your mouth and exhaling into it--or by pushing the purge button to clear water out of it.

A face mask (without an exhaust valve) can be cleared of water by putting it back onto your face properly (right side up and in place), then tilting your head backward slightly--as in looking up at the water's surface, pressing the top part of the mask firmly against your forehead, and gently blowing through your nose into the face mask, as shown in Figure 6-28. By blowing into your mask in this way, the water inside the mask will be forced out the bottom edge. With a little practice at this, you will be able to remove every bit of water from the inside of your mask--all while underwater.

Some masks come with an exhaust valve on the face of the mask itself, which is designed so water can be flushed out of the mask simply by exhaling into the mask through your nose. Some divers like this kind of mask and some don't; it's a matter of preference.

If you are new to underwater diving, the first thing you should do before you start dredging is practice removing your mask and putting it back on while underwater. Keep on practicing this until you feel very confident about doing it and are certain that you can keep your cool should your mask become swept off your face unexpectedly. You should also practice with the regulator in the same way. If something happens to cause both your mask and your regulator to become filled with water or swept off your face at the same time, the first thing to set right is your regulator. That's your air supply. Remember, you'll have lead weights attached to your body keeping you on the bottom. So you'll want to get this down well; your life could depend on it. Practice makes perfect!

The buddy system: One safe system of dredging is in having two divers team up on the same dredge. One person controls the nozzle (*''nozzle man''*), and the other person helps him by moving the larger sized rocks out of the way (*''rock man''*). The rock man in an operation such as this is usually put in charge of safety--it being his or her duty to keep an eye out for any possible rocks or boulders that could be hazardous. He should see that all dangers are removed before they can cause any trouble. When a team such as this is working a larger-sized dredge, in which it takes a lot more effort to move the suction hose and nozzle around, the two men or women can switch jobs back and forth so that neither person gets overworked.

Another good way to team up is to have two individuals with separate dredges working in the same hole together. Each can keep an eye on the other--for safety reasons. And also, each can help the other to move the larger sized boulders.

WINCHING BOULDERS

One tool commonly used among gold dredgers is the *''come-along''* (see Figure 6-29). The come-along is basically a hand operated winch. These come in various sizes, the larger and better ones being able to move larger and heavier loads.

When a gold dredger runs across a boulder too large to be moved by hand, or with the use of a steel pry bar, and he feels that it is worth the effort to move the boulder so that he can dredge underneath, he can sometimes do the job by himself or with the help of another by using a come-along.

The boulder itself can be attached with a harness, which is available on the market--as shown in Figure 6-29; or a tow chain can be used, as shown in Figure 6-30.

Many dredgers prefer a heavy tow chain to sling boulders when winching, because they are quickly adjustable. They also hold up well, despite the pounding they are subjected to when being used to pull large boulders around.

Sometimes the come-along is best positioned next to the boulder, rather than next to the anchoring object. In this way, the dredger can observe what is happening with the boulder while he

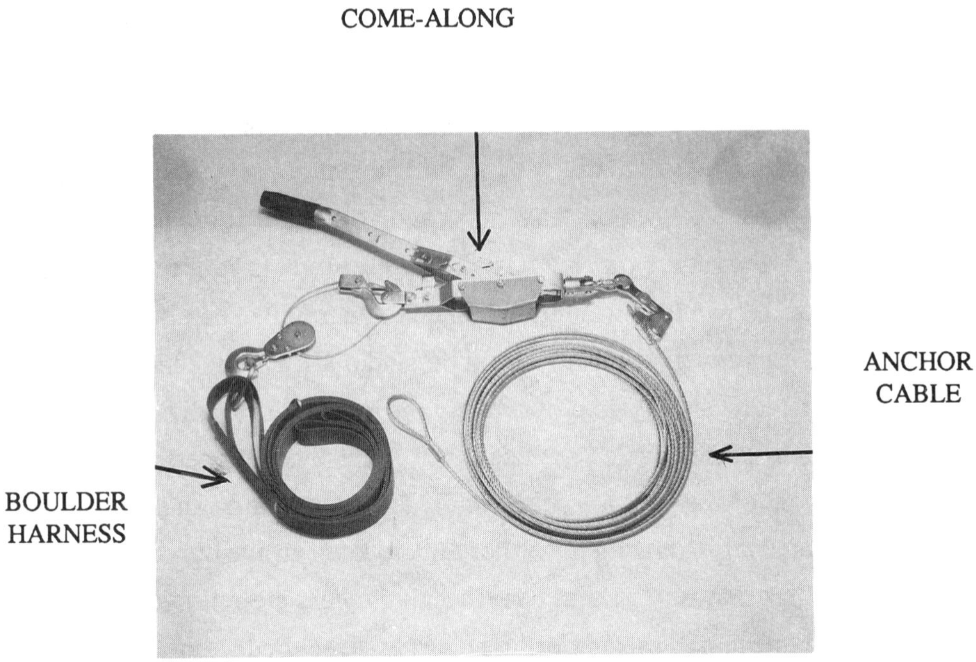

Fig. 6-29. The "come-along" is a common tool for a gold dredger to move larger sized boulders.

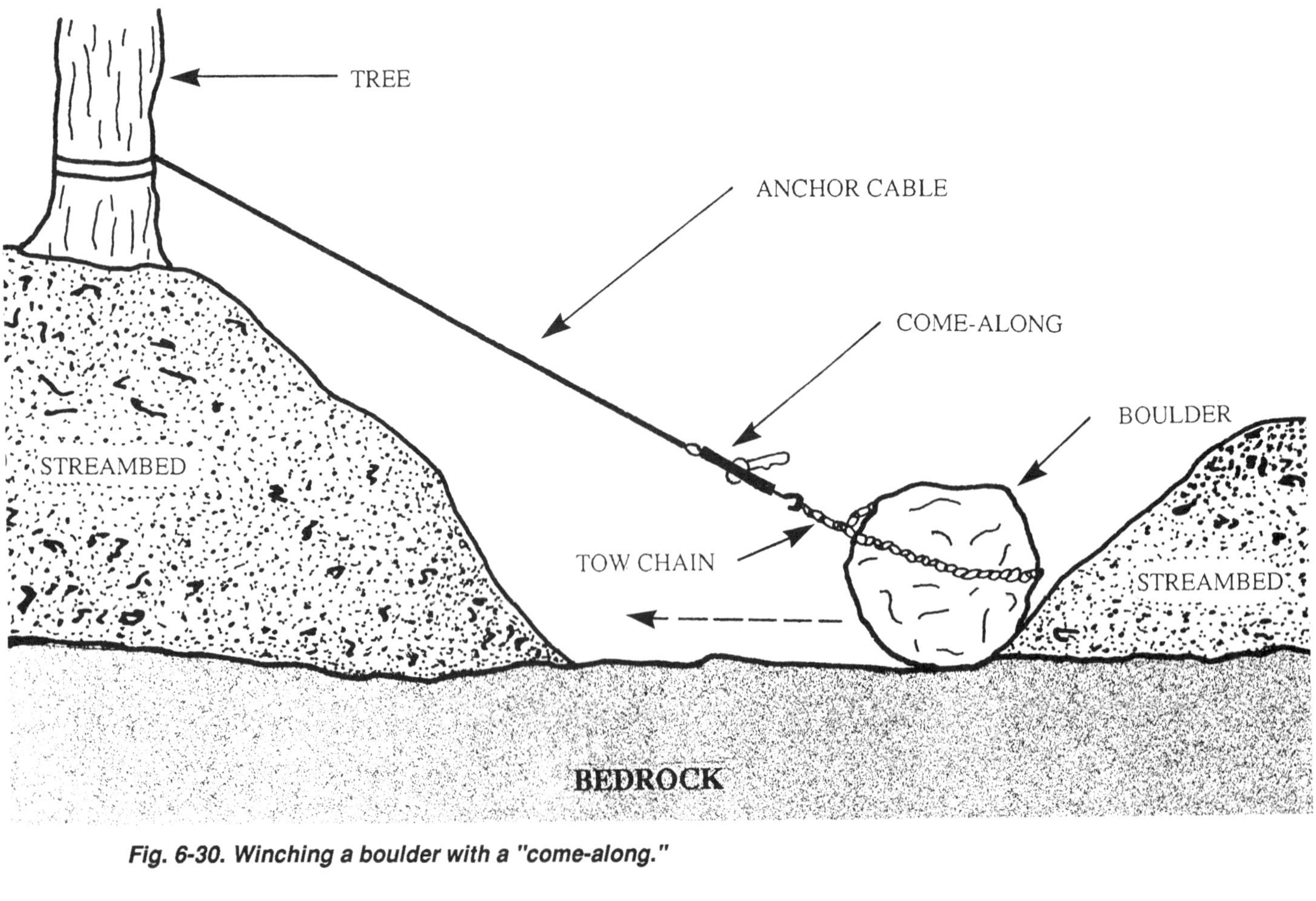

Fig. 6-30. Winching a boulder with a "come-along."

winches it along. That is, unless the boulder is being winched in a downhill direction where it could possibly roll on top of the diver.

Steel cable is best for attaching the come-along to the anchoring object. This is because steel cable is very strong and won't stretch like rope when being used to pull heavy loads.

It's necessary to be extra careful when winching boulders around while underwater--especially if you are doing it alone.

The idea is to move the boulders to the rear of your hole where bedrock has already been cleaned up by dredging.

CAUTION: If the come-along (or cable or boulder sling) you are using to move boulders is damaged in any way, have it repaired or replaced before you use it further. If a person is using a faulty come-along to move boulders, and he puts a lot of strain on it, and it comes apart in his hands, it can cause serious injury. Keep a close eye on the winching equipment to make sure that it's fit for use.

EXTRA TOOLS AND PARTS

It's a good idea to have a small toolbox on hand as part of your standard field dredging gear. Put in it the basic tools you will need to assemble the dredge and make minor field repairs. You might also want to include an extra spark plug, extra hose clamps, engine manual, pump seal, etc. If your dredge floats on innertubes, throw in a small tire patch kit and bring along a small tire pump--just in case.

HOW FAST TO RUN YOUR DREDGE

Some mini-dredges are powered by two-stroke engines, which give them their optimum power in the high RPM range. These types of dredges usually need to be run at full throttle.

Some of the dredges being powered by four-stroke engines have a little extra power in the high RPM's and might not need to be run at full speed all the time, depending on the altitude above sea level and the depths at which you are dredging.

The proper speed to run your dredge is the speed which gives you enough suction power to move material as fast as you are able to feed it into the nozzle. Your recovery system should be adjusted to recover well at this same rate of operation.

When you are dredging along in a paystreak and suddenly uncover a lot of gold, it's good practice to let clean water run through the dredge for a few seconds to allow the riffles to clean out the lighter material. Then suck up the gold. Most experienced dredgers, when into a heavy paystreak, will suck up the gold a little more slowly than they normally suck up other material. This is probably not necessary if the dredge you are using recovers gold properly. But it is safer to slow down just a bit when sucking large quantities of gold into the nozzle, to ensure that the riffles are not loaded up. This ensures that all of the gold will be recovered.

DREDGING PROCEDURE

Dredging basically consists of sucking up the streambed material off of the underlying bedrock and tossing or rolling the larger-sized rocks out of the way.

It's always a good idea to dredge in an upstream direction, when possible, and place the dredge behind you so the tailings are dumped downstream from where you are dredging. This way they will not be washed into your hole by the water's current. If you place the larger rocks and boulders behind you as you move forward, they will only have to be moved by hand one time. As you move your hole forward, the tailings off the back end of your dredge will drop into the hole behind you and fill it in.

If you are dredging in a paystreak which is paying a pennyweight in gold or more per hour (1/20th of an ounce), you will be seeing plenty of gold while you dredge. Gold usually shows up very bright and shiny and easy to see when uncovered underwater.

When moving your hole forward, be careful to clean out all the cracks, crevices and other bedrock irregularities. This is sometimes where a large portion of the gold is trapped--especially if you are into a bedrock paystreak, as opposed to a flood gold paystreak.

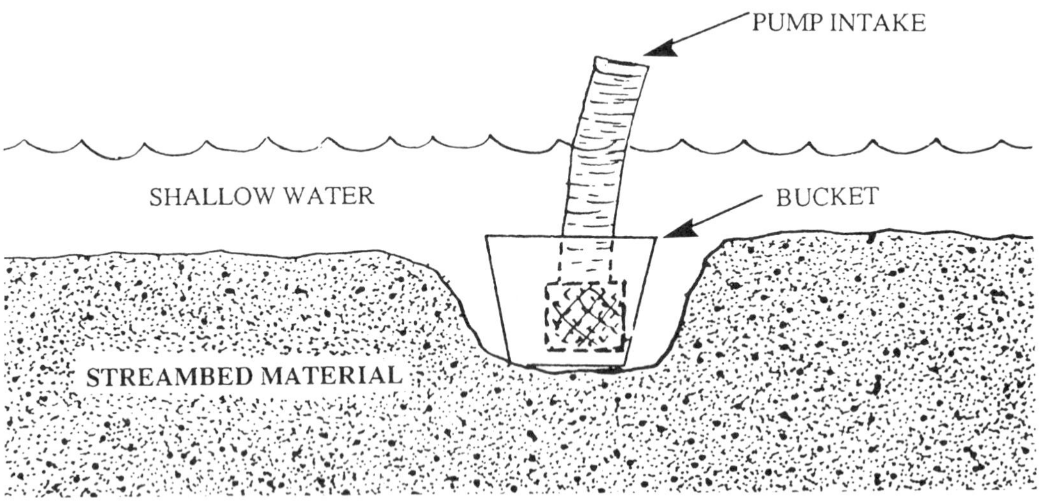

Fig. 6-31. In shallow water, place a bucket around the pump intake to prevent sand and gravel from being sucked into the pump.

It you are dredging in shallow water, be sure to keep an eye on your dredge's pump intake to make sure that it is not allowed to rest on the bottom where it can suck sand and small rocks into the pump. Otherwise, this can cause a great deal of unnecessary wear on your equipment. If the water is so shallow that this can't be avoided, dredge a small hole for the pump intake to rest in, and place a small bucket around the intake, as shown in Figure 6-31.

LOSS OF POWER: TROUBLE SHOOTING

If you notice that your gold dredge has a loss of suction power at the intake nozzle, the following points should be checked and handled accordingly:

Plugged-up pump intake: Sometimes leaves, grass or even pebbles can accumulate on the outside screen of the pump intake. This will prevent a full force of water from being pumped into the jet and cause a power loss at the suction nozzle. When this happens, shut down the engine and clean the intake screen thoroughly.

Clogged pump impeller: It is possible for some pebbles to be sucked up and become lodged in the pump's impeller. This usually only happens after the pump intake has been close to, or lying on, the river bottom while operating. It can also happen when another dredge has been dumping its tailings in the water just upstream of your dredge's pump intake--or if you have been dredging backwards in the river.

A clogged impeller can be detected by pulling the pump intake off the pump and by looking at the impeller. Any rocks or other foreign material lodged in the impeller should be removed.

Air leak in the primer: Occasionally, an air leak will develop in the upper section of a primer on the pump intake of a dredge. When this happens, the amount of air being sucked into and through the pump subtracts from the amount of water and pressure to the jet. The result is a loss of suction power. An air leak can be recognized by the air bubbles being passed through the pressure hose. There is no place underwater for air to come from. So when air is seen pumping through the pressure hose, you know there is some kind of an air leak on the pump intake or primer. Every effort should be made to find the source and plug it. Sometimes silicone rubber glue comes in handy in repairing such leaks.

Water leak in the pressure hose: Once in awhile a pressure hose will spring a leak. This will subtract from the amount of water and pressure being pumped to the jet and cause a suction loss at the nozzle. This usually only occurs after the pressure hose has had lots of wear. For temporary field repair, these holes can normally be patched with the use of silicone rubber glue and duct tape. A pressure hose having such leaks should be replaced at the earliest convenience.

Holes in the suction hose: Holes in the suction hose itself can subtract from the amount of suction at the nozzle. Also, it is possible that gold can be lost out of such holes. These too can be patched with the use of duct tape.

Plug-ups: The most common cause of a sudden loss of suction while dredging is when a rock, or a combination of rocks, lodges in the suction hose or the jet (see Figure 6-32).

When a plug-up occurs in the suction hose, it can usually be quickly freed by gently tapping around the obstruction with a heavy rubber hammer or with the smooth surface of a hand-sized

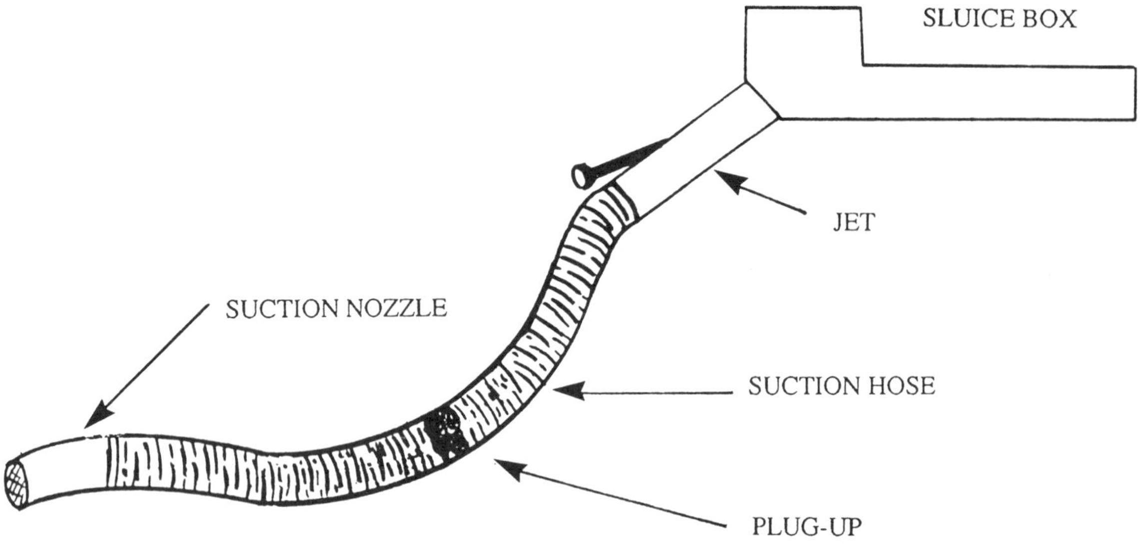

Fig. 6-32. A plug-up in the hose is a major cause of sudden loss of suction power at the nozzle.

rock. If the obstruction is inside the jet, it can usually be freed up easily with the use of a plugger pole from the surface, as shown in Figure 6-33. Most store-bought dredges come with a plugger pole designed for the purpose of tapping free jet obstructions from the surface.

Sometimes a plug-up in the upper section of a jet can cause the water to reverse the direction of water flow in the suction hose. If you are dredging along and the water suddenly starts to pour

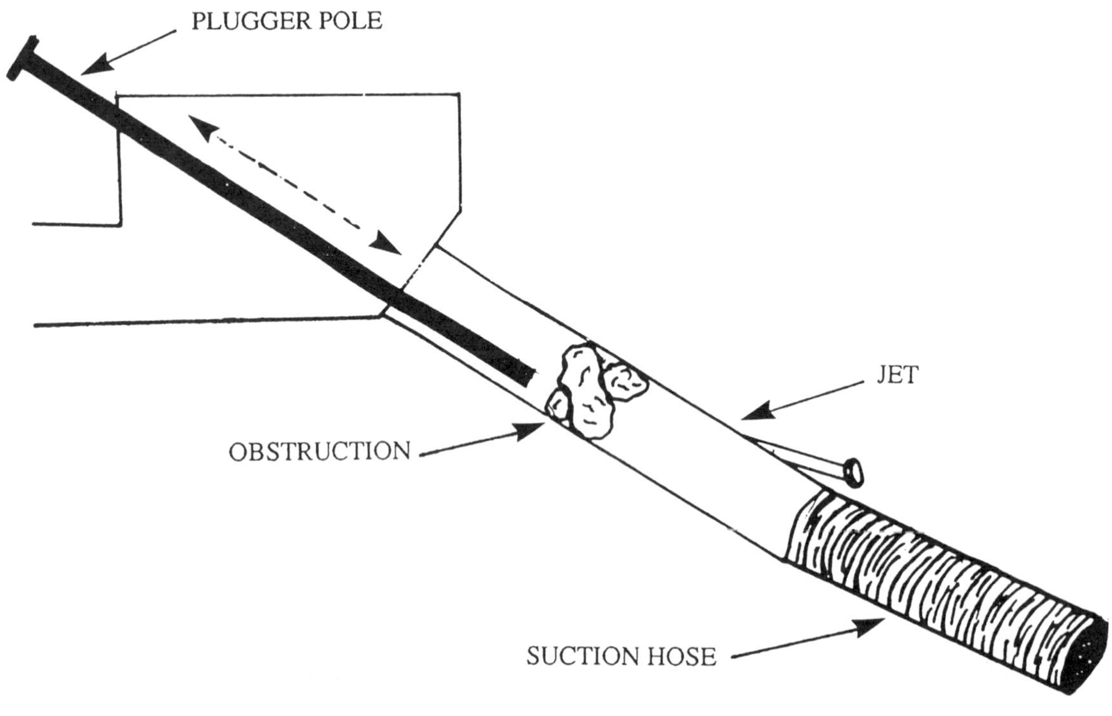

Fig. 6-33. Obstructions in the jet can be knocked out from the surface with the plugger pole.

out of the suction nozzle, instead of into it like it is supposed to, it's almost a sure thing that you'll find an obstruction in the jet. The only other time water will reverse flow like this is when the dredge quits running--out of gas or whatever, or when running at low idle.

When your dredge undergoes a power loss, always check thoroughly for a plug-up before looking into the other possible causes. If you do not have a plug-up and none of the other earlier mentioned causes seem to be the reason for your loss of suction power, you should pull the spark plug and clean it, or replace it if needed.

There's a lot more to learn about gold dredging: Gold dredging is a subject with much technology--it is today's fastest progressing field in small-scale gold mining. In this chapter, I have attempted to cover all the data you will need to know about what suction dredging is, what dredges are, how they are used, how to dredge safely, and the fundamentals of how to run a successful dredging operation. In no way have I covered everything there is to know about dredging.

I have written a series of how-to books on advanced dredging techniques, giving a professional viewpoint on how to succeed in gold dredging. These books cover many of the ''do's and don'ts'' and ''tricks of the trade'' a gold dredger normally picks up after lots of actual field activity and experience in dredging. I picked up most of the data by hanging around and working with some of the most successful professional gold dredgers in the field--and also by doing many things the wrong way the first time, or first several times. I highly recommend my dredging books as good reading material to anyone who wants to do better at gold dredging--on any scale of operation. I have also produced several how-to videos on the subject of gold mining and dredging. If these educational materials are not available at your local mining supply store, you can order by mail--turn to the first pages of this book for more information.

FINAL CLEANUP PROCEDURES

CHAPTER VII

After you have finished panning off a set of concentrates from a production period, there are usually some impurities remaining which need to be separated from your gold. These impurities most often consist of a few pieces of lead, some small iron rocks, and a little black sand.

There are probably many different ways to go about separating the impurities from your gold in the final cleanup steps. Here follows a workable procedure which has proven fast and effective.

These final cleanup steps can be done at camp, preferably in a dry environment, where the wind is not blowing much and where there is a table top or some other flat surface available.

STEP 1: First dry out your final concentrates. This can be accomplished by pouring them into a metal pan and slowly heating them over an open fire or stove--whichever is at hand, as shown in Figure 7-1.

CAUTION: Heating of concentrates should not be done inside a closed environment. Heating should be done outside and/or in a well-ventilated location, where vapors given off by the various steps will be swept away from you and other bystanders.

Fig. 7-1. Final clean-up Step 1: Dry out concentrates.

You don't want to heat the concentrates too much at this stage. This is because there may still be some lead in them. Excessive heat at this point has a tendency to melt the lead onto some of the gold within the concentrates. Pay attention to heat just enough to thoroughly dry out your concentrates.

STEP 2: Once the concentrates have cooled enough that they can be handled, they should be screened through a piece of window screen. A small piece of window screen, about 6-inch square, is handy to use for this purpose, as shown in Figure 7-2.

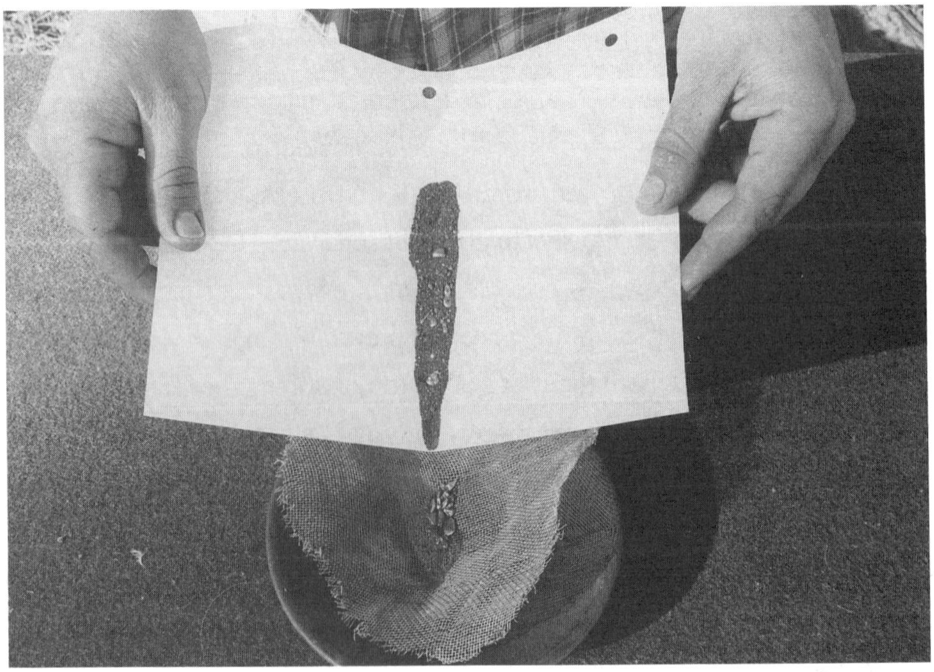

Fig. 7-2. Final clean-up Step 2: Classify concentrates through window screen.

STEP 3: Take the larger-sized concentrates, the material which would not pass through the window screen, and pour them onto a clean piece of paper. If there is a lot of this size concentrate, this step will have to be done in stages, handling a little at a time. Once the concentrates are poured onto the paper, it's easy to separate the pieces of gold from the impurities. The impurities should be swept off the paper and the gold should be poured into a gold sample bottle. This is where a funnel comes in handy (see Figure 7-3).

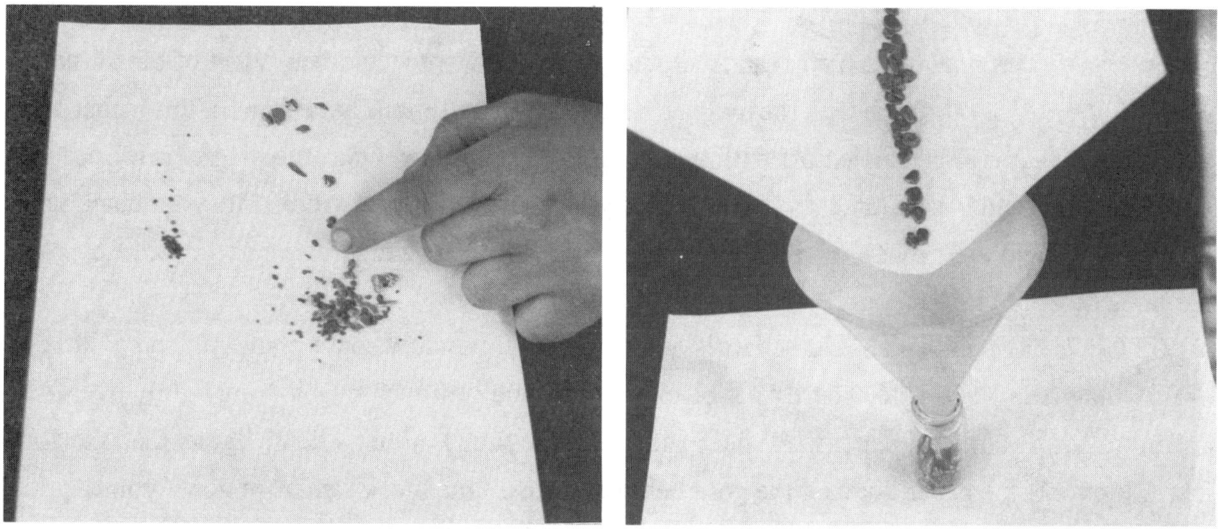

Fig. 7-3. Final clean-up Step 3: Separate gold from the largest classification of waste material and pour into gold sample bottle.

STEP 4: Once the larger-sized concentrates have been handled as shown above, the remaining concentrates can be classified through a finer mesh screen. A stainless steel, fine tea strainer works well for this. One can be found in just about any grocery store (see Figure 7-4).

Fig. 7-4. Final clean-up Step 4: Classify remaining concentrates through finer mesh screen.

STEP 5: Take the larger classification of concentrates from the second screening, pour them onto a clean sheet of paper, and separate the gold from the impurities in the same way that it was done with the larger material in Step 3 above. Some prefer to use a fine painter's brush to do this step. It can also be done by using your fingers. This step goes faster if you only do small amounts of concentrate at a time. Pour the gold recovered in this step into the gold sample bottle.

STEP 6: Take the fine concentrates which passed through the final screening and spread them out over a clean sheet of paper. Use a magnet to separate the magnetic black sands from these final concentrates. The magnetic black sands should be dropped onto another sheet of clean paper, spread out, and gone through with the magnet at least one more time. The reason for this is that some gold always seems to be carried off with the magnetic black sands. Once the magnetic black sands have been thoroughly separated from the gold to your satisfaction, pour them into your black sands collection. There are probably still some values left with them which can be recovered by other methods (covered later).

STEP 7: Now all that should be left is your fine gold, possibly some platinum, and a little bit of non-magnetic black sand. These final black sands can be removed by blowing lightly over them and vibrating the sheet of paper while doing so. Since the sand is about 4 times lighter than the gold, it will blow off the paper, leaving the gold behind. Once all the black sands are gone, you can pick out the pieces of platinum, if present, and separate them from the gold. Pour the gold into the same gold sample jar used in the earlier steps.

This process (Steps 1-7) goes very quickly if an effort was made during the final panning stages to get as much black sand and other waste material as possible separated from the gold. In other words, the less impurities you have to deal with in the final cleanup, the faster the steps go and the easier they are to complete.

CLEANING GOLD

Sometimes placer gold, just out of a streambed, is very clean and shiny. If this is the case with your gold, after the final cleanup procedure is completed, your gold is ready to be weighed and sold or displayed or stored away in a safe place.

Once in a while gold will come out of a streambed with a coating of mercury or some other impurity stuck to it, and it will be necessary to go a few steps further to make the gold's natural beauty stand out.

If your gold is not clean and shiny, and you want to get it that way, place it in a small water-tight jar about half full of water and add a little dishwashing liquid. Fasten the top on the jar and shake the contents vigorously until the gold takes on an unnatural glittery color. Sometimes this happens quickly and sometimes it takes a little longer. It depends on how much gold is in the jar. The more gold, the faster the process, because it is the friction of gold against gold which does the cleaning. Once the gold is glittery, rinse the soapy water out of the jar, pour the gold into a metal pan, and heat it up (outside and down wind) until the gold takes on a deep, natural, shiny luster.

Sometimes the gold will be deeply coated with a thin layer of mercury and will need to be heated to vaporize off the mercury or immersed in a solution of nitric acid first, before washing in soap and water (covered later in this chapter).

Gold has a tendency to turn a dull color after having been stored in an airtight container for an extended period of time. For this reason, some gold miners and dealers store their gold in water-filled jars, and dry it out just before displaying it or making a sale. In this way the gold keeps its beauty all the time.

If you should happen to store your gold in an airtight container and notice that its color does not seem to be as bright as it once was, wash it with soap and water and re-heat it, as in the above steps.

The best time to weigh your gold to get the most accurate measurement is after you have completed all of the final cleanup steps.

AMALGAMATION

Mercury (*"quicksilver"*) is a silver-colored liquid metal which has a tremendous affinity for some other metals. Clean mercury tends to attract to itself and ball up into a single mass. A ball of mercury also attracts pieces of gold to itself, swallowing them up into its mass. A droplet of mercury will continue to attract more pieces of gold into its body until it becomes packed so full, that it cannot hold together as a single mass any longer (see figure 7-5).

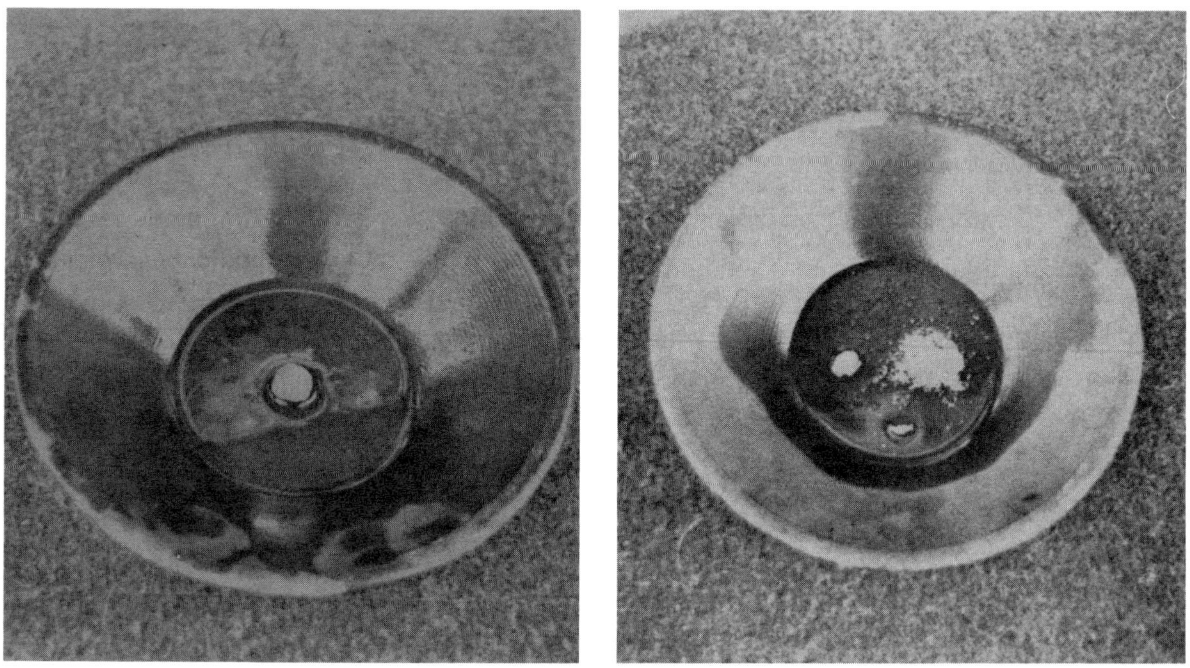

Fig. 7-5. Right-hand picture shows gold and mercury separately. Left-hand picture shows them mixed together.

Mercury is often used in mining to separate the fine pieces of gold and silver from other heavily concentrated materials. The process of mixing mercury with metals is called *"amalgamation."* The resulting mass of gold and mercury mixed together is called *"amalgam."* Amalgamation is one of the oldest gold refining techniques in existence. It is still widely used in the field of mining today.

CAUTION: Mercury is a poison. Care must be taken to avoid breathing its vapors, or allowing mercury to be absorbed into your body through open cuts, or perhaps the pores of your skin. When working with mercury, it is best to use rubber gloves. Goggles are a good idea. The process should be outside and downwind of yourself and nearby habitation.

Mercury is very heavy, having a specific gravity of about 13.5. Some old-timers used to place mercury in their sluice boxes in order to catch more of the fine particles of gold, which otherwise would have been swept out of their boxes because of their minute size.

Gold must be clean for mercury to be able to attach itself. Sometimes placer gold will be covered with a thin layer of oil. This will prevent the gold from being amalgamated unless the oil is cleaned off first. If you are going to use mercury to amalgamate the gold and silver values out of a set of concentrates, it's always a good idea to bathe the concentrates first in a 10-to-1 solution of nitric acid--meaning 10 parts water to 1 part acid. This process should not be done in a metal pan, because the acid solution will attack the metal of the pan itself. Plastic gold pans and glass jars work well for cleaning concentrates with a nitric acid solution.

CAUTION: Working with nitric acid can also be dangerous! Be extra careful to avoid spilling it onto yourself, splashing it into your eyes, or breathing in its fumes. If contact is made, fresh water will help to dilute the acid. Always pour the nitric acid into water when mixing up a solution. This helps prevent the full strength acid from making contact with impurities which can cause the full strength acid to spatter out of the container and onto you or your equipment.

All work with nitric acid and mercury should be outside and downwind of habitation and/or under a well-ventilated hood.

Nitric acid can be neutralized by adding baking soda until no more foaming takes place.

When a solution of nitric acid is poured onto a dirty set of concentrates, the effect will be a bubbly reaction. A set of concentrates, when being cleaned with an acid solution, should be allowed to bathe until all such visible reaction has stopped. Then the concentrates should be rinsed with fresh water so the acid is diluted and washed away. Once this is done, the concentrates are properly set up for amalgamation.

A small amount of concentrates can be amalgamated (after the nitric acid bath) in a gold pan--either metal or plastic. Approximately the same quantity of mercury should be used as the estimated amount of gold in the concentrates. Too much mercury should be avoided, because it becomes unwieldy to work with in the pan. If anything, attempt to pour in a little less than the estimated amount that you will need. More can be added if it is needed. A little water should also be used in the pan during amalgamation.

Fig. 7-6. Amalgamating in a gold pan.

Take the pan in your hands and jiggle it around until all of the visible gold is attached to the ball of mercury (see Figure 7-6).

Mercury will not grab onto black sands. In essence, what you are doing is getting the mercury to grab all of the free gold out of the black sands. Then you will pan the black sands and be left with a mercury ball filled with gold.

Once all of the visible gold is gone, pan off the black sands into a tub of water. The reason for the tub during this step is in case you goof and toss your amalgam ball, or part of it, out of the pan--especially if you are using too much mercury. If you pan into a tub, and some of the amalgam is lost from your pan, you can retrieve it from the tub and try again at no loss. During this final panning, it is handy to have two gold pans. The amalgam can be poured from one to the other, while the remaining black sands can be washed from the pan which doesn't contain the amalgam. In this way, all the black sands can quickly be separated from the amalgam with no loss.

Mercury in general use will not grab onto platinum either. Care must be taken to watch for platinum during this final pan-off if you wish to save it. Platinum is heavier than black sand so it will be at the bottom of the gold pan. Platinum can be picked out of the pan after most of the black sands have been washed out.

During amalgamation, if you do not have enough mercury in your pan to accommodate all of the gold present, you will notice the amalgam starting to break apart into separate pieces. In this case, add more mercury so the entire ball of amalgam can hold together and still collect the gold out of the pan.

A thoroughly packed ball of amalgam will have about 50% gold and 50% mercury.

After mercury has been used a number of times in an amalgamation process, it may itself become dirty. When mercury becomes dirty, it tends to break down into smaller separate bubbles, rather than come together into a single mass. The way to clean dirty mercury is to bathe it in a nitric acid solution of 30 parts water to 1 part acid. This process must be done outside and downwind of any habitation, because the vapors given off are potentially harmful. This will clean the impurities out and allow the mercury to amalgamate properly again. Mercury can be used over and over to amalgamate, and cleaned when necessary in this way.

If you have a larger set of concentrates, a rock tumbler can be used to amalgamate the gold by placing the concentrates and a correct estimation of mercury into the tumbler and allowing them to turn for a few hours. Some larger-scale operations use portable cement mixers to amalgamate their concentrates.

Mixing lye (Drano) with your tumbling mixture will help keep the mercury alive--that is, keep the mercury grabbing gold well.

Once all the gold has been amalgamated, and the amalgam has been separated from the black sands, it's necessary to squeeze off the excess mercury from the amalgam ball. This is sometimes done by tightly squeezing the amalgam ball within a wet chamois until all of the excess mercury is forced through the pores. This should be done underwater to keep the mercury from jetting through the pores (like a squirt gun) and onto the floor or ground. A thick cloth, piece of canvas, or a nylon stocking can also be used for this, but a thin leather chamois is preferred by experienced miners. The squeezing should be done over a water-filled gold pan, or some other open container, to catch the excess mercury as it is squeezed through. If the catching container is filled with water, the

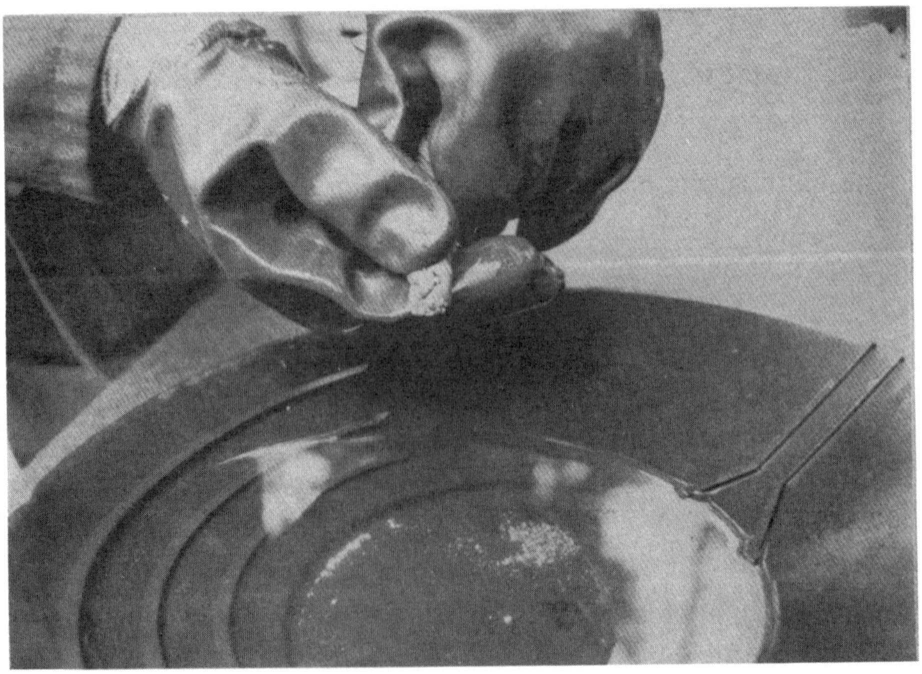

Fig. 7-7. Squeezing the excess mercury out of amalgam with a chamois.

mercury will be prevented from splashing or bouncing out as it drops into the receptacle.

It's always a good idea to use rubber gloves when doing the squeezing step. This will prevent any mercury from becoming absorbed into the cuts or bruises that may be on your hands.

A hypodermic syringe (without a needle) also works very well to remove the excess mercury from amalgam. It's best to find a larger syringe with a strong plunger--with the syringe being made of flexible plastic. Veterinary supply stores generally will sell you these syringes. Pliers can be used to squeeze the hole into as tight an opening as possible. This will prevent significant amounts of gold from being sucked out of the amalgam with the mercury. This syringe method is cleaner and easier than using a chamois, and no gold is lost in the process. Any gold sucked out of the amalgam ends up in your supply of mercury for a bonus at some later time.

Mercury which has been removed from an amalgamation process will be found to hold onto some of the extra fine gold. These remaining gold residues cause the mercury to have even more affinity for gold when used in later amalgamation processes.

Once all of the excess mercury is squeezed out of the amalgam ball, the next step is to separate the mercury from the gold. This can be accomplished in two different ways. The first way is by heating the amalgam until all of the mercury has been vaporized off the gold. The other way (covered later) is to burn off the mercury with nitric acid.

VAPORIZING SMALL AMOUNTS OF MERCURY

CAUTION: Mercury vaporizes at 675 degrees Fahrenheit, which can be obtained over any open fire and over most propane stoves. Mercury vapors are extremely poisonous and can cause death if enough of them are inhaled. ***NEVER VAPORIZE MERCURY INSIDE AN ENCLOSED STRUCTURE!*** Mercury can give off dangerous vapors even at room temperature. So any time mercury is being heated at all, it should be done outside and from a position where the wind will blow the vapors away from you and anyone else in the vicinity. Mercury can attach itself to gold in small amounts and it's not uncommon to have some present, even if it is not visible in any great amount. It is for this reason I suggest, when you heat up your gold during the final cleanup steps, you do it outside and downwind. Mercury should be stored inside water-filled containers.

Aluminum pans do not work well when working with mercury, because the aluminum reacts to the mercury in an amalgamation process of its own--which tends to make the cleanup procedure much more difficult. The small steel gold pans, about six or eight inches in diameter, work very well as all-around clean-up pans for heating and amalgamation purposes.

When heating up an amalgam ball in a steel pan, a good effort should be made beforehand to remove as much excess mercury as possible from the amalgam. Also, it should be heated slowly at first, to prevent the mercury from boiling and spattering values out of the pan. Once enough mercury has vaporized off so that there is no longer any danger of this, the heat can be increased to make the job go faster. Best results with this method can be obtained by placing a carved-out half of an uncooked potato over the amalgam, to prevent the spattering and loss of values, and to trap most of the mercury vapors (instructions later in chapter). If your gold just has a small amount of mercury attached to it, you won't have to worry about spattering, and just heating your gold will suffice. Don't forget: Those vapors are dangerous--do it outside and downwind!

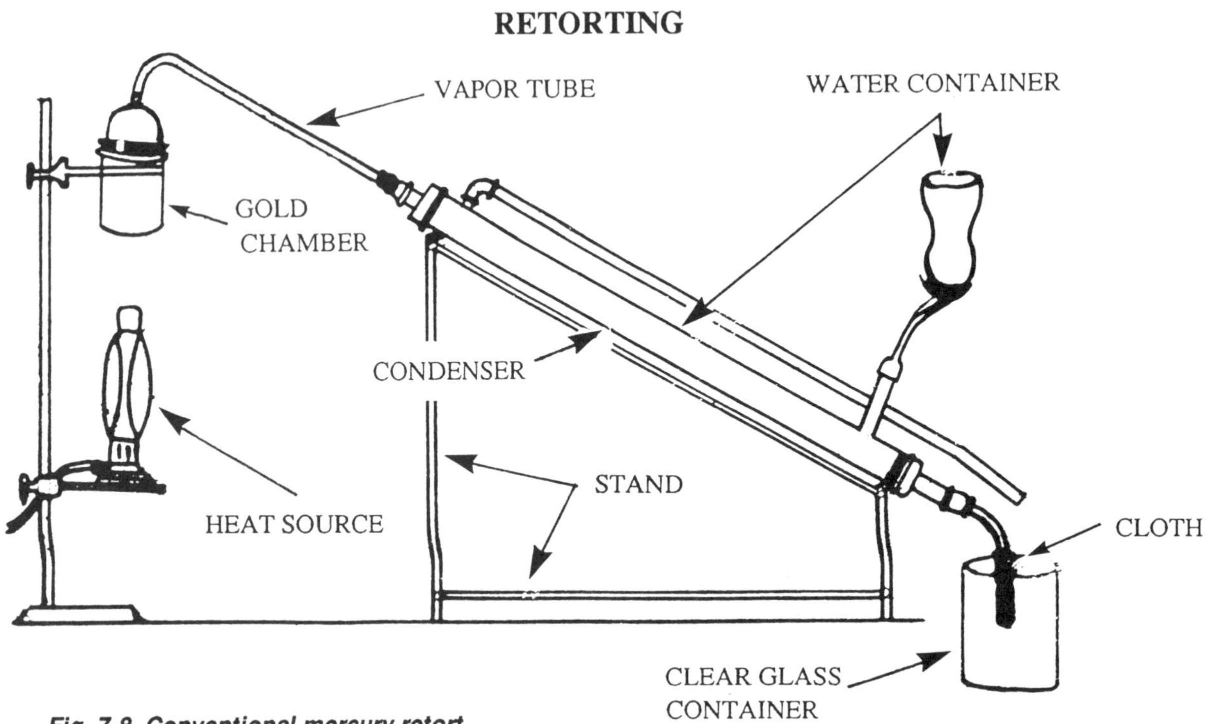

Fig. 7-8. Conventional mercury retort.

A *"retort"* is an apparatus that is used to vaporize the mercury off of gold. The device condenses the vapors back into native mercury, and drops it into a separate container so that none is lost. A standard retort (Figure 7-8 above) consists of a gold chamber, a vapor tube, and a condenser, as shown in the above diagram. The retort is usually held in place with a stand. Sometimes a propane torch is used as the source of heat. The condenser is basically a water jacket that surrounds the vapor tube with cool water. As the vapors are cooled, they condense back into liquid form again and run down the tube to fall into a container at the bottom (see Figure 7-9).

The procedure commonly used in setting up a mercury retort is as follows:

STEP 1: First, look over the vapor tube to ensure that there are no obstructions inside it. Sometimes it's necessary to blow through it to make sure. Be careful not to get any remaining mercury residues in your mouth!

STEP 2: Coat the inside of the gold chamber with chalk or graphite and allow to dry. This will prevent gold and silver spatterings from adhering to the cast-iron interior of the gold chamber. A double thickness of newspaper lining on the inside of the chamber will also prevent spatterings from sticking to the inside walls of the chamber; but if used repeatedly, will cause an undesirable residue

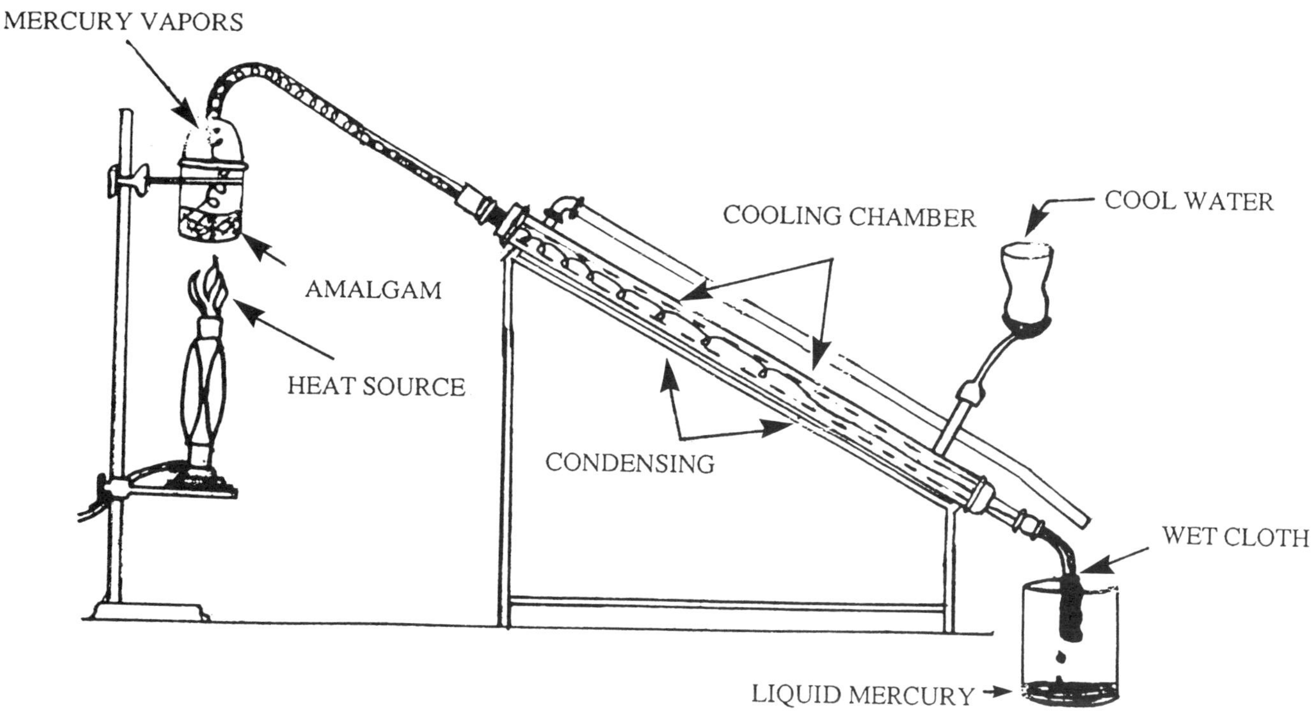

Fig. 7-9. Retort condenses the mercury vapors back into liquid form again.

to collect inside the vapor tube.

STEP 3: Place the amalgam inside the gold chamber. The amalgam should be broken apart and placed loosely at the bottom of the chamber for best results. When using amalgam, the chamber should not be filled more than two-thirds of the way to the top. When retorting liquid mercury, the chamber should not be filled more than 1/3 to 1/2 way to the top. This is to prevent spatterings of amalgam from getting into and possibly blocking the vapor tube passage--or from simply boiling bubbling mercury through the system instead of vaporizing it.

STEP 4: The cap should be sealed (''*luted*'') with clay or sealing compound. A water/flour mixture works well for this. Once the compound is placed on the upper outer edge of the gold chamber, the cap should be tightly screwed shut.

STEP 5: Check the seal on the gold chamber by blowing into the vapor tube. No air should escape from the seal around the upper edge of the chamber. If any does, re-seat the chamber and check again until you are certain of a good seal.

STEP 6: Securely fasten the retort to its stand so there is no danger of it falling or being disturbed while retorting is being done.

STEP 7: Place a small, clear, water-filled jar just under the vapor tube opening so the mercury will fall into the jar as it drops out of the vapor tube. Ensure that the end of the tube is close to the water, but not submerged. **Important:** Also make sure the jar is filled with water to the brim. This way water will overflow from the jar as the mercury drops in. This is to prevent the water from rising up in the jar, as it is displaced by the mercury, and submerging the end of the vapor tube.

STEP 8: Tie (or rubber band) a piece of cloth securely around the lower end of the vapor tube so that it surrounds the end of the tube and enters the water in the collecting jar (see Figure 7-8 for example).

STEP 9: Fill the cooling chamber with cool water, or turn on the cool water, depending on the make and model of your retort.

STEP 10: Start with a low heat on the gold chamber for the first 10 or 15 minutes. This prevents excessive boiling at first, which could, in theory, cause pieces of gold to become lodged in the vapor tube and possibly block the exit of further vapors. This could cause an explosion.

STEP 11: Slowly increase the heat on the gold chamber until mercury starts to flow from the vapor tube into the collecting jar. Continue with just enough heat to keep a steady flow of mercury into the jar.

STEP 12: When mercury stops flowing from the vapor tube, continue the heat on the gold chamber for a few minutes.

STEP 13: Turn off the heat and allow the retort to cool. **Important:** Ensure that the end of the vapor tube is not submerged under water in the collecting jar during cooling.

STEP 14: Return the mercury from the collecting jar to its proper container.

STEP 15: Once the retort has cooled, unseal the chamber and remove the gold. **CAUTION:** Some mercury vapors always remain in the gold chamber just after retorting. Be careful not to breathe them when you unseal and take off the lid.

If retorting is completed, you will be left with yellow sponge-gold as the final product.

PRECAUTIONS: There are several things to keep in mind when setting up a retort to prevent the possibility of an explosion. One is to never fill the gold chamber with amalgam or liquid mercury. Retorts come in a wide range of sizes, from the smaller ones designed to retort about 2 ounces of amalgam, to the largest ones able to handle up to two hundred pounds or more of amalgam at once. The idea is to acquire one a little larger than is necessary to service your needs. In this way, there is less chance of your needing to overfill it.

Another precaution is to make sure you start with a slow heat on the gold chamber to ensure that excessive boiling does not occur.

Always check the vapor tube, before using the retort, to make certain it is not blocked and will pass vapors freely.

And finally, the lower end of the vapor tube should not be allowed to submerge under the water in the collecting jar. The reason for this is that heating the gold chamber causes the gases there to expand and be forced out of the vapor tube. When the chamber is allowed to cool, a vacuum is created within which pulls air back up the vapor tube and into the chamber. If the end of the vapor tube is submerged when cooling is started, water can be sucked up into the extremely hot chamber, where it will suddenly expand into super-heated steam and have nowhere to go. This could possibly cause an explosion. It has occurred in the past, and is the reason for the cloth at the end of the vapor tube. A damp cloth is enough to ensure that no vapors are allowed to escape during retorting, yet will also allow air to pass back into the vapor tube during cooling.

Retorting should be done outside and downwind of any nearby occupancy. Even though the retort is supposed to recover all of the mercury, you can never be too safe. Plus, there are a certain amount of vapors released when the gold chamber is opened after retorting has been done. *So be safe and do it outside, eh?*

POTATO RETORTING

Another form of retorting can be done with the use of a large uncooked potato. This method of separating mercury from gold is not used much these days, even though the old-timers used it as a regular routine. It is, however, something you can do if you find yourself in a bind out in the field, working with enough mercury that you want to save most of it, but don't have a conventional retort on hand. The procedure is as follows:

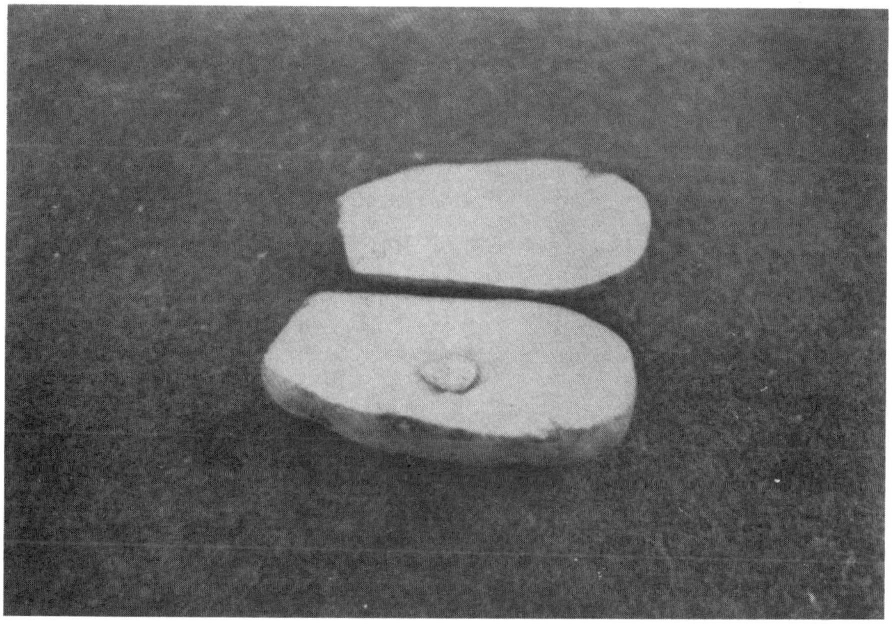

Fig. 7-10. Potato retorting Step 1: cut an uncooked potato in half and carve out a cavity slightly arger than the size of your amalgam ball.

STEP 1: Cut the potato in half lengthwise and carve a cavity in the center of one of the halves, as shown in Figure 7-10. The cavity should be made a bit larger than the size of your amalgam ball.

STEP 2: Place the amalgam in the cavity, put the other half of the potato on top, wire the potato tightly together, and wrap the entire potato in aluminum foil.

STEP 3: Place the potato as wrapped in the hot coals of a campfire for a couple of hours, or a little longer. Some vapors may escape in using this method of retorting, so make sure that you stay upwind. Also, there is the possibility a potato may explode; so keep your distance.

STEP 4: Remove the potato from the fire and allow it to cool.

STEP 5: Open up the potato and remove the gold. If retorting is complete it should be yellow and spongy.

STEP 6: Thoroughly crush the potato in your gold pan and pan it off. This should leave you with most of the mercury that you started out with.

This is a rather easy and effective way of retorting mercury when out in the field. The main precautions to keep in mind when retorting in this way are to stay upwind and DON'T EAT THE POTATO! Don't laugh, it's happened.

HALF POTATO METHOD

Fig. 7-11. Half potato retorting Step 1: Cut an uncooked potato in half in a crosswise direction and carve out a cavity slightly larger than the size of your amalgam ball.

Here's another method of vaporizing off the mercury from gold. The method can be used when a miner wants to save as much of his mercury as possible, yet he doesn't have a conventional retort, and it's inconvenient to start up a camp fire.

STEP 1: Cut a large-sized uncooked potato in half in the crosswise direction (see Figure 7-11 for example).

STEP 2: Cut a cavity for your amalgam ball into one of the potato halves.

STEP 3: Put the amalgam in the center of your small steel clean-up pan, and place the carved-out half of the potato over top of the amalgam. It should be covered by the potato as shown in Figure 7-11.

STEP 4: Place the pan over a source of heat for about 30 minutes for each ounce of amalgam being retorted, and stay upwind.

STEP 5: Allow to cool, remove potato, and extract the gold.

STEP 6: Crush the potato thoroughly in your gold pan and pan off the waste material so that the mercury can be recovered.

BURNING OFF MERCURY WITH NITRIC ACID

CAUTION: Nitric acid is a highly corrosive substance which can generally be obtained from mining supplies stores, chemical suppliers, or sometimes your local pharmacy. Anytime you work with nitric acid, you should have a fresh supply of clean water right in front of you. This way, if the acid should splash or spatter onto you, the acid can be quickly diluted with fresh water.

Acid spilled onto your skin will cause a burn if it is not diluted immediately. Acid spilled onto your clothes is just about guaranteed to cause a burn; you must remove the affected clothes immediately and dilute the acid on your skin.

Avoid breathing the vapors given off by nitric acid; they can attack the membranes inside your lungs. And the primary precaution is to avoid getting nitric acid into your eyes. And if you should, get your eyes into water to dilute the acid with no delay. Then get to a doctor. Goggles are a good idea!

Nitric acid will attack most metals--including your car. Be careful not to spill it on your valuable possessions! It should be stored in a glass jar, the right kind of plastic sealed containers, or stainless steel containers. Keep nitric acid out of the direct sunlight to maintain its potency.

Nitric acid will attack mercury, whereas it has no effect on gold. A solution of 6 parts water to 1 part acid (or stronger solution) will eat up the mercury that has attached itself to your gold. If you are dealing with a small amount of amalgam (several ounces or less), or some gold which has a small amount of mercury attached to it, nitric acid can be used to dispose of the mercury.

This process is perhaps the most often used method of separating mercury from gold today by small-scale miners.

The process can be done in a plastic gold pan, although it is more efficiently done in a small glass beaker or jar.

STEP 1: If working with amalgam, make sure that all excess mercury is removed before starting--and that all black sands and other free impurities have been separated.

STEP 2: Place the gold or amalgam in a small glass jar and set it in a safe place, downwind of any presently populated area in the vicinity.

STEP 3: Pour in a solution of 6 to 1 nitric acid (or stronger) and allow it to boil until there is no visible reaction occurring. **CAUTION:** BE CAREFUL NOT TO BREATHE THE FUMES GIVEN OFF BY THE CHEMICAL REACTION! Also, do not make skin contact with the solution, even when diluted.

STEP 4: Carefully flush the jar with fresh water to dilute and wash out the nitric acid into a separate container.

STEP 5: If all of the mercury is not yet dissolved out of the jar, with the gold back into its natural flake and powder form, use a probe to poke it around and break up the remaining amalgam. Pour the water out of the jar and pour in another dose of nitric acid solution. Sometimes it's necessary to poke at the gold just a bit to help break it up while it's being worked on by the acid. An old screwdriver works well for this.

STEP 6: When the reaction stops, flush again with fresh water. If the gold is still not back into its natural form, increase the strength of the acid solution and do the above steps again.

When dealing with small amounts of mercury, usually the gold will be thoroughly cleaned after the first bath in the acid solution. Sometimes when working with larger amounts of mercury, it is necessary to do the steps a few times, as covered above.

When the mercury is dissolved out of an amalgam mixture with the use of nitric acid, usually the gold is left with an unnatural dark film covering it afterwards. This film can be removed easily by putting the gold in a jar with some vinegar and salt, placing the top securely on the jar, and vigorously shaking until the gold is clean of the dark residue. The vinegar solution can then be flushed out of the jar with fresh water. For best results, the gold should then be washed with dish soap and water and then heated, as outlined earlier in this chapter.

CHEMICAL RETORTING

If you are burning off larger amounts of mercury with the use of nitric acid, and if you want to save most of the mercury in the process, it can be done by pouring the diluted acid solution into a separate glass jar. The acid dissolves the mercury off the gold and into the solution. Once the solution is in a separate jar, some aluminum foil should be dropped into the acid solution. When this happens, a chemical reaction takes place and the acid solution will drop the mercury in order to attack the aluminum. This causes the mercury to revert back to its natural liquid metal form at the bottom of the jar. The acid solution can then be rinsed out and you will be left with most of your original mercury. If the process is not working fast enough, it can be warmed over low heat to make it go faster. The remaining acid solution can be further neutralized by pouring in some baking soda until all foaming action is stopped.

OTHER INTERESTING INFORMATION

A solution of sodium cyanide will dissolve gold into the liquid state of the solution. Some refiners use cyanide solution as a method (called *"leaching"*) of removing all of the gold out of a set of concentrates--or out of raw ore. The solution is then removed and placed in a container from which there are various methods of extracting the gold. One way is by adding powdered zinc to the solution, which will form a compound. The compound is then heated in a furnace until the entire content is melted, which allows the zinc to float to the surface where it can be skimmed off, leaving pure gold behind.

Another advanced refining method generally used in hardrock recovery of gold particles that are extremely fine and difficult to recover by other methods is called the *''flotation process.''* Chemicals are used which cause the particles of gold to attach themselves to air bubbles being blown into the chemical solution. The fine particles of gold then float to the surface where they form a froth, which can be skimmed off or allowed to float off, leaving the useless waste material behind.

SELLING GOLD

There are numerous markets where you can sell your gold. Refineries will pay you for the fineness (purity) of the gold itself and subtract a percent, or a few percent, for refining charges. In this case you will receive a little less than the actual value of the gold itself. Generally, refineries will not pay you for the silver and platinum contained in your placer gold unless you are delivering it in large quantities. The refineries prefer that you bring your gold to them in large amounts and will often charge less for refining and sometimes pay just a bit more for the gold when it is brought to them in larger quantities.

Flakes of gold and nuggets have jewelry value on a different market. If marketed properly, flakes and nuggets can bring in twice as much as a refinery will pay--or sometimes even more.

If you are in gold country and ask around, you can always find someone who is buying placer gold from the local miners. These individuals usually pay cash. Unless the fineness of the gold within the area is extremely low, there is no reason to settle for less than 70% of the market value of the gold for that day. This means that the gold is weighed and he pays you for the weight of what you deliver. Impurities are never calculated into this type of deal. There is always someone in gold country who will pay you at least 70% for your gold--as is. If you look around, you can usually find someone who is willing to pay 75%. Sometimes you can even find an 80% straight out-buyer--which is good.

There are also plenty of guys out there who are ready to gyp you out of your gold if they can get away with it. It's always well to bring your own pocket calculator along when going to deal with a new buyer. If you go to a dealer who starts figuring a certain percentage of the fineness etc., and his final figures end up lower than a straight out 70% of the bulk weight of your gold as it is, go find another dealer. This is not to say that 70% is the going rate. You can do better if you look around. You should never have to accept less than 70% for your gold. If a dealer starts to tell you all sorts of reasons why your gold is not worth what you want for it, go find someone else. There are plenty of gold buyers around who will at least validate your gold, so there's no reason to hang around and listen to someone who's trying to steal it from you.

Local miners will know who pays the most!

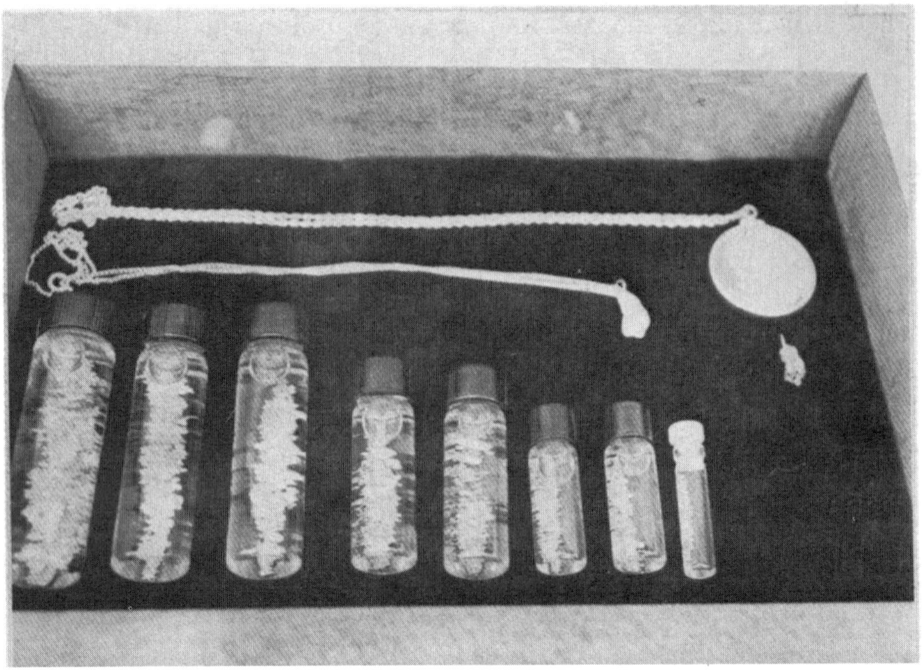

Fig. 7-12. Top dollar can be had for your gold by selling it out of jewelry boxes placed in tourist areas.

Always clean your gold well before you take it somewhere to be sold--it helps a lot.

Sometimes dentists will give you a good price for your gold, and a phone call or two can pay off. Also, some lawyers and businessmen like to invest in gold. Sometimes you can get up to 100% of spot for your fines (fine gold) when dealing with them.

Some jewelers will pay well for your flakes when they have a demand for them. It's not uncommon to get as much as 90% or better when you make such contacts.

One good way to market your placer gold is to build some small jewelry boxes and place your gold in different sized gold bottles and lockets and display them in the well-frequented stores and restaurants in your area (see Figure 7-12). The price can be marked up to about double the spot market value of gold, plus the store commission can be tacked on. There is enough gold in the gold bottles and lockets to make them nice keepsakes, gifts or souvenirs; and they do sell well to the tourists. You can fill up the gold bottles with clean water (distilled water is best). This tends to magnify the gold to a degree, making it sell even better. A jeweler can place bails on your larger pieces of gold--or you can easily learn to do it. Those can be put on a chain and sometimes bring in as much as three times the current market value--or more, depending on the individual characteristics of the various pieces. Smaller pieces can be made into earrings with the same result. Even your fine gold can be placed in gold bottles with water and sold for double spot. This is a good way to sell your gold when you don't need the money immediately. Nine or ten boxes as such, placed in well-populated areas, can bring you in a good income, as long as the boxes are kept up and moved when the sales drop off.

The best way to get top dollar for your gold is to do a lot of inquiring, always with the intention to find more and better markets. Then when you need some cash, you sell to the buyer who pays the most.

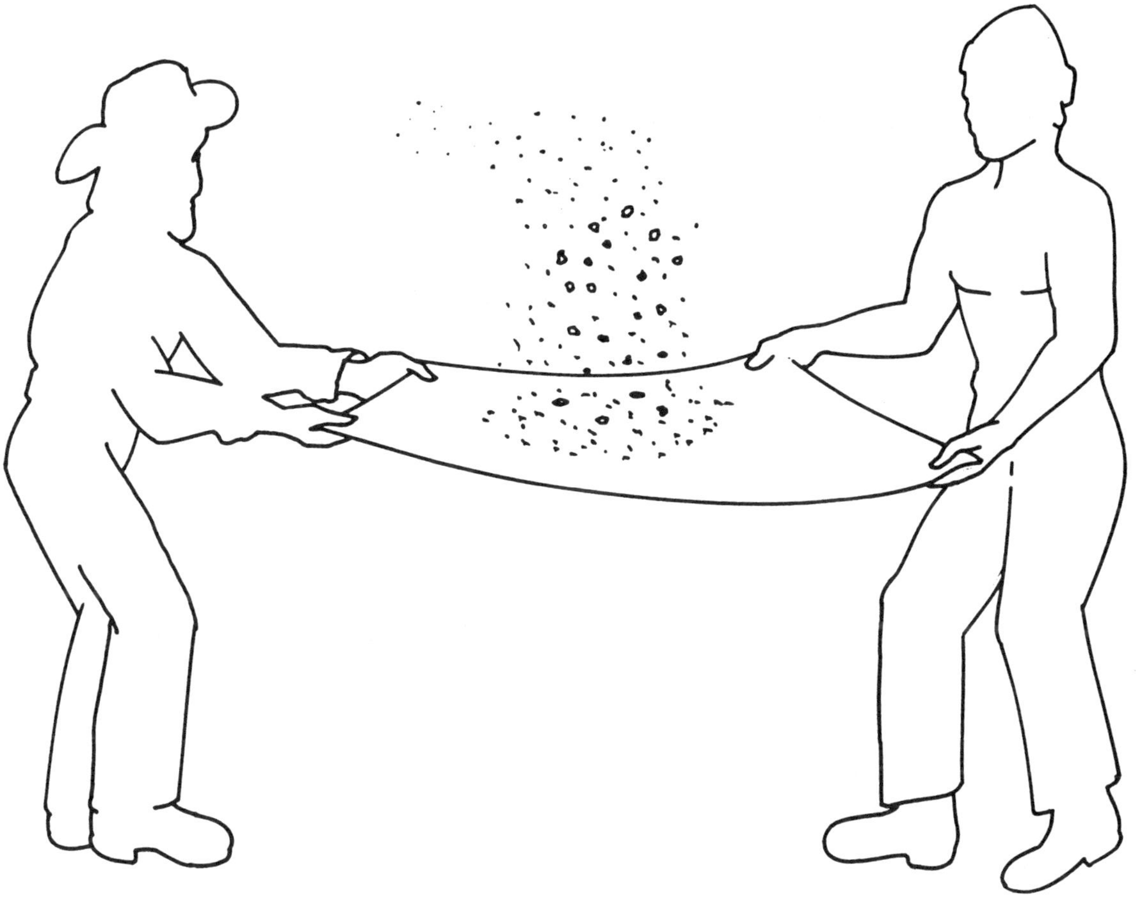

DRY WASHING FOR GOLD

CHAPTER VIII

The deserts consist of huge deposits of sedimentary material which have been affected by ancient ocean tides, ancient rivers, glaciers, floods, gully washers, and windstorms; they are literally a gold mine of placer deposits.

There is also an enormous amount of gold-bearing mountainous dry placer ground which has remained relatively untouched by large-scale gold mining activity because of the scarcity of water in the area needed for wet recovery methods.

Generally speaking, dry methods of gold recovery are not as effective or as fast as wet recovery methods. Yet, dry methods do work well enough to make some ground pay in sufficient amount to be worked in this manner at a good profit. Plus, recent developments in dry washing equipment--namely the *"electrostatic dry concentrator"* (shown later in Figure 8-8)--have made it possible for a one- or two-man operation to work larger volumes of dry placer ground without water, and obtain good results in gold recovery too.

Dry processing methods use air currents to do the same job water does in wet recovery methods. Under controlled conditions, air currents can be made to effectively wash off (blow off) the lighter worthless materials.

WINNOWING--OLDEST FORM OF DRY WASHING

''Winnowing'' is one of the oldest methods of working dry placer materials. This procedure was first developed by the early Spanish miners. The use of this method on a very broad scale for a long time is evidence that the procedure works well enough to pay, and is also further evidence of the extensiveness of well-paying placer deposits out in the desert regions. It is done with a heavy duty, large wool blanket, two men, and a mild amount of wind (see previous page).

The procedure basically consists of placing a conservative amount of dry paydirt in the center of the wool blanket. Each man then grabs hold of two corners of the blanket, and they toss the material up into the air, where the breeze can blow off the lighter waste materials. During this process, the two men should be positioned in such a way that the waste materials do not blow on top of them or back onto the work site. Best results are obtained when the material being worked has already been screened through 1/8 inch mesh screen (or smaller). Otherwise, it takes substantially more wind to blow off the larger materials. When you have that much wind, you will start losing a higher percentage of the fine-sized gold values. How far down in size the material needs to be classified to get the most efficiency out of winnowing depends a great deal upon how much wind is blowing. A very mild current requires that the materials be classified to a smaller size for winnowing to work effectively. So it's best to have a few different classifications of mesh screens on hand for changing wind conditions. With a little trial and error, the operators will get the idea of what classification is necessary to get the most out of the breeze blowing at the time.

The wool of the blanket will tend to latch onto and trap some of the smaller-sized pieces of gold during the winnowing process.

The optimum gold recovery in winnowing can be obtained by working materials down to the heaviest concentrates, picking out the larger pieces of gold, and then pouring the concentrates into a container to be worked later by wet processing methods. In this way, a large amount of material can be worked into a much smaller amount of concentrated gold-bearing materials, which can then be more easily packed to a water site. To get the concentrates from the wool blanket into a bucket without losing any of the values in the process, a strong sheet of plastic or a tarp can be used. The concentrates can be poured from the blanket onto the plastic, and then from the plastic to the bucket. When this method is used, care should be taken to winnow the material down to the heaviest of materials before pouring into the bucket. This way, less material needs to be packed and worked again at a later time. Winnowing, when being done properly, is effective--more so than dry panning, and perhaps even more so than some of the dry washing plants available on today's market. You'd be surprised after trying it for a while! The secret is in screening the material properly and in not trying to winnow too much material at once. It doesn't take long to get the hang of it. The main disadvantage is that you are depending on a steady mild breeze to continue.

Winnowing is probably not best used as a production activity anymore. This is because of the limited amount of material that can be effectively processed, and because of the dry production plants now available, at rather low cost, which can produce so much more (covered later).

Probably the best use for winnowing is in a two-man sampling activity, in an effort to locate some good production ground. A winnowing operation has great accessibility, the only tools needed being: wool blanket, bucket, shovel, classification screens, a few sheets of plastic, and a gold sample bottle. This isn't very much for a two-man team to carry around. Also, winnowing is very fast to set up. Actually, it's just a matter of laying the blanket down and screening the sample material onto it. A dedicated two-man winnowing team can get a lot of area sampled in a day's time. And the samples would be pretty accurate too, if the concentrates were packed back and worked later by wet processing methods. This is one way good paying ground could be located which could then be mined with larger, more productive equipment.

DRY PANNING

Panning dry placer deposits without the use of water can be done with some success after just a little practice. The method is really not much more difficult to do than wet panning. However, no matter how well dry panning is done, it won't be as effective as wet panning. This is especially true when it comes to fine gold recovery. Also, wet panning methods are much faster than dry panning. But, when no water is available, and you are in a potentially high producing gold location, you have to use whatever methods are available to you that work. And the gold pan can be used to process dry material to some results.

Dry panning is not best used as a production activity, although records show that it has been used for this purpose in the past with success. Today, there are simply too many economical machines that can do a better job in a dry production operation. Dry panning can be used rather effectively in a sampling activity, especially when used in conjunction with wet methods to process the heavier concentrates recovered by dry panning.

The procedure is as follows:

STEP 1: Fill your gold pan about half way to the top, or perhaps just a little less, with sample material. Ensure that all of the material you put into the pan is broken up thoroughly into separate individual parts, instead of clumps of conglomerated material (see Figure 8-1 on the next page).

STEP 2: Classification of material through 1/8th-inch screen before it is panned is best. If such a screen is not available, use your fingers as a rake to sweep the larger sized stones and pebbles out of the gold pan. Do this until most or all of them are out (see Figure 8-2 on the next page). Be extra careful to look for the larger pieces of gold while doing this step so that you don't throw any away along with the larger pieces of waste material. Keep in mind that pieces of gold in

Fig. 8-1. Dry panning Step 1: Fill pan a little less than half full of thoroughly broken up dry material.

dry material do not show themselves nearly as well as when uncovered in water.

STEP 3: Toss three pieces of lead shot or BB's into the pan of material. These will help you see what is happening to the heavier materials in your pan while you are working it. Lead is best for this because it is closer to the specific gravity of gold than BB's are and will act more like the gold in your pan (see Figure 8-3). Once you have learned how to dry pan and have some experience in knowing what to look for, you can skip using the lead shot. Many prospectors prefer to use them

Fig. 8-2. Dry panning Step 2: Rake the larger stones and pebbles out of the gold pan.

Fig. 8-3. Dry panning Step 3: Drop three pieces of lead shot into the material within the pan.

though, because they consistently show what is happening with the heavy material in the gold pan.

STEP 4: Vigorously shake the pan back and forth to start the heavier materials working down towards the bottom of the gold pan (see Figure 8-4). This is the time to crumble up any conglomerated material which has not yet been broken up. Be careful not to overdo the shaking so much that some material is spilled from the pan.

STEP 5: Now, take the pan in one hand, holding it in a level position, and start tapping the side of the pan with your other hand. The tapping should cause quick jerks to the pan. This should make the material inside jump slightly (see Figure 8-5).

Fig. 8-4. Dry panning Step 4: Vigorously shake contents.

Fig. 8-5. Dry panning Step 5: Start tapping pan to the side to cause the material to start jumping slightly.

STEP 6: While continuing to tap the pan on the side, slowly tilt it forward just enough to regulate a small amount of lighter materials to flow over the forward edge of the gold pan as the tapping action vibrates them forward (see Figure 8-6). Watch closely for the location of the pieces of lead in your pan while doing this. If the pieces of lead start moving towards the forward edge of your pan, tilt the pan back to the level, start a new tapping action and then tilt forward again to vibrate the lighter materials out of the pan.

STEP 7: Every once in a while, or whenever the pieces of lead show themselves to be moving towards the forward edge of the gold pan, tilt the pan back to the level position and reshake the materials. Then continue to tap the lighter materials off the forward edge of the pan. This is to

Fig. 8-6. Dry panning Step 6: Tapping the lighter materials out of the forward edge of your pan.

prevent any heavier material (gold) from being worked up the forward surface by the tapping action.

The tapping of the pan is the main key to successful dry panning. It's this action which vibrates the material in the pan and shakes it into a somewhat suspended state. The suspended state will cause the heavier materials (gold) to work their way downward in the pan. So how the pan is tapped can make all the difference in the results of this activity. You can see the results of this tapping action by watching what happens to the lead shot in the pan, while trying different kinds of tapping actions. When the tapping action makes the lead shot stray down toward the lower end of the pan, while the lighter materials are vibrated out, you have got it right. You'll catch on with a little practice at this.

Dry panning can be continued until there is nothing left in your pan except the lead shot and gold. Yet, I recommend that instead of panning all the way down to gold, you only dry pan down to the heavier concentrates. Then take them, once enough are accumulated, to a water site and finish them off by wet panning. The reason for this is that dry panning is not very efficient. It becomes less efficient after you have worked your way down to the heaviest concentrates within the pan and are attempting to vibrate the lightest of these out. Fine gold and small flakes can be lost in the process anyway. This becomes increasingly true once you are down into the heavier materials within the pan. If you are sampling around in an attempt to locate a paystreak, often these fine flakes of gold are the only sign that rich paydirt is present. So you don't want to lose any more of them than you can possibly prevent--especially in a sampling activity. Otherwise, you can waste your time and effort passing over paystreaks and testing them, and never know of their existence.

One of the best plans when sampling out dry areas with a gold pan is to bring along a container. After the sample material is worked down to the heaviest concentrated material, pick out the lead shot and larger pieces of gold and pour the concentrates into the container for more efficient processing methods to be done later. In this way, you can get a somewhat accurate index of how rich certain areas are by dry panning. A large coffee can works well as a container for dry samples. Sometimes a larger container is needed, depending on the amount of sampling to be done on a single trip.

One thing you will want to remember in the field is not to mix the concentrates taken from different sample locations. Ziplock baggies are useful to keep samples separate when being carried in the same container.

One thing worth mentioning is: if you pan down to the concentrates, and find pick-out sized pieces of gold in your pan, consider that you have found a likely hot spot and start figuring out how to best work the location production-wise. Smaller pieces of gold do not show themselves readily when using dry methods. So if you are seeing it, it's definitely a good sign.

One of the first things to do when you've found a hot spot in the desert is to take note of its location. It is *very easy* to lose a spot in the desert, once you have left it. On top of that, a single wind or rain storm can change everything around and make it look different. The best way to mark a spot in the desert is by checking its position with regard to the large fixed objects in the near vicinity--or those further away if none are close. Here's where your magnetic compass comes in handy. With a compass you can note the exact bearing to three or four permanent objects in different directions and write them down. Then you will always be able to get back to that exact position, no matter what happens to the face of the desert. And believe me, if you find a rich deposit out in the desert, you can never be too careful about not losing track of it.

DRY PAN SAMPLING PROCEDURE

Here's an effective dry panning procedure which can be used successfully to locate paying placer ground in dry areas:

STEP 0: Tools needed: gold pan, ziplock baggies, bucket, 1/8 inch mesh screen--to fit over the mouth of the bucket, shovel, concentrates container (coffee can), and some black plastic sheets.

STEP 1: Uncover the material you want to take a sample of.

STEP 2: Screen the prime sample material into the bucket until you feel you have a sizable enough sample.

STEP 3: If the sample material is not perfectly dry, spread it out thinly over your plastic sheet in the direct sun until it dries. This usually doesn't take long. If for some reason this step is going to take a while, it's good to have extra plastic sheets on hand and move on to set up your next few sample locations in the same way. Then come back later to complete each sample.

STEP 4: Dry pan the screened sample material down to the heavier concentrates and pour the separate sample results in separate baggies. Mark each baggie properly and place in the sample container.

STEP 5: Once you have a full load of concentrate samples, bring them back to your car or truck and wet pan them into a washtub to get an idea of how much gold is present in each sample.

If you have a large car or a truck, you can haul plenty of extra gallons of water and several washtubs out into the field with you to assist in your sampling activities.

Sometimes you can drive your vehicle directly to the sites where you want to sample. In this case, you can skip dry panning and wet pan your samples directly into your washtub. A second washtub can be used to pour the water into from the first tub, once it starts getting too much material accumulated inside from the panning. When panning raw material like this directly into a washtub, the water becomes muddy very fast and it is a little more difficult to see what you are doing. But you get used to it, and it's still much more effective and a lot faster than dry panning.

If you can't get your vehicle out to the sampling site, then you will have to settle for hauling the dry concentrates back to the vehicle to be wet panned. If you do not have room in your vehicle for wet panning supplies (filled water jugs and washtubs), then you really don't have the proper vehicle for the job. But it can still be done; you'll just have to bring more sample containers along with you and take them somewhere else to be wet panned later.

If you plan to take more than one sample on a single trip out into the field, it's always good practice to bring along a magic marker and be sure that you mark each sample separately. This way you can keep track of which sample came out of which hole. You can also bring along some paper and make a map if necessary.

Never mix the samples from different locations together in the same container before you have finished testing them. If you do so, then you no longer have an accurate index of the material value from each separate location you have sampled. For example, let's say you sample six different locations. One is very rich and the others carry little or no amount of gold. If you mix all of the samples together, the result will probably show a poor average of values. You might never realize that one of the samples was very rich. And if you do realize that one of them might have been rich, you'll have to go back and re-sample each one separately to find out which. But that's what you are supposed to do in the first place. Get the idea? This holds true of any kind of sampling activity in mining, no matter what it is that you are doing.

If you are able to haul water in your vehicle, yet you can't quite get your vehicle out to your sampling site, you might consider the idea of also bringing along a wheelbarrow, if there's room. Of course, it depends on the terrain, but a wheelbarrow can be a big help in either hauling sample material to water, or in hauling water-filled jugs to your sample site. Sometimes you can get your vehicle to a good generalized location where you'll want to do lots of sampling. Then the wheelbarrow can be used to haul different samples to the water so they can be wet panned. This can be lots faster and more effective than the combination wet panning/dry panning method. It depends on how far the samples will have to be hauled and what the terrain is like. It's something that you'll have to decide. If a wheelbarrow is to be used for hauling sample material, you should consider the idea of building a screening device that will rest over top of the wheelbarrow--to speed up the classification procedure. The Chinese miners were well known for hauling material to water sites this way.

Needless to say at this point, sampling for gold deposits in the dry regions is a lot of work. This is true of sampling in any area, but more so in the dry areas; because the process often entails hauling material a distance, or setting it out to dry. However, do keep in mind that the dry regions, the deserts especially, are almost entirely virgin of earlier mining activity. The gold deposits, which were once there thousands of years ago, just about all still remain there today. The deserts of today remain as virgin as the Mother Lode area in California when the 49'ers first found gold there. Some extremely rich deposits have been found in the deserts during the past. Many still remain in place. To find them requires sampling activity. Sampling activity means hard work and persistence to make it pay off. The sampling activities as laid out above do work.

One other thing to keep in mind about the dry regions is that you don't have to get off the main beaten trail in order to locate virgin ground. It is often necessary to get off the beaten trail in order to find entirely virgin ground when you are prospecting for gold in the gold bearing mountainous areas where water is available. But in the dry regions, especially in the deserts, this is almost never the case; because these areas are mostly unworked by large scale mining activity. This is not because of the lack of gold, but because of the lack of water being immediately available for large-scale processing methods. So you don't need to get way out in the middle of nowhere in the deserts to find likely areas to carry bonanza-sized deposits. You may be just as likely to find them in the washes located just off the highway. It's just within the last 60 years or so that the deserts have been accessible to the interested miners, because of the availability of transportation that will get you, your equipment, and water to the promising locations. It's just within the past few years that interest has picked up again in mining gold. Taking all of this into account, very little effective gold mining activity has been done in the dry regions. Yet, with the equipment that is available today, these areas probably show more potential for the small-scale miner than any other single area, possibly with the exception of gold dredging. Which area shows more promise? That is a question which will be answered truthfully within the next few years. I'm hoping that the desert areas prove themselves out--simply because of the vast size of the totally virgin ground that is presently available.

WORKING DRY GROUND ON A PRODUCTION SCALE

Once you have located ground in a dry area which is paying well enough to be worked on a production scale, in order to recover paying amounts of gold, you will need to figure out the best way to work the ground.

Sometimes a road can be bulldozed to your spot. Sometimes you can drive right in with a 2- or 4-wheel drive truck. In these cases, you might consider the idea of screening the paydirt into the back of a truck and hauling it to a wash plant to be processed elsewhere. Actually, this is just slightly more difficult than shoveling directly into a wash plant. The hardest part is breaking the material away from the streambed and classifying it. It takes a little more time to haul the material to the wash plant--but that depends on the distance and the condition of the road. It's also more difficult to shovel up into a truck. Some small operations use a conveyor belt to lift the material into their truck. Feeding the material from a truck into a wash plant is not difficult, because it's usually downhill. A one or two-man team could probably move the equivalent of a pickup-sized load of screened paydirt, plus run it through the wash plant at another location, in the period of a full day's work--perhaps two truckloads. And if the material was paying well, they could do well at it, too. This is just one possibility of working dry paydirt on a small production scale that should be considered.

Fig. 8-7. Lightweight, hand operated, drywashing plant--used for sampling or production purposes.

DRY WASHING PLANTS

If conditions do not allow you to truck the paydirt to a nearby water site to be processed by wet methods, you will have to consider processing the rich material by dry production methods.

While dry panning and winnowing do work, and have been broadly used as a means of production in the past, they are not nearly as effective as some of the modern drywashing plants which are available on today's market.

Dry washing machines use an air blowing fan or bellows-type device to blow a controlled amount of air current up through the dry material being processed. The air current will blow off the lighter materials and allow the heaviest particles, gold included, to collect.

Dry washing plants are available which can either be operated by hand, or by lightweight engine/air fan assemblies.

The hand-operated dry washing plant, as shown in Figure 8-7, has its own classification screen

Fig. 8-8. Electrostatic concentrating drywashing production plant, which uses high frequency vibration to improve recovery.

as part of the unit. Raw material can be shoveled directly onto it. The bellows air blower is worked by turning a hand crank, which is conveniently located so that one man can both shovel and alternately work the bellows at the same time . Two men working together can process up to a ton of gravel per hour by taking turns at one man shoveling while the other works the bellows. The particular machine as shown in Figure 8-7 weighs 32 pounds and so makes for an excellent sampling tool, too, as well as a small-scale production device.

There is also a 12-volt electric conversion kit, weighing an additional 10 pounds, available for this unit. This gives you the option to either hand crank it in the field, or connect it to the 12-volt battery on your vehicle for automatic bellows operation.

Figure 8-8 shows one of the latest developments in dry washing equipment available on today's market. It is a gasoline motor driven dry concentrating unit, which uses static electricity and high frequency vibration to help with gold recovery. In taking an overall look at this machine, first, there is a high-powered air fan, which pumps air through a discharge hose to the concentrator's recovery system. The air currents which pass through the recovery system are adjustable so that the proper amount of flow of lighter materials through the recovery system can be obtained--similar to a sluice. The purpose of the steady airflow is to "float off" the lighter materials through the box. The heavier materials, gold especially, will have too much weight to be swept through the recovery system by the airflows.

The bottom matting in this concentrator is made of a specialized material which sets up an electrostatic charge as high velocity air is passed through from the air discharge hose. Fine pieces of gold, while not magnetic, attract to surfaces which have been electrostatically charged, similarly to the way iron particles are attracted to a magnet. So the bottom matting in this type of concentrator attracts the fine gold to itself and tends to hold them there. The hand-operated dry washing plant shown in Figure 8-7 also sets up an electrostatic charge when being operated. But this motor-driven unit keeps pumping all the time. So the electrostatic charge remains in full force all the time. The result is improved gold recovery.

This type of concentrator also uses a high frequency vibrating unit to keep the entire recovery system in continuous vibration while in operation. The purpose of this is the same as tapping during dry panning, or shaking during wet panning. The way to get gold particles to settle quickly down through other lighter materials is to put the materials into a state of suspension--either in water or in air. This is the purpose of the vibrating unit on this concentrator. It vibrates the fine particles of gold down through the lighter materials being suspended by air.

This machine is excellent for the production demands of a one- or two-man sized operation. It is able to process up to two tons of raw material per hour, which is the equivalent of what a medium-sized wet sluicing operation can produce. This is as much or more than one or two men can shovel at production speed when in streambed material. All that's necessary on the part of the operator is to set it up, start the engine, and to shovel into its classification unit. The concentrator does its own screening of materials and everything else automatically. This leaves the operator free to produce at his or her own comfortable speed.

This vibrating electrostatic concentrator is one of the best gold recovering dry washing production plants that has ever been developed. Some large-scale professional dredging operations use this concentrator for their daily cleanups, and they seem to be very happy with the results.

Total weight is about 75 pounds, but the unit breaks down into separate pieces which can be carried by a single person. So the electrostatic concentrator can be carried to a hot spot if it's worth a few trips to do so. It also gets excellent gas mileage, about 3 hours to the gallon. Therefore, it can be used as an effective sampler if larger and more accurate samples are wanted.

SETTING UP A DRY WASHER

There is no fixed formula for setting up the proper air flows and downward pitch of a dry washer recovery system. Much depends upon the nature of the material you are processing, how heavy it is, whether or not the material is angular or water worn,and the purity (specific gravity) and size of the gold being recovered.

You want to remember the purpose is to separate the gold from the lighter waste materials. If you have a small amount of air flow, naturally, you need more pitch on the recovery system--and you may need to feed the material slower. Too much air flow can also be a problem and would require less pitch on the riffle board.

Watch how the material flows over the riffle board. You should see the dirt rise up in an orderly fashion and flow over top of each riffle. It looks an awful lot like water. Keep a steady feed of material going through a dry washer at all times. The riffles should be filled about half to three-quarters, with a steady flow moving from one riffle to the next. The material in the riffles should have a fluid look to them; they should not be packed solid.

It's a good idea to shovel lower-grade material into your dry washer while adjusting for the proper air flows and pitch. Once set, you can shovel in the paydirt.

One thing about dry washing is that because it is slower and more difficult, the paydirt must have more gold. High grade areas in the deserts certainly do exist! All the more reason to make sure your recovery system is set properly before processing paydirt. Chances are you will not see the gold lost to the tailing pile through a dry washer recovery system.

Another thing, which goes without mentioning, is that you always set up a dry washer downwind of where you are working!

Once gold falls into the dead air space in the riffles, it will stay there. The air flows are not strong enough to move it anymore. There is a limit to this, however. Just like a water recovery system, a dry washer will concentrate the heaviest materials which it processes. After some time, the heavier concentrates may require stronger air flows or a steeper pitch to keep them moving. At this point, it is time to clean up the recovery system. If you are keeping a close eye on your recovery system, you can see when it is time to clean up; the fluidity of the material inside the riffles becomes more concentrated and slows down.

DRY WASHING AND CLAY-LIKE MATERIALS

Again, material must be thoroughly dry to get the best results out of any dry washing plant. Sometimes, when out in the dry regions, you will run into moist clays--just like you do in the wet streambed areas. It is also possible to find a pay layer along with the clay. Clays make dry washing procedure more difficult, because they must be thoroughly dried out and crushed before being processed effectively by dry methods. Sometimes this means the material needs to be set out in the sun to dry for a full day or more before anything further can be done with it. Sometimes it's necessary to dig clay a couple of days ahead of the processing stage. You can alternate spending a day digging and laying out material to dry, and then a day processing dried material. This takes longer and the procedure is more involved, but if a good paystreak is involved, you do what is necessary to recover the gold out of it. And if you are using a dry recovery process, it's necessary to dry the material out and fully break it up first.

Sometimes clay material will dry into very hard clumps, and to break it up and crush it down on any kind of a production scale, it is necessary to use rock crushing machinery--the same kind which is used in lode mining operations (see Figure 9-6 in the following chapter).

The cleanup of concentrates from a dry washing plant is also best done by wet-panning. Usually, if you have room enough to haul around a dry washing plant in your vehicle, you'll also have room for enough water to pan off your final concentrates, too. However, if for some reason water is not available to you, the cleanup of your dry concentrates can be done quite effectively by running them through your plant several times. The final cleanup steps, as laid out in Chapter 7, can then be done to separate the gold from the last bit of remaining waste material.

DESERT PLACER GEOLOGY

The chances of finding a hot spot out in the desert, or in the other dry regions, are probably just as good or better than your chances of finding a hot spot in the watersheds of the gold bearing mountainous areas. These chances are pretty good, providing that you are willing to spend the time, energy, and study required to find such locations.

Probably your best bet is to start off with the ***"Where To Find Gold"*** books and study the geological reports that apply. There has been some small-scale mining activity out in the dry regions. Much of it was lode mining, but some placer activity took place also. A good portion of prior activity is recorded information today. It can be of great value to you to know where gold has already been found. It's almost a sure thing that the areas which were once worked for gold at a profit were not entirely worked out. They might be worked again with today's modern equipment at a profit. An area which has once proven to pay in gold values is a good generalized area to do some sampling activity to see if any new paydirt can be found. The desert areas were pretty much left alone by the large-scale mining activities of earlier times because of the accessibility problem. Often, there was not enough water to sustain life, much less to process gold-bearing material.

But the desert prospector should not limit himself to only the once proven areas. Most of the desert regions have gone pretty well untouched by effective sampling activity because of the accessibility problem, lack of water, and proper equipment to do the job up until recent years. So the desert rat has got a lot of ground at his or her disposal and very few competitors to worry about.

A single large rain or wind storm can change the entire face of the desert in just a few hours' time. There is very little undergrowth in these areas to prevent a good-sized rain storm from causing an incredible amount of erosion. And so you hear all the old-timers' stories of finding bonanza-sized gold deposits, marking their position, going out after tools and supplies, and then returning to find the desert entirely changed and the bonanza apparently gone. Undoubtedly, some of these treasure stories are true. After all, many of those old-timers had gold to go along with their stories. Many of them spent the rest of their lives looking for their ''lost mine.'' All indications point to the desert having the greatest placer deposits left on earth--other than the oceans. Many of the deserts were once the ocean floor, not to mention that a lot of ancient streambed still remains there. The same ancient streambeds have proven to be extremely rich when encountered in the mountainous areas.

All of the placer geology which applies to streambeds, as laid out in Chapter 2, also applies to desert placer deposits. The same remains true of eluvial deposits--which is the gold that has weathered from a lode and which has been swept some distance away by the forces of nature. Eluvial deposits in the deserts (called *''Bajada placers''*) tend to spread out much more widely, and in different directions, because they are usually not eroding down the side of a steep mountainous slope. Because of this, they are sometimes a little more difficult to trace back to their original lodes. But it can be done, the answer is to do lots of sampling.

History has shown that one of the best locations to look for gold is where the hills meet the desert and fan out. This is where the water slows down during floodstorms and drops gold in the gullies and washes. There also may be more gold traps further up the hillside.

When doing generalized sampling in the desert, concentrate much of your activities in the washed-out areas, where natural erosion has cut through the other sediments and created a concentration of heavier materials. Dry washes, dry streambeds and canyons are good for this. Get an eye for the terrain, and in looking over the high points and the low points, you'll get an idea of where the water runs during the large storms. Areas where the greatest amount of erosion has taken place are areas where the highest concentration of gold values are most likely to be. Remember that we're looking at many thousands of years of erosive effects--not hundreds. Most of these areas are untouched. In some low areas, as in canyons and dry washes, bedrock will be exposed. These are ideal places for you to get into the lowest strata of material--where the largest concentrations of gold values are most likely to be. Look over the various canyons, large and small alike. These have been formed by many years of erosion and are likely spots to find paying quantities of gold.

The desert and dry areas also commonly have a *''false bedrock''* layer specifically called *''caliche.''* Sometimes (often), this caliche layer is only a foot or two deep (see Figure 8-9). In some areas, gold is concentrated along the caliche, just like on bedrock. After a storm in the desert, in some places you can find small pockets of gold in the caliche gravel traps, under rocks and under boulders. Sometimes the gold is pounded directly into the caliche and needs to be removed with a pick or crevice tool. Caliche layers close to the surface allow small-scale dry washing operations to be economically feasible, because of the lesser amount of gravel and material needing to be shoveled off the gold deposits.

Streambed material can be recognized by the smooth waterworn rocks. Where actual streambed material is present, it is a prime area to be sampling. Such material indicates that it has been exposed to a great deal of running water. This means concentrating activity took place with

Fig. 8-9. Caliche is a cement-like false bedrock which is commonly found in desert placer areas.

those same materials. It's possible that the material was once washed out of an ancient river, and a good-sized gold deposit may be present, too--or nearby.

But gravel and material does not need to be water worn in order to carry gold in the desert areas. Rough and angular gravel, which has not been greatly affected by water, also sometimes carries gold in volume amounts. Testing is the key.

Sometimes it's worthwhile to do some sampling in different strata of desert material when they are present and exposed. Flood gold layers apply even more so in the desert because of the flash floods which can occur there.

Sometimes substantial deposits are found just beneath boulders that lie on bedrock, or up in a layer above bedrock.

When you find a gold deposit in a dry area, whether on bedrock or the caliche, you will want to thoroughly clean the underlying surface upon which the gold is resting. Seldom will you see gold in dry placer--even when there is a lot of gold present. Use a whisk broom to sweep up all of the loose material--or better yet, a Mack-Vack. Sometimes it is also productive to break up the caliche or bedrock with a pick or other crevicing tool.

Sometimes, in dry washes, you can actually see stringers of black sand along the bedrock or caliche--especially directly after a storm.By following these stringers, and digging out the concentrated gravel traps, you can sometimes do exceptionally well. Don't forget to test the roots in such areas. Vegetation requires mineralization to grow. Roots can grow in and around high-grade gold deposits. I have heard of single roots which have produced as much as three ounces of gold!

Some electronic prospectors use their metal detectors to trace concentrations of black sands, and follow up by testing the areas showing the strongest reads from their detectors.

Some desert areas, like Quartzsite, Arizona, also have gold just lying around anywhere--even on top of the ground. Such places are excellent for electronic prospecting and dry washing. The deserts of Australia are famous for this. I have friends who have been very successful in the Nevada deserts, using metal detectors to recover large numbers of nuggets, some very large, directly off the surface of dry desert ground.

One thing to know about this is that if you find a piece of gold on the surface of a dry placer area, it is likely that there are more in the immediate area. Electronic prospectors call these areas *''patches.''* Gold generally does not travel alone--unless it was dropped there by mistake.

Sand dunes in the desert are not usually productive because they mainly consist of lighter-weight sands that were deposited there by the wind. However, sometimes the wind can blow off the lighter-weight sands, leaving the heavier materials behind and in an exposed state, similar to the gold beaches. This is something that should be watched for.

OTHER TIPS

When prospecting around in the dry areas, when you encounter tailing piles from past dry washing operations, it is worthwhile to do some raking of the tailings, and scanning with a metal detector. Sometimes these old tailing piles can be very productive; sometimes worth running through a modern dry washer.

DESERT SAFETY

The desert can be very dangerous if you go out into it unprepared for its potential of extreme change. Those people who have spent a lot of time in the desert respect it in the same way that experienced seamen respect the mighty ways of the ocean. In the desert, the extreme heat during the day, and the bitter cold nights, can be equally hazardous if you are unprepared for them.

Here are a few safety tips for you to take note of, if you should decide to venture out into the desert in your quest for gold and you are unfamiliar with desert ways:

1) Be sure that your vehicle is in good running condition with plenty of gas in the tank and water in the radiator.

2) Bring a lot of extra drinking water along--GALLONS, even if you don't plan to stay long. Also, some extra nutrition is a good idea. It's possible to get stranded, in which case you will get hungry and thirsty.

3) When traveling out on the desert trails and roads, keep track of your route and of where you are. You'd be surprised at how easy it is to get lost when you are wandering around out there. Mark your trail as necessary. Bring along a map of the area and a compass, and know how to use them to determine your exact location if necessary.

4) Bring along a snake bite kit, and know how to use it, too. These are not expensive, and you never know; it could save your life. When walking around in snake country, watch where you step. Never step over a rock or log into its shadow. That's where the snakes take shelter from the hot sun. Avoid such places, or step up onto the rock and jump over the shadow area out of striking distance.

5) It's also not too difficult to get your vehicle stuck in the desert sand. Bring along a good sized piece of heavy rug or carpet. It will be a big help to get you some extra needed traction. Sometimes, you can let a little air out of your tires and get extra traction when stuck in the sand. However, be careful not to overdo it. A few pieces of 2 X 4 wood, about 3 or 4 feet long, can also be handy to have along when you get stuck. Be sure that your jack is in good operating condition before venturing out into the desert. Sometimes it can be your most useful tool in getting your vehicle out of the sand.

6) Always camp on high ground, never down in the dry washes or arroyos, even if it's nicer down inside of one because of a stream or scenic view, etc. The reason is that when a good-sized rainstorm hits the desert, the various washes and erosive waterways flood very quickly. If you happen to be in one when the flood hits, it's goodbye to you and your gear. These floods can and have occurred just as readily from a rainstorm which is happening elsewhere.

7) Bring extra warm clothes, even if you don't plan to stay the night--you never know.

8) And finally, don't overexert yourself when working in the heat. It's better to work steadily and stop for a short rest when needed, if the heat starts getting to you. Have salt tablets on hand. Know the symptoms of heat exhaustion and know how to handle it, should you feel it coming on.

LODE MINING BASICS

CHAPTER IX

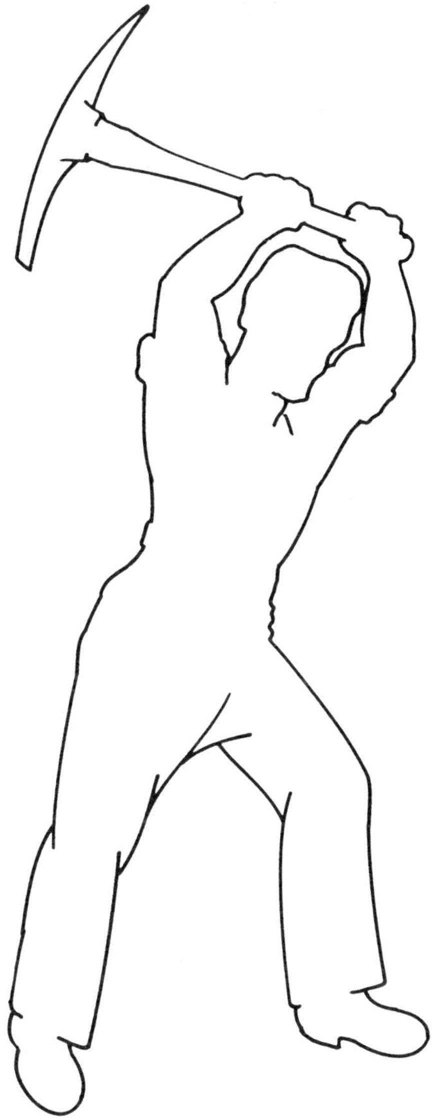

Prospecting for lode mines and developing them is a highly specialized field. It often requires a substantial financial investment and application of a very large and sophisticated technology. The entire subject is far more involved than the scope of this simple chapter. However, anyone engaged in the field of placer mining should have a general idea of what lode mining is and how lodes are found. Therefore, I will outline the basics of lode prospecting in this chapter.

FINDING LODES

Probably the oldest and most often used method of finding a lode is looking for and tracing quartz float back to its original source--the lode itself. *''Float''* is the large and small pieces of quartz which have weathered off the vein, and have been moved down the mountainside as an eluvial deposit, as covered in Chapter 2.

Quartz float is usually some shade of white in color. Consequently, it is often rather easy to see in contrast to the predominantly darker colors of the surrounding rock and debris on the mountainside.

A good pair of binoculars is helpful when looking for, and tracing, float deposits.

Gravity, and the other forces of nature, tend to move quartz float downward at an ever-widening distance from the lode. The further such an eluvial deposit is moved from its source, the wider and more dispersed it usually becomes--much like in a triangle, as shown in Figure 9-1.

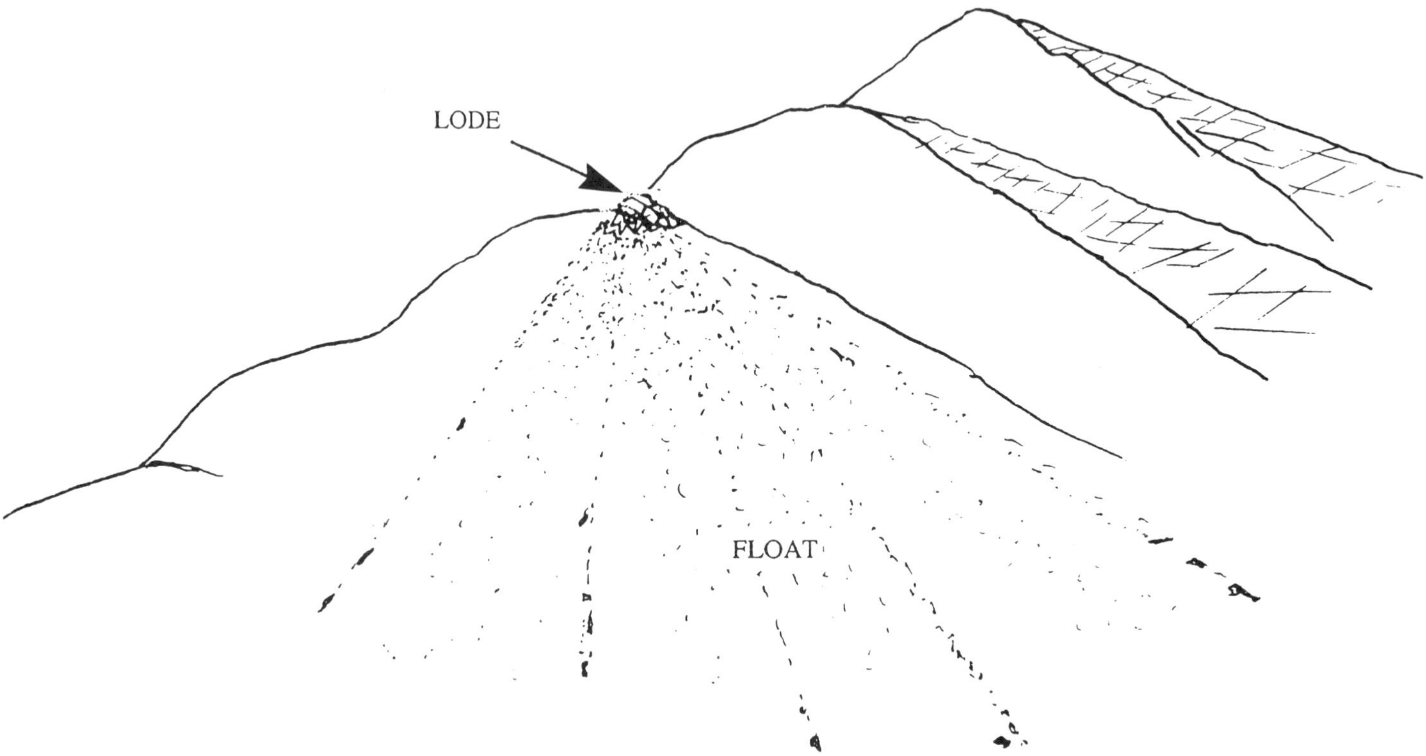

Fig. 9-1. Float usually becomes more widely dispersed, as it is swept further away from the lode.

When promising looking float is located on a mountainside, the general direction to follow in order to find more, is upward.

In a stream of water, when a placer deposit is located where all or most of the pieces of gold are very coarse and jagged, and perhaps a good deal of quartz is still attached to some of the gold, it's a sign that the lode might be nearby. The gold is possibly being swept into the streambed a short distance upstream. This is more apt to occur in the tributaries to the larger rivers, because the gold in most streams and creeks has been supplied by such lodes. That is, unless such waterways cross an area where an ancient river once ran. Many of the larger rivers have received a large portion of their gold from the ancient riverbed deposits. Regardless of the type of stream, if a placer deposit of very rough gold is located, it generally warrants an investigation of the banks in an upstream direction to see if any gold-bearing quartz can be found.

Float, much like gold, tends to round out and acquire smoother edges as it is washed further away from the lode. So the roughness of a piece of quartz float can be an indication of how far from the lode it has traveled.

How much work should go into the searching for a lode depends on the characteristics of the float being found. A piece of float showing visible signs of gold will prove to be a very rich specimen. It indicates the likelihood of a very rich lode too, if it can be located. Prospectors have been known to search for years in an effort to find a single lode from which they have located one or more pieces of promising-looking float. In fact, many very rich mines were located in just this manner, after a long period of systematic searching done by a persistent prospector.

Float does not have to show visible signs of gold or silver content in order to be rich. The main indicator showing the possibility of gold being present, and possibly in paying quantities, is a large amount of mineralization showing in the quartz itself. Mineralization in quartz float often shows itself as a rusty-red color. Any red spongy quartz float located during prospecting should be followed to its lode, if possible.

To find a single piece of promising float on a mountainside is encouraging. But it's not a sure index that the lode is nearby or will be easily found. There are many different ways a single piece of gold-bearing quartz could arrive at a specific location. Perhaps an Indian liked the looks of it and picked it up, only to discard it later. Maybe a bird picked it up because it was shiny and alluring. Or, perhaps running water or a glacier once picked it up from somewhere else millions of years ago and helped to deposit it where you found it. It's also possible that it is the single specimen once dropped out of some earlier prospector's backpack.

Two pieces of similar quartz being found near each other enormously increases the chances of a lode being present somewhere up above. Three pieces of similar float found in a near vicinity is even that much better, and so on.

Once a prospector has found some promising-looking float, his or her aim is to trace it back to the lode by looking for more pieces and following their path upwards toward the source of the deposit.

Float can sometimes be visibly followed up a hillside. Sometimes, most of the float is covered by other alluvial material. In some situations, modern metal detectors can be used to trace float--if the float is highly mineralized or contains specimen gold (covered in Chapter 10).

When following the pieces of float up the side of a mountain, it sometimes happens that you run across a larger piece of quartz which has also broken away (or has been blasted away) from the lode. Sometimes a very large piece can be partially buried in the sediment and take on the look of a lode outcropping. You can be fooled into believing it is the actual lode. If this is the case, you will usually find more float in evidence above. This is not as likely to happen if the ''*false outcropping*'' were an actual lode. Of course, there is the possibility of multiple outcroppings.

Sometimes sediments have been washed down over the mountainside to cover the actual outcropping and much of its float. In this case it becomes more difficult to find the lode. When this occurs, short of using electronic searching procedures, the primary method of continuing is to dig sample holes to find the pieces of float, and continuing to move upward as long as you are finding traces. Because a float deposit becomes more widely dispersed as it descends downward, by looking at how wide a float deposit is, you can get a better idea of how much further up the lode is likely to be. The basic idea is to follow the float up the triangle to its apex in order to locate the lode. This can be done by using a systematic set of sample holes when the float is buried by sedimentary material.

Abnormalities in bush terrain were always investigated by the old-time prospectors. Such abnormalities signify the possibility of a change in the hardness of the underlying rock, meaning a potential mineral vein. The exposed walls in canyons and gulches should be carefully looked over. In their case, there is no sedimentary material to cover mineral lodes that might be present.

Any highly mineralized areas are worth the effort to check out. The richest mines have been located in areas where the surrounding country rock is highly mineralized. Such ground is indicated by quartz veins and narrow quartz stringers being widely dispersed throughout the county rock--or the country rock being multicolored.

One thing to keep in mind about lode prospecting is that a substantial amount of erosion has taken place within the last hundred years or so since the old-timers prospected the countryside. There is a good possibility that some outcroppings, once buried by sediments, have since been uncovered by the erosive effects on the countryside. Valuable lodes which were once hidden to view, or overlooked, might now be easier to locate.

RESEARCHING MINING RECORDS

Fig. 9-2. Old mines can often be spotted by recognizing their old tailing dumps.

GOLD, LODE—Continued

Map no.	Name of claim, mine, or group	Location	Owner (Name, address)	Geology	Remarks and references
	Armstrong	Sec. 6, T 37 N, R 12 W, MDB&M	Undetermined	9-in. quartz vein in quartz porphyry strikes NW, dips 20° NE.	About 13 mi. NE of Denny. Located in 1890 and developed by adit 210 ft. long; some high-grade ore was mined from surface workings. Idle. (Crawford 96:437; Brown 16:884.)
	Bailey	Sec. 15, 16, 21, T 35 N, R 10 W, MDB&M			No. 2 level of Globe Mine driven through mountain from Stuarts Fork side passed through this claim. Adit partly caved. (Miller 90:711-712; Averill 41:25.) See Globe also.
	Bateman and Mouldin	Sec. 14, 15, T 33 N, R 8 W, MDB&M	Undetermined	Quartz vein at contact of diorite porphyry and slate varies in width from a few inches to 5 ft.	Vein exposed in two old tunnels 100 ft. apart vertically. A small amalgamation plant was installed May 1932. Idle. (Averill 33:9.)
33	Bear Tooth (Fischer)	Sec. 33, T 7 N, R 7 E, HB&M		Quartz vein. Gold in oxidized sulfides of pyrrhotite, pyrite and chalcopyrite.	An old property. 70 tons of sorted sulfides shipped to Tacoma in 1947 yielded $55 per ton in gold, silver and copper. (Aubury 08:144; Averill 41:25; herein.)
	Bell				See Golden Chest.
34	Belli	Sec. 32, T 38 N, R 7 W, MDB&M		In gabbroic rock near contact with serpentine.	Old pocket mine 7 mi. N of Trinity Center. (MacDonald 13: Plate 1.)
	Bette				See Golden Chest.
	Big Chief	Sec. 16, T 8 N, R 8 E, HB&M	Undetermined	14-in. quartz vein in schist.	Two claims near Old Denny located in 1908. Tunnel on vein for 100 ft; ore assayed $15 to $60 per ton in gold. (Brown 16:884.)
	Bigelow	Sec. 15, 16, T 34 N, R 10 W, MDB&M	Undetermined	A short ore shoot in hornblende schist and granodiorite.	6 mi. NW of Weaverville. 70-ft. shaft. A little high-grade produced. Idle. (Brown 16:884.)
	Bismark	Sec. 12, 13, T 33 N, R 8 W, MDB&M			Patented quartz claim in Deadwood district 4 mi. E of Lewiston. Idle.
35	Black Cloud Group	Sec. 31, 32, T 34 N, R 9 W, MDB&M	Undetermined	Dark and light colored, fine grained, porphyritic dikes penetrate mica schist and interbedded siliceous slate and metavolcanic rock.	3 mi. NE of Lewiston. 9 unpatented claims prospected by shallow surface cuts; ore estimated to assay $4 to $6 per ton, based on 3 surface samples and 40-ton mill run. (Averill 41:26-27.)
	Black Diamond				See Mason and Thayer.
36	Blagrave	Sec. 10, T 37 N, R 7 W, MDB&M	Undetermined	Meta-andesite country rock	3 mi. NE of Carrville. No published description. (MacDonald 13: Plate 1.)
37	Blue Jacket	Sec. 17, 18, T 37 N, R 7 W, MDB&M	Undetermined	Ore shoot of low-grade quartz with rich spots, 2½ ft. wide, 200 ft. long. Strikes N 40° E, dips 40° SE; vein cuts granodiorite, aplite, and serpentine. Some free gold in aplite and granodiorite.	Near Carrville. Attempt was made to develop it on large tonnage basis; has adit 820 ft. long with drifts, raises, and stopes. Producer prior to 1896. (Crawford 96:439; MacDonald 13:25-27; Brown 16:884-885; Averill 31:29.)

Fig.9-3. Here's a sample of a single page taken from the "Tabulation of Mines" section in a county geological report. Note: The monetary figures concerning the ore values from the mines are based on $35 per ounce figures, so must be computed accordingly.

There are several other ways of locating lode deposits and valuable mines. One is to talk with the locals and old-timers in the general area of your interest. They usually have a wealth of data, and will have plenty of interesting stories to tell about such matters.

Probably the best means of locating a proven lode mine today is by studying the various geological reports concerning the county, or counties, that you are interested in. Many of the notable lode mines within a county are listed in its various geological reports. Sometimes the reports contain the old production statistics for the mines in question, not to mention other valuable data which would be of interest (see Figure 9-3 on previous page).

By taking the information located within a geological report concerning the various hardrock mines within a county, and cross-referencing to the data contained in the County Recorders Office, you can find the exact location of any mine that you are interested in. You can also find out whether or not the mine is presently owned (claimed) by anyone--and if so, by whom. The geological report will tell you on what scale the mine was being operated, the consistency and general make-up of the ore, as well as how the ore was being processed and how much it was yielding in gold and silver values. This is a means of prospecting for mines which for the most part have already proven themselves. The preliminary hard work of finding them, and their initial development, has already been done.

There are thousands of abandoned lode mines in gold country that might be profitably worked at today's gold exchange market value. The majority of these were old mines shut down and abandoned years ago when they became impossible to work because of rising inflation and the fixed market exchange value for gold. Many such mines were shut down during the Second World War and were never reopened afterwards.

CAUTION: Old abandoned mines can be very dangerous! Timbers used to shore-up the ceilings in old mines may now be dry rotted or disintegrated entirely. Gas pockets and/stale air inside old mines could be fatal if encountered. Vertical shafts could be entirely hidden by what looks to be several inches of water on the floor of the shaft. Sometimes, dirt and dust might cover a thin layer of water over top of vertical shafts. Snakes, bears and all sorts of other potential varmints might make old mine shafts unsafe. Never go alone. Always leave someone outside. Always keep others informed of your plans in case you don't make it back. And, the best bet, if you are going to explore old mine shafts, is to take an experienced lode miner along.

It takes some time and study to locate old mines this way. It's also necessary to go out and take samples for assay to see if the mine or mines are presently of value. A portion of them are, and so it's an excellent way to find good gold mines; because you are studying records concerning mines which were once successfully operated. Actually, now is the time to be locating mines in this way, because very little has been done on this so far--but there is more and more interest all the time. When you consider the amount of work and expense that can be bypassed in locating good mines by this method, it's surprising that all the good ones have not already been snapped up--which they have not.

SAMPLING A LODE

Once a promising-looking lode is found, whether it is an already developed mine or an outcropping that you have discovered, you will want to take samples so you can have an assay done. An *''assay''* report will give you a good idea of how much gold and silver is contained in the ore which you are having tested. At this writing, there are many assayers who will give you an accurate report on your hardrock samples. The going rate, per fire assay, is about $10 for gold and $12.50 for a report on both gold and silver. Some outfits charge a little more. The more prominent assayers advertise their services in the monthly mining publications. Samples can be sent to them through the mail. Most serious hardrock prospectors do not generally rely on a single assay report, but send their most promising or most important samples off to three different assayers and take the average for the final analysis. The reason for this is that the actual assay is done on only a fraction of the sample which you have taken. The sample's accuracy depends on the entire sample being crushed and thoroughly mixed first. Even with this properly done, the assay can still be off. One speck or flake of gold, or the lack of one, can create a misleading report. There's also a chance of a technician making a mistake. So in having three separate assays done, by different individuals, and discarding any one of the reports if it greatly contradicts the other two, and in taking the resulting average, a prospector is pretty well assured of getting a fairly accurate report on his samples. Some prospectors learn to do their own assays; it's faster and far less expensive to do so (about 20c per assay), when a lot of testing is being done. ''Do it yourself'' assay kits are available, and are also advertised regularly in the monthly mining publications.

When testing for the valuable mineral content in an outcropping or a mine, even the most precisely done assay can only give you as accurate a report as your sample of the lode will allow. To get an accurate sample of a lode for assay purposes, you don't just take samples of the best looking ore-part of the lode. You want to get, as precisely as possible, a sample which shows the overall average of the ore you will be processing if you should decide to mine the lode.

The reason you need the assay done in the first place is to have a reasonable idea of how much gold and silver is present in the average ore that will be processed if you mine the lode. Taking samples of only the highest-grade parts of the entire ore body will give you a better report, but it's also misleading. This is perhaps one of the most important things to know about hardrock prospecting. The assay reports resulting from the sample that you take off the face of the ore body will be used to determine if it is worthwhile to undergo any further activity or investment in that lode. So you want to do as good a job as possible in taking a sample showing the overall average of the ore body. This is usually done by taking a number of smaller samples from different parts of the entire face of the lode, being careful to take equal proportions of both high- and low-grade ore to the degree that they are present on the face of the ore body. This can be done by systematically taking small samples at regular intervals across the face of the exposed ore body. These smaller samples are then all put in a single body, crushed down, and mixed together to form the overall

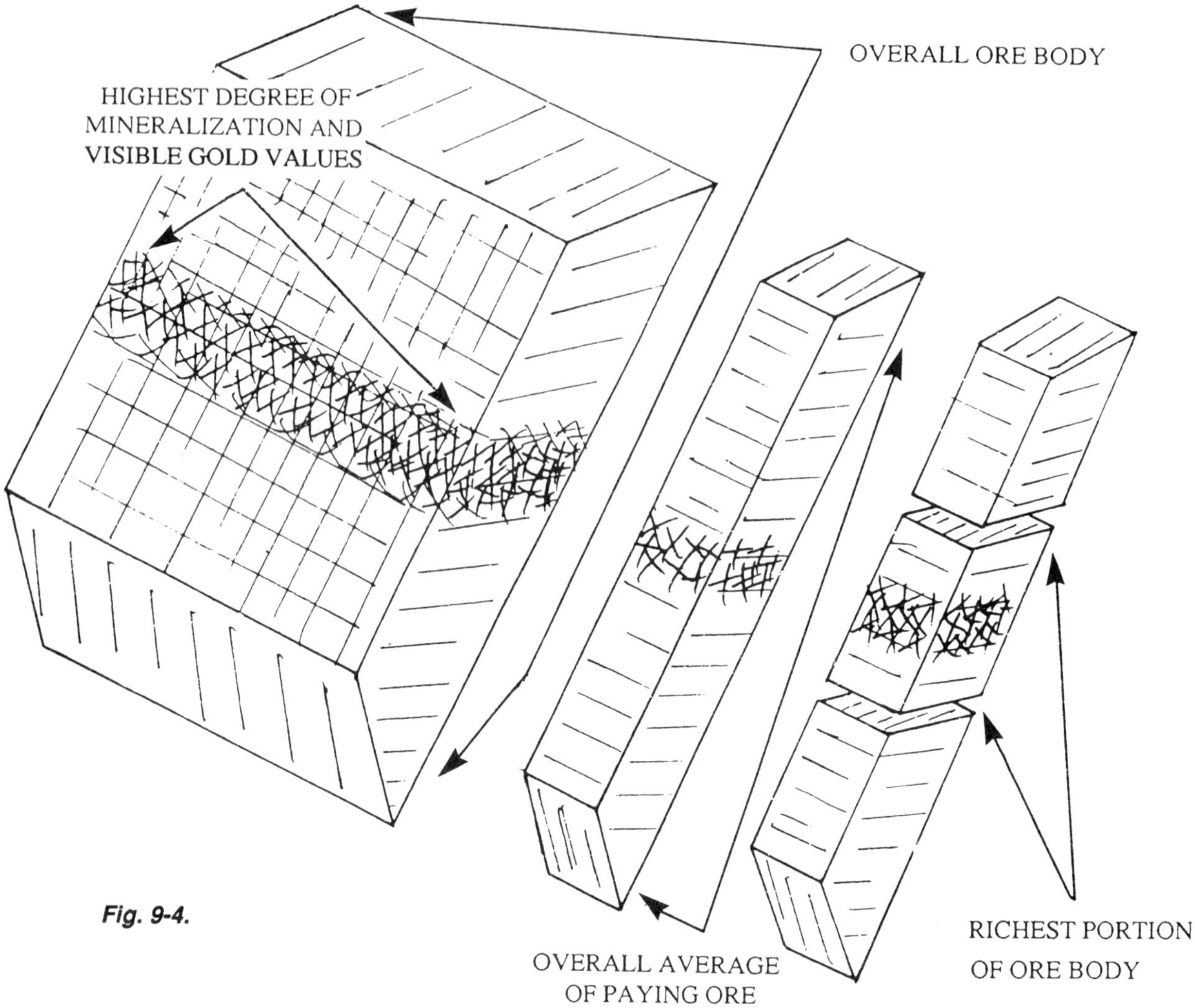

Fig. 9-4.

sample which will be assayed. It's also worthy of note that unless you are planning to send the entire sample off to a single assayer, you should attempt to crush the sample down as finely as you can, and mix it together as thoroughly as possible before separating portions of it to be sent off to separate assayers. This will improve your chances of getting an accurate portion of your sample to each assayer.

If there is a particularly rich portion of the lode that you would like to know about, there is certainly nothing wrong with having a sample of it assayed separately, just as long as it doesn't get mixed up with the overall sample. As an example, it's interesting to know that one portion of the ore body is paying 10 ounces to the ton of ore, but it's more important to know what the overall average of ore will pay only one ounce to the ton (see Figure 9-5).

One thing to keep in mind is that once a mine is put into production, it is not necessary to process all of the rock which is blasted. Your tunnel may need to be five feet wide and seven feet tall to have working room. But your vein may only be 2-1/2 feet wide. This means you might be blasting away more than you are processing, which you would have to consider when figuring out

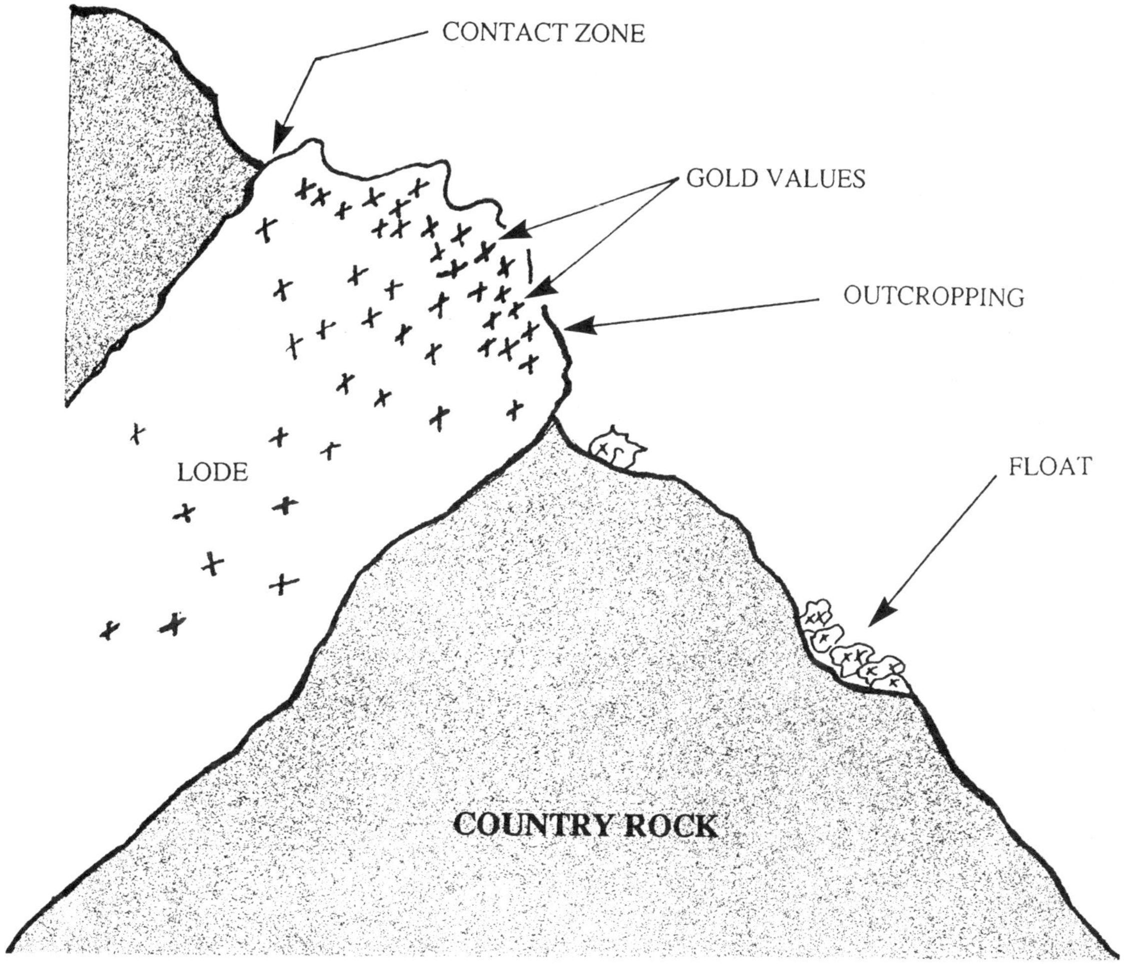

Fig. 9-5. Sometimes the outcropping can be richer than the underlying ore body.

the costs of operating the mine. The ideal scene is to have a vein that is wide enough and pays consistently enough so all of the blasted rock is also processed. However, this is not always the case. It should be pointed out that the narrower a lode vein is, the richer it will need to be in order to mine at a profit. This is because, in a narrower vein, there will be just that much more worthless rock *(''gangue'')* which must be blasted out and hauled out of the way to get at the lesser amounts of high-grade ore.

When taking samples for assay, don't disregard taking a portion of the *''contact zone''* which lies between the lode and the country rock. Sometimes the contact zone contains the most values.

When sampling an undeveloped lode, one thing to keep in mind is that the outcrop itself is often richer or poorer in mineral values than the underlying ore body. This can give a false picture of what the rest of the ore body contains in values. Years upon years of natural erosion may have caused the surface quartz to be washed off an outcrop, leaving much of the mineral values behind to be absorbed *(''leached'')* into the remaining outcrop. This can make the outcrop itself richer in mineral content than the underlying ore body, as shown in Figure 9-5.

The opposite of this may also be the case, where mineral values have been leached out of the ore body's face--leaving a lower grade ore directly at the surface of the vein.

So in order to get an accurate index of the mineral content in an undeveloped outcrop, it is best to get down into the vein as deep as possible before taking samples. It is for these same reasons that quartz float sometimes assays out to have more or less mineral values than the lode from which it came.

An ore body does not need to have visible gold present to be rich in gold content. However, if visible gold is present, it is likely to be a very rich lode. Sometimes, pyrites can look a lot like gold when visible in hard rock form. The solution to this is to get an assay done, or try the hardness test to see if the golden material is brittle.

Most developed mines have no visible gold whatsoever. An ore body can be so saturated with microscopic particles of gold that it can be very rich indeed. Yet, you may never see any gold at all until after it's processed out of the ore. Some ores can pay as much as five or six ounces in gold, or more, to the ton of ore being processed, and yet not show any visible sign of gold.

Probably the best index of whether or not gold may be present in an ore body is how much mineralization is in evidence, just like in the float. The presence of iron pyrites can be an index that gold might be there, too. Iron will oxidize (rust) when exposed to air and moisture, which turns into a rusty red-brown color. The presence of a lot of rusty-red and brown mineralization within the quartz rock of the lode is a good indicator of potential gold, too. Down into the ore body itself, where the minerals have not yet been exposed to the effects of oxygen and moisture, the minerals will take on a darker color, not a rusty red color like they do after being oxidized.

Mica, by itself, is not the same as iron pyrites and is not an indication that gold is likely to be present in the ore.

DEVELOPMENT

At today's market value of gold, and with the equipment readily available on the market, it would be reasonably safe to say that a small-scale hardrocking operation could be run at a profit in ore which is paying a half ounce or more in recoverable gold values to the ton of ore being processed. A cubic yard of hardrock ore weighs on the average of about two to three tons. There are other factors which can come into play and make it necessary for the ore to be higher grade in order to run a viable operation. One example would be having a narrow vein where much more rock needs to be blasted than that which is being processed for its gold and silver values. Another example is having ore which needs an involved and expensive processing procedure in order to release and recover the gold and silver values from the ore.

Also, an assay report which says there is a half ounce of gold to the ton of ore is stating that there is a half ounce of gold present. However, it is not necessarily true that a half ounce of gold can be recovered out of the ore by using ordinary processing methods. Sometimes, you will run into rich ore bodies where a large percentage of gold is chemically locked in so tightly with various sulfides that it is all but impossible to recover some or most of the values without using the most advanced (and expensive to set up) chemical extraction processes.

So, after the fact of a favorable assay report on a lode, it's a good idea to take a larger sample, like a couple hundred pounds, or a ton or two if possible, of the average ore. Then have a professional refinery or processing plant process the sample and do an analysis for you. In doing so, you can obtain a much more accurate report on the recoverable values, a flow chart of how to process your ore to recover the most values at the least expense, and a list of the equipment you will need to use. This is not an inexpensive process to have done, and if you are planning to enter operations on a small scale with portable equipment, perhaps it is not necessary. On the other hand, if you are planning to invest a substantial amount of money to open a single mine, it is highly recommended that you have a large sample treated as such before you invest in a processing plant.

One thing that should be mentioned at this point is that it is always a good idea to keep quiet about the exact location of the source of your samples during the initial testing stages, at least until you have obtained positive results and have filed the necessary paperwork to claim for yourself the more valuable mines found.

Once you know the recoverable values contained in an ore body you have sampled, you can calculate the values against the costs of operating a plant and developing the mine, and determine if an operation within your limits can be run at a profit. Sometimes a road needs to be put into the area. This is another factor which would need to be entered into the initial investment figures. If you are considering refurbishing an already developed older mine, sometimes much of the shoring (tunnel supports) needs to be strengthened. That would be another factor to enter in.

How much initial investment it will take to start up a mine is largely determined by the size of the operation you are planning. A normal flow plan of a plant, from start to finish, would be something like this: 1) Breaking away the ore from the ore body. 2) Transporting ore to the processing plant. 3) Breaking large pieces of ore into medium-sized pieces. 4) Breaking medium-sized pieces of ore down into small pieces. 5) Crushing small pieces of ore down into fine powder. 6) Concentrating the heaviest powder (contains gold and silver). 7) Amalgamating the gold and silver values out of the concentrates. 8) Separating the mercury from the gold and silver by use of a retort to give a final product ready for sale. This process can be done in a multitude of varying systems and sizes of operation.

A small operation might consist of breaking the ore away with the use of a pick, pounding it into powder with the use of a mortar and pestle, panning the gold, and amalgamating as necessary. Many of the earliest mines were run in just this manner, and at a profit, too.

On a larger scale, a portable impact hammer could be used to break away ore from the ore body and to break it down into medium-sized pieces, which could then be placed into a portable rock crusher, as shown in figure 9-6. This particular machine is a combination *''rock crusher''* and

Fig. 9.6. Portable "Rock crusher--Roller mill."

''roller mill,'' so it carries out two functions. First it takes medium-sized pieces of ore of about 4-inch to 6-inch size and quickly breaks them down into 1/4-inch pieces. These then drop down into the second stage (roller mill), which crushes the ore down into fine powder of about 150 mesh. Depending on the type of ore being processed, the machine is capable of crushing up to one ton of medium-sized ore down to 150-mesh per hour. This is more than sufficient to meet the needs of a small hardrocking operation of this size. The powder from the mill can then be fed into a portable recirculating concentrating plant. The concentrates can be amalgamated to recover the gold and silver values.

An operation of this size can be effectively run by one or two men and can process up to four or five tons of ore per day at little expense, under ideal conditions. This size operation is becoming very popular today with the small-scale operator, because it doesn't require a huge investment to get set up with equipment. Also, there are plenty of abandoned mines around that potentially will make a small operation such as this pay well. And finally, the equipment is all portable. It can easily be moved from site to site for sampling purposes in an effort to locate richer lodes. Some prefer to use this scale of operation to sample lodes with the idea of investing in the development of a good lode once it is found, or to sell the proven lodes to large mining companies.

A small professional operation, on a slightly larger scale, would probably use a portable power drill to make holes in the face of the ore body so that a few charges of explosive can be set to blast out larger portions of ore. The ore could then be run through a series of rock crushers and a mill. The resulting powder could then perhaps be run over a concentrating table. The gold could then be amalgamated or extracted chemically. The entire setup would use conveyor belts and sand screws so the ore could be placed in the first rock crusher and end up as the final heavily concentrated material and gold, with little or no further action from the operator of the plant.

An operation such as this could probably move about 10 tons of ore per day under favorable conditions, and could be effectively run by a three-man team: one drilling, one shoveling ore into a cart *(''mucking'')* and hauling it to the processing plant, and one man overlooking the plant's operation.

The larger an operation is, meaning the more volume of ore it is able to process, then the lesser grade ore it can process at a profit.

Also, the larger the size of an operation, the larger the size of the initial investment and the operating costs.

If you are considering investing a large amount of money into the development of a single ore body, it's a good idea to consider having the ore body *''core drilled''* beforehand to ensure there is enough valuable ore present to warrant the size of your investment. A core drill will drill deep into the ore body and bring out samples from different depths. These samples can be tested to see what kind of values the ore contains as it stretches deeper into the earth.

There is always the possibility of an ore body crossing a fault zone, being sheared off, and causing the rich ore body to suddenly end, with the other section being somewhere else not to be found. This is something that would be discovered by having the ore body core drilled. Core drilling is a rather expensive procedure, but so is the development of a lode mine on a professional scale. If large amounts of capital are being invested, it's usually worthwhile to know that the ore body is large enough to warrant the expense.

Once the ore body has been sampled by core drilling, and it has been established that the values continue for hundreds or thousands of feet into the mountainside, and that you have lots of valuable ore to process, well, then you have got yourself a real gold mine and can be pretty certain of good returns on your investment--providing, of course, that you have a skilled manager to run the show and develop the mine.

Full development procedures are out of the scope of this single chapter on basics. If you have gone far enough along to be interested in developing a lode mine, I highly recommend that you acquire the technical manuals covering the subject and study them well. It's also a good idea to bring on a proven successful and competent mining consultant to help you get started.

CAUTION: Don't ever run an internal combustion engine inside a mine tunnel. Carbon monoxide poisoning can creep up on you quite fast in such places; it can be very dangerous! If electric lighting is needed in a tunnel, run the generator outside the shaft and stretch an extension

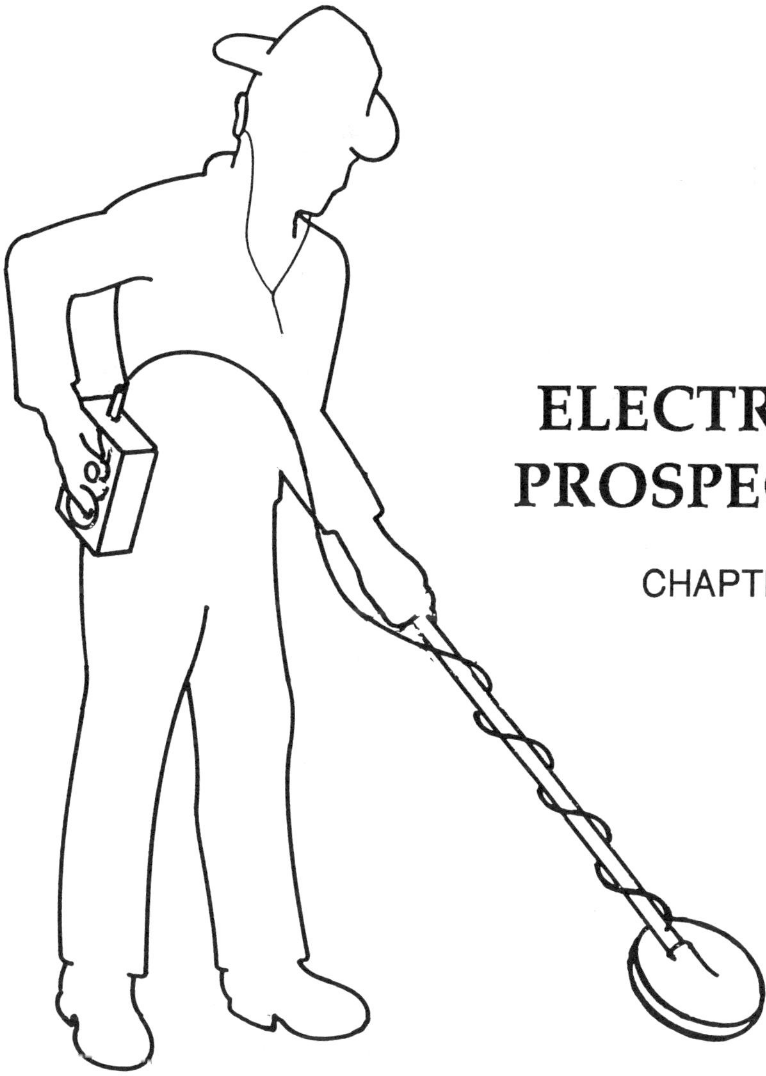

ELECTRONIC PROSPECTING

CHAPTER X

There are many different kinds and models of electronic metal/mineral detectors to be found on today's market. This chapter is set up as a guideline to give you the basic knowledge you will need to choose the proper detector for your prospecting needs, and to show you how to use the one that you do choose as an effective prospecting tool.

There is also a tremendous amount of electronic prospecting and gold nugget hunting activity going on at the time of this writing. Consequently, there are different tools being used--and several different popular approaches in how to tune detectors properly, and how to search in order to achieve the best results. One thing I would point out is that no two gold bearing areas are exactly alike. An approach which might work the best in one area, might not work very well in a different area. So, the purpose of this chapter is not to tell you what I think the "best" approach is. It is to give you as much information as possible on the different approaches; so you can gain a larger bag of tools to use when confronted with different problems in the field.

The first point to remember is that the type of electronic detector used in finding gold and other precious metals is not a *''Geiger counter.''* A Geiger counter is an entirely different electronic tool used to detect radioactive elements.

The type of electronic device used to prospect for gold is called a metal/mineral detector (*"metal detector,''* for short). Metal detectors are quite simple to use, and can be helpful in assisting a prospector to locate gold or silver deposits or specimens once he or she has experience in using one properly. While they are rather simple to use, it does take some practice with a metal detector before a person can use one proficiently in gold prospecting activities.

There are many different models being offered on today's market, most which are more useful to the treasure hunter than the gold prospector (two entirely different fields of detecting activity and procedure). Those detectors of most use to gold and silver prospectors generally fall under two separate categories: Beat Frequency Oscillator *(''BFO'')*, and Very Low Frequency *(''VLF'')*.

BEAT FREQUENCY OSCILLATOR

First we'll take up the BFO, which is the simpler of the two--but is less often found these days due to the substantial electronic advancements of VLF detectors.

The BFO detector usually has two main settings, which are *''metal''* and *''mineral.''* As far as electronic detectors are concerned, the difference between the two is that ''metals'' are targets which are conductive of electricity--such as copper, gold, silver or iron. And, ''minerals'' are targets, or target areas, consisting of magnetic non-conductive materials such as magnetic black sands (Fe_30_4). These are also known to prospectors as *''black sand concentrates.''* An iron object which has been in the earth for an extended period of time, and having thoroughly oxidized, will usually read out on a metal detector as a mineral instead of a metal object--which it no longer is.

So the two basic settings on a BFO detector are *''metal,''* electrically conductive targets (gold and silver), and *''mineral''* --non-conductive magnetic particles (magnetic black sands).

The various models of detectors have different ways of sounding out on reading targets. Some detectors have a light which turns on and off. Some have a meter with a needle on a dial--which will also give you an idea of the intensity of the signal given off by various targets. Other detectors have a tone which changes in volume or pitch when passed over a specific target. Some newer-model detectors have an LED display which spells out the different types of targets being encountered. Some detectors have a combination of these features. Probably the best type of metal detector for prospecting purposes is the type which includes an audio tone in which the audio pitch changes when the search coil is passed over a reading target, and which also allows a set of headphones to be connected.

The advantage to using headphones while prospecting is that you can shut out the background noises from the surrounding environment and concentrate fully on any possible change in audio while searching.

On most tone-changing BFO detectors, the tone will not only raise in pitch when the search coil is passed over a target for which it is set to sound, but it will also lower in pitch when the search coil is passed over a target of the opposite setting. For example, if a BFO detector is on the metal setting and is passed over a large gold nugget, the detector's audio tone should rise in pitch. If the detector on the same metal setting is passed over the top of a high concentration of magnetic black sand, the audio tone should lower in pitch. The same thing holds true in the opposite for the BFO detector which is adjusted to the mineral setting.

One other interesting fact about BFO detectors is they tend to read out on the most dominant element, either "metal" or "mineral," whichever is most present in the ground which the detector is being passed over. For example, if you are passing the search coil over ground which contains gold (this would read as a metal), yet there is a large amount of magnetic black sand in the same ground, it is likely that the BFO detector will read out on the black sand as a mineral while ignoring the gold. Equal reading amounts of both metal and mineral elements in a section of ground, in any quantity, will prevent the BFO detector from sounding out on either element.

Because BFO detectors read out so well on highly mineralized ground, the presence of highly mineralized ground tends to block out reading traces of gold which lie in or under. This is known as *"interference"* in the electronic detecting field. Magnetite (magnetic black sands) has such a strong affect on metal detectors that a concentration of only one percent magnetite in the ground may create an imbalance signal which is hundreds of times stronger than the signal which might be given off by a small gold nugget.

So, a mineral reading on a BFO detector does not mean there is no gold present, only that there is heavily mineralized ground--which may be blocking out gold readings.

One of the problems in electronic prospecting is that gold targets are often associated with highly mineralized ground. Therefore, as a tool, the BFO has its advantages and limitations. It can be very helpful to have a device which is good at pinpointing areas of concentrated heavy mineralization, and the BFO does this exceptionally well.

Also, some places where nuggets and larger flakes of gold become trapped do not allow heavy concentrations of black sand. One example would be a location (rapids) where the water runs fast over top of exposed bedrock.

GOLD TARGETS

Unfortunately for the gold prospector, gold, as a metal, is generally not picked up very well by metal detectors. This is a comparative statement. Gold does not sound off on a metal detector nearly as well as an iron object of the same size and shape. Specialized metal detectors will detect gold well enough that they will sound off on nuggets, deposits of smaller pieces of gold, or even very small individual flakes of gold. At the time of this writing, no metal detectors are able to detect particles of gold dust. This is good, however, because there is so much super fine gold spread throughout gold country that it would probably create additional interference problems on a more sensitive detector. We are looking for flakes, nuggets and deposits; targets which will add up more quickly to something of good value.

It is important to understand that different makes and models of metal detectors are not equal in their ability to detect gold objects. Some detectors will just barely sound out on gold objects and others will not sound out at all--despite claims made by the manufacturer.

I highly recommend any person buying a metal detector for gold prospecting bring along some samples of natural gold to test the various detectors BEFORE deciding which one to buy. This is to be sure that the one you do buy will sing out well when it is passed over natural gold objects--even very small gold targets. If a specific detector will not sound out on gold held in the air, it will never detect gold targets located in the ground. It is better to use natural gold samples--like nuggets, flakes or a sample bottle filled with fine gold, when testing out the various detectors. Some detectors will, and some won't, sound off on bottles filled with fine gold. Using natural gold targets is better than using a gold ring or some other type of jewelry. Jewelry is usually made of gold which has been alloyed with other metals (like copper)--which are likely to read out on a metal detector better than natural gold objects.

The best detectors for finding gold are not necessarily the most expensive. Varying costs in detectors are often in proportion to the amount of additional electronic circuitry put into the detector for extra features. These usually have little or nothing to do with the operation of the detector for the purpose of locating gold targets.

Gold targets give a solid, mellow sound on a metal detector--similar to lead or brass. Pieces of steel wire and bigger nails usually give a stronger beep--or often a double beep.

The capability of a metal detector to sound off on a natural gold target will partly depend upon what other metals the gold is alloyed with. Silver and copper make natural gold targets sound out stronger. Nickel, mercury and platinum alloys make natural gold targets more difficult to find.

Metal detectors read out on gold better as the pieces become larger. As an example, an average gold detector might sound out very well when its search coil is passed over an eighth-ounce nugget from several inches away, yet not sound out at all when passed over three times as much fine gold accumulated in a glass jar at the same depth or distance from the search coil. Actually, it is not just the size of the target which counts. Its shape makes a difference, and also the direction which a target is pointing. A larger, more solid surface area of gold will sound out stronger. For example, a flake-shaped nugget is likely to sound out better on a metal detector than a round nugget of the same weight--provided that the flat surface area of the flake is pointed in the direction of the metal detector. Also, coarse and irregular-shaped nuggets, as commonly found in dry placer areas, residual and eluvial deposits, do not generally sound out as well as nuggets which have been worked over and pounded by a streambed (more dense and solid).

How tightly a gold deposit is concentrated also makes a difference in how well it will cause a metal detector to sound out. Whereas a quarter ounce of flake gold in a jar might sound out well on a particular detector, perhaps two ounces of the same flake gold spread out over a slightly larger area might not read out at all with the same detector when the targets are at the same depth beneath the surface. This is one factor which is important for the gold prospector to realize. Any detector will read out on tighter concentrations of gold better than larger amounts of gold which are more widely dispersed. They will also read out on nuggets (larger solid pieces of gold) best of all.

DEPTH CAPABILITIES

How deep into the ground a specific metal detector will sound out on an object depends on various conditions. Surprising to many, how much a detector costs has little to do with its depth-sounding capability. In fact, some of the less expensive models are able to probe deeper, and pick up on gold better, than some of the more expensive detectors. The Federal Communication Commission has put a maximum limit on the signal strength which can be used in metal detectors. So the idea that a more expensive model puts out a stronger signal to probe deeper is simply not true.

The type of object itself has much to do with how deep into the ground it can be located with a metal detector. Different kinds of objects have varying amounts of magnetic and electrically conductive properties. Therefore, they affect electronic detectors differently. Also, some detectors will sound out on some kinds of objects better than others. As mentioned earlier, gold is not one of the better reading metals, so cannot be picked up with a metal detector as deeply as an iron object of similar size and shape.

Another factor which determines how deep an object will be picked up by any detector is the size of the object itself. Whereas a two pennyweight nugget (1/10th ounce) might be picked up five inches deep into the ground with a certain metal detector, a five pennyweight nugget (1/4 ounce) might be picked up eight inches deep into the same ground with the same detector.

How much an object has deteriorated and has been absorbed into the soil is another factor in how deep the object will be picked up. Iron objects tend to oxidize and become slowly absorbed into the surrounding material. This causes the target to appear larger and read out more strongly, so it will be picked up at greater depth with a metal detector. Once such a target has thoroughly deteriorated as an object, it will stop reading as a metal and start reading as highly mineralized ground. As mentioned in Chapter 1, gold does not oxidize or deteriorate, so this factor does not apply to natural gold targets.

The size of a search coil on a metal detector is also a factor in how deeply the detector will pick out objects. Larger coils generally are able to pick up objects at greater depth than smaller coils. But they do not have as much sensitivity in detecting smaller gold targets. Smaller search coils have greater sensitivity to small objects, yet do not have the depth-probing capability that larger coils do. Medium-sized coils, from five to eight inches in diameter, often combine the features of having both a reasonable amount of sensitivity for the smaller objects, and acceptable depth-scanning ability. One thing to keep in mind is that a larger coil will also increase the size of the area being covered by each sweep.

Many nugget hunters prefer to have the smaller search coils handy because they have the greatest small object sensitivity (gold flakes) and because the smaller coils can get into tighter spots--like in and around tree roots and inside of exposed crevices, where nuggets are most likely to be found with a metal detector.

Almost all detectors today are made so that various sized coils can be attached, depending on what they are to be used for. When testing a detector, don't make the mistake of assuming if the device sounds out well on a gold sample when using a coil of one size, it will also sound out well when using a coil of a different size. Your best bet is to test the detector with the various sized coils to see which works best for your particular needs.

One of the most important factors determining how deep a metal detector will sound out on a gold object is how much mineralization (interference) is present in the ground being scanned. More minerals equals less depth. This is especially true of BFO detectors. Because black sands are usually in evidence in the same streambeds or soils where gold deposits are located, metal detectors are not usually used to directly detect gold in streambeds or material of substantial depth. They are more often used to scan places where there is a very shallow amount of gravel or material present over top or the gold, or none at all.

One excellent use of the BFO detector as a prospecting tool is to locate concentrations of black sands in a streambed. Black sands often accumulate in the same locations that gold does. From your fundamental knowledge of placer geology, after potential paystreak locations have been pin-pointed, those specific locations can sometimes be scanned with a BFO detector to locate the increases in other heavy elements. Specific sites which sound out heavily on the ''mineral'' setting can then be sampled by conventional gold mining techniques.

VERY LOW FREQUENCY DETECTORS (VLF)

The VLF detector is a more recent development in the field of electronic prospecting. Very Low Frequency detectors may come under other names or descriptive abbreviations such as: VLF, GEB, MF, GCD, and others. These are designed with circuitry which is able to cancel out the effects highly mineralized ground has on a BFO detector. VLF detectors have the ability to look through or past highly mineralized ground and pick up metal objects (gold) that may not read at all on a BFO metal detector.

The VLF, being able to cancel out interference caused by mineralized ground, is more suited for locating gold deposits and gold specimens directly. However, it still remains true that gold targets will have to be large enough, or located close enough to the surface, or deposits will have to be tightly concentrated enough, to sound out on a VLF, just as with the BFO detector.

Just because a particular detector is of VLF design does not mean it will sound out well on gold. In fact, there are some VLF detectors which have difficulty in sounding out on gold samples at all. So this type of detector must be just as thoroughly tested before buying.

The VLF detector, being a mineralization-cancelling device, sometimes does not have the ability to detect the heavy black sand concentrations the way a BFO detector is able to. Consequently, the VLF is best suited for scanning directly for gold, whereas the BFO is generally best suited in helping the prospector locate gold deposits in an indirect sort of way--by finding the highly mineralized ground in a gold-bearing area.

MULTI-PURPOSE DETECTORS/SPECIALIZED GOLD DETECTORS

VLF detectors are sometimes also made with discrimination circuits designed to cancel specific types of targets--like bottle tops, aluminum foil, pop-tops, etc. For the most part, this type of electronic circuitry is best suited for treasure and coin hunters. When used in prospecting for gold targets, discrimination circuitry sometimes has a tendency to also reduce the detector's depth probing capability, especially in highly mineralized soil or streambed material. Since gold targets are already difficult to locate, it is better to not utilize additional circuitry which could hamper sensitivity towards gold. However, some conditions probably do exist in which discrimination circuitry may assist a gold prospector. If using such a detector, always test against a nugget sample planted in the ground you are testing to determine whether or not you can trust the discriminating circuit.

Even some of the newer specialized gold VLF detectors are utilizing specific discrimination circuitry called *"Iron Identifiers."* This works at moderate depths, but does not necessarily reduce the total depth capability in the detection of gold targets. In other words, the circuitry will identify iron objects which it is certain to be iron. If the target is too deep or too small, it will still identify it as gold. However, if the ground is highly mineralized, the accuracy of iron identifying circuitry is likely to be reduced.

Some experienced electronic prospectors utilize discrimination circuitry only after a target has been located. This way, depth and sensitivity is not forfeited during searching. Other experienced prospectors insist that no discrimination circuitry is needed. Once you are familiar with the area you are searching, and know the specific audio tone changes of gold and/or trash targets, you form a judgment of which targets to dig and which targets to leave alone. Different prospectors have different methods. Also, different locations often require different methods. Some experienced electronic prospectors simply dig every target.

Some VLF detectors are made with circuits designed to analyze targets. This means they are able to tell you if the target is a nail, bottle top, nickel, silver dime or a piece of gold. Such circuitry has only limited accuracy in electronic prospecting; because highly mineralized ground tends to interfere with the signal and can give a false reading in the analyzer.

None of these circuits are a problem with multipurpose detectors providing they can be shut off or bypassed--and/or providing the additional circuitry does not hamper the detector's efficiency in locating gold and silver targets.

Some VLF detectors are designed with manual ground balancing controls, and others are designed with automatic ground balancing circuitry. Some prospectors prefer the manual controls. Others prefer automatic ground balancing. There is nothing wrong with automatic ground balancing circuitry as long as it is fast enough to keep up with the rapidly changing mineralized conditions of the different areas you intend to prospect--and providing the additional circuitry does not hamper the detector's ability to locate gold and silver targets.

Some VLF detectors have been specifically designed as gold prospecting tools. Since most specialized gold detectors operate at a higher transmitting frequency, have extensive ground balancing capabilities, and have special circuitry to avoid sensitivity overload in highly mineralized ground, they definitely do have some advantage over most multipurpose detectors in their ability to locate small gold targets--or gold targets which are deeper. The high performance of some of today's gold detectors even make pinhead-sized gold targets recoverable. Which detector you choose to buy will depend on what you plan to use the detector for. If you plan to only use it for prospecting purposes, a special gold machine is probably best for you. If you intend to search for coins, caches, artifacts and lost articles, as well as prospecting for gold and silver, perhaps a multipurpose detector is best--or two separate detectors.

I would suggest you buy your detector from a dealer located in the general area where you plan to prospect for gold. The local dealer will know which detectors are performing best in that area. Local dealers will also introduce you to other local prospectors, and perhaps the local prospecting and/or treasure hunting club or association. Communication with local miners will be a huge help in determining which detectors are best for specific areas. You can also get tips on productive places to prospect with your detector.

No detector made is the best for all locations. Some machines work better than others in wet or dry conditions. Some work better in hot or cold climates. Some detectors are affected by alkali "salts" in the soil or gravel more than others. When any of these examples, or others, are the case, a smaller coil might handle adverse conditions better than a larger coil. Each area is different.

Another reason to purchase your detector from a local dealer is the help and support you will receive. Success in the field comes from understanding the workings of your detector, and perhaps receiving inside information on good places to hunt. The money saved by buying from a discount mail-order house is not worth the loss of support you would receive from a local dealer--especially when just beginning.

When buying a detector which you intend to use for prospecting purposes, keep in mind that probably the most important feature is the detector's capability of cancelling heavy mineralization found in most gold-bearing areas.

Practice makes perfect. You must start with good equipment. The rest will be up to you.

ELECTRONIC PROSPECTING DRILLS

The following is a set of drills put together to give the new (or old) owner of a metal detector some practice with his tool and to allow him (or her) to get a good grasp of what the detector's gold finding capabilities are.

DRILL No.1: Take a file to a piece of iron or steel (like a nail), and allow the pieces of metal to fall into a container. Pour some filings onto a piece of paper and pour some glue over the filings to hold them intact. Pour more filings on top of the glue and then pour on more glue. Continue this until the conglomerate is giving off a strong mineral reading on your detector. Make three different sheets of mineralization, one giving off a very mild mineral reading, one causing a medium signal, and one which gives off a strong signal.

If you are already an experienced gold miner, and have some black sand concentrates lying around somewhere, use a magnet to collect some magnetic black sand and use these instead of iron filings. Sometimes you can get prospecting supply outlets to send you a small package of black sand concentrates.

These different mineralized conglomerates will give you a good index of how your detector will react to different degrees of mineralized ground.

DRILL No.2: Acquire at least a half-ounce of placer gold, preferably more, with a variety of fine, flake, and nuggets so a wide range of testing can be done.

Carefully place the gold in a pile on a clean sheet of paper in a location where there is no other metallic object reading on your detector. Scan the gold with your detector from varying distances to get an idea of your distance capabilities on a large compacted gold deposit.

Now spread the gold out over a slightly wider space on the paper and scan again to check distance. Continue to spread the gold out wider and wider until it no longer reads on your detector--or until you are picking up on individual flakes of gold. This drill will give you a good idea of what size pieces and accumulations of gold will read at what distances. Try different coil sizes to see what their capabilities are.

Pay particular attention to the type of sound readings that you get when scanning over gold targets. These drills should be done with headphones. After a while, you'll start to be able to tell the difference between gold and other metallic readouts by the difference in the strength and tone of the read. Stronger reading metals will give a sharper and louder change in tone, whereas gold tends to cause a softer and more indistinct read--especially when located in smaller amounts or at a distance. Do the drill and see for yourself.

Do this same drill while using various sized gold nuggets and flakes, if available. Check out the different readings caused by the assorted pieces of gold and accumulations from various distances.

After this drill, you will have a good idea of your detector's gold finding abilities and know the proper sound that your detector makes when it is reading out on a natural gold target.

DRILL No.3: Using the flake gold and nuggets in different accumulations, as done in drill No. 2, place the different sheets of mineralization over the top of the gold and note the responses on your metal detector. If you have a VLF, practice cancelling out the mineralized sheets, and test to see what size accumulations of gold can be picked up while doing so. Try more and more mineralization, combining the sheets together if necessary to see how much mineralization your VLF detector will look through and still have sensitivity to gold targets. Notice how even a larger piece of gold puts out only an inkling of a reading when covered by heavy mineralization and scanned from a distance. Recognizing these light signals can make the difference of success or failure in the field.

If you are doing these drills with a BFO detector, try combining different amounts of mineralization with the various sized accumulations of gold. Determine for yourself on your own detector how much mineralization it takes to block out the different accumulations of natural gold.

I am certainly aware that sometimes it is difficult to come by a collection of gold flakes and nuggets if you don't already have a collection of your own. However, the time spent in locating one, or in talking a friend into lending you his collection--or in talking him into doing these drills with you--will be worth many times as much time spent out in the field with your detector. These drills will not teach you how to prospect for gold deposits. Only practice and experience out in the field will do that. However, these drills thoroughly done, will familiarize you with your detector and give you certainty on the use of it, and also give you the basics that you will need to learn to prospect with a detector.

HELPFUL TIPS ON TUNING

Each model of detector has its own set of operating and tuning instructions which you should follow. And, I highly suggest you familiarize yourself with every aspect of the manufacturer's instructions. In addition, here are a few pointers which have proven successful in the prospecting field.

Some manufacturers recommend that their volume-changing detectors be tuned to just below the hearing range. The purpose of this is so that the slightest reading will make a sound-- which can be easily distinguished from the silence. For prospecting purposes, it works better if you tune these detectors so the audio signal is always within hearing range. This will run down the batteries just a bit faster, but it's much better to be able to hear the signal at all times. The audio threshold (''*threshold*'') of a tone-difference sounding detector should also be set just in the hearing range. When looking for natural gold targets, the slightest change can mean a gold target. Changes in volume and/or audio tone also are an indication of changes in ground mineralization and let you know when adjustments are needed to ground balance. Also, the detector's audio signal will sometimes drift off to a lower volume range due to temperature changes or loss of battery life. If the audio signal is tuned into the non-hearing zone and drifts into an even lower range, you might be scanning for several minutes without having the detector tuned properly. In the case of electronic prospecting, this can be an expensive lesson to learn.

Sometimes a warming coil will cause the threshold sound to drift upwards. A cooling coil might cause the threshold to drift downward. Hunting in and out of water environments, while scanning the banks of a stream, might cause threshold changes.

The main cause for a detector's tuning to drift is loss of battery life. When this occurs, it's time to replace the batteries with a new set in order to get the best performance out of your detector--which is needed when hunting directly for gold.

It's always a good idea to bring along an extra set of batteries into the field when prospecting; because when they quit, you are finished until new batteries are installed. Extra batteries should be kept cool and dry--Ziplock baggies work well.

Prospecting for gold targets directly with a VLF detector should always be done in the *''all-metal''* mode.

Setting Sensitivity: It is important to stress that you do not want to set the sensitivity too high on your VLF detector while prospecting in a heavily mineralized area. A high sensitivity setting while testing a nugget in the air will show improved perception--and therefore can give you a false impression of the detector's scanning ability for gold targets in the field.

Turning the sensitivity up too high in mineralized ground is similar to using high-beam headlights in the fog. You get lots of flashback and irregular sounds, false targets, etc. If your sensitivity is set too high, your detector will operate in an erratic manner; there will be many false signals which do not repeat themselves *("flashback").*

Consequently, less sensitivity can give you more depth penetration in mineralized ground. There is actually a middle ground, depending on ground mineralization, which will give you optimum sensitivity, without too many ''ground noises'' which are confusing and prevent you from selecting the true signals. Try and run with the sensitivity as high as possible--until the steady tone of the threshold begins to give off an uneven, wobbly sound.

I don't recommend using factory preset marks on your detector controls. Such settings are for average conditions. Prospecting for gold targets requires regular adjustment to ground balancing, and the threshold and sensitivity need to be set as accurately as possible to changing conditions. You need to get the most possible out of your detector to avoid missing gold targets.

''Peak Performance'' on a metal detector for nugget hunting purposes, in most cases, is maximum volume on detector, threshold set in minimum audio hearing range, maximum sensitivity without receiving too much flashback, and ground balance to the average ground being scanned. When you accomplish peak performance on your detector, the rest is up to you!

Ground Balancing: Setting the proper ground balance on your detector, and keeping it properly adjusted while you search, is perhaps the most important factor in successful nugget hunting. I cannot overstate this point; because without proper ground balance, you simply cannot find natural gold targets--unless you just get lucky. All of the skills we will talk about in this chapter, skills and methods which will make you good at finding gold targets, all depend on your detector being properly ground balanced.

Always set your ground balance to the average soil or material which you are searching. You will find the majority of gold nuggets in average ground. If you ground balance to specialized heavier mineralized zones which are not the average matrix, you may forfeit some depth probing capability or sensitivity to smaller or deeper gold targets.

Detectors which come with permanently set, predetermined ground balance are not particularly good for electronic prospecting.

When your detector is on properly, you should hear a low hum (threshold sound). As the detector is raised or lowered from the ground, the threshold hum will get louder or softer. This tells you what needs to be done in order to ground balance. Handling the ground balance knob on your detector is similar to handling the volume control of a radio. If the threshold hum is disappearing as you lower the coil to the ground, turn the knob up. If the hum gets louder as you lower the coil, turn the knob down. The basic idea is to adjust the ground balance knob until raising and lowering the coil to the ground creates little or no change in the threshold hum. When ground balancing, make certain that you are balancing to average ground, rather than making the error of ground balancing over top of a metal target or heavier mineralized zone.

Ground balancing has to be redone on a regular basis while prospecting. The reason for this is because placer deposits do not contain uniform amounts of magnetic mineralization. Water flows create low pressure zones and high pressure zones from one place to the next. Often, you can see changes in mineralization just by noting changes in the color of the ground you are scanning. Also, changing from gravel-like material to bedrock surfaces almost always changes the amount of ground mineralization. Get into the habit of re-ground balancing about every 15 or 20 feet, or about every five minutes. This changes from one location to the next. Your detector will tell you what is going on. If the threshold hum is getting louder, it usually means there is less mineralization in the ground you are now searching. If the hum goes lighter, the mineralization is increasing. With experience, you will form a perception of when it is time to re-ground balance.

It's almost never a good idea to balance a detector over top of a piece of metal in the ground. Move around until you find a non-reading area to ground balance.

When ground balancing, move your coil all the way down to touch the ground if possible. I say "if possible," because you occasionally run across areas with so much mineralization that you are not able to put the coil within a few inches of the ground! Alkali "salts" in damp soil can sometimes also create so much interference that the coil of your detector needs to be raised several inches above the ground to search for targets. Naturally, by doing this, depth penetration is lost. But sometimes you have no other choice. Sometimes you can also get around this problem by making adjustments to your detector's sensitivity. This will allow you to search with your coil closer to the ground; but the reduced sensitivity will likely eliminate some perception of smaller or deeper gold targets.

Sometimes, you can also obtain better results by ground balancing your detector a little on the positive side. A slight positive ground balance increases the detector's sensitivity to smaller gold targets when hunting in an area of lighter mineralization. This means that the threshold makes a slightly louder hum as the coil is lowered to the ground. When operating this way, be sure to keep the threshold in the audio hearing range. You don't accomplish this by adjusting the threshold; reset the ground balance as necessary to remain in the audio hearing range when lowering the coil to the ground. Just a slight positive ground balance boost is all that is needed. Some experienced prospectors like to operate in a positive range all the time. However, you may find instances when working around highly irregular ground, vegetation and/or rocks when a slight positive ground balance creates a problem. Lifting the coil up and down and around with a positive ground balance setting can create a similar situation as with too much sensitivity.

In highly mineralized ground, when there are too many flashback signals which could be real targets, you can try and ground balance your detector to a slightly negative setting with the coil on the ground. This may reduce your sensitivity to some of the smaller gold targets. But, it is likely to settle out your machine, and it might make it possible to locate targets which otherwise would not be accessible.

Always bring along your small sample natural gold target (about the size of a match head). This should be glued to a bright colored poker chip, or something similar, to keep it from being lost. Some prospectors go so far as to drill a hole and tie a string to the poker chip to avoid losing valuable time searching for lost poker chips! When in doubt about your tuning, toss down the sample, cover it over with the ground in question, and see how your detector reacts.

One thing which should be mentioned is that while you are searching around, your threshold hum is likely to change. The answer is not to reset the threshold; it is to adjust ground balance and sensitivity as necessary to challenge the changes in ground mineralization. Your sample gold target will be the final test of whether or not your adjustments are proper. If you don't have a small natural gold nugget, get one! Otherwise, a small piece of lead will create a similar target.

Other Tips on Tuning and Setting Up a Metal Detector for Prospecting: When you are operating a metal detector, it's good practice to remove all rings, bracelets, watches and other jewelry from your hands and arms. They can give a false read on the detector. This is especially true when you are testing a detector before buying, or when you are tuning your detector to sound out properly on a special metal target while passing it over or under the search coil with your hand. Sometimes belt buckles, canteens, knives and other digging tools or large metal objects carried on a belt can create false signals when using the more sensitive specialized gold detectors. Even metallic eyelets on boots have been known to cause problems when scanning too close to your feet! It doesn't take much practice to figure out how to solve these problems.

Make sure to adjust the shaft length on your detector to a comfortable position. Bending over too far will create uncomfortable back strain when hunting for extended periods. Also, when the angle of the search coil on the shaft is changed to fit a new set of conditions, the detector must always be re-tuned to correspond with the new relationship between the coil and the metal shaft.

A lot of prospectors prefer to mount the control box of their detector on their belt or hip. This lightens the arm load during longer periods of prospecting activity.

It's also a good idea to wind the coil connecting cable firmly to the shaft. This way, it is not flopping around, giving false signals, getting caught on objects and vegetation, etc. Be careful not to pull the cable so tight as to break inner wires and create irregular operation of the detector.

OTHER IMPORTANT FACTORS TO CONSIDER WHEN BUYING . . .

If you are looking over a metal detector you are interested in buying, test it to make sure that its tuning doesn't drift on its own. This test can be done by placing a good set of batteries into the device, turning it on, allowing it to warm up for a minute, tuning it in, and allowing it to sit and run for 5 to 10 minutes. If the audio tone drifts during this time, you ought to look around for a similar detector which has better electronic stability.

CAUTION: Wetness and dampness are not good for the control box of any type of electronic detector. Be careful to avoid getting yours wet when working around water. If you intend to use a detector out in the field on a damp or rainy day, you can cover the control box with a clear, loose-fitting plastic bag and secure it to the shaft of the detector. The bag should be loose enough so you can work the various control knobs without having to untie the bag and take it off to setup or re-tune the detector.

HEADPHONES

It is important to use quality headphones. This point cannot be over emphasized. Some detectors work best with the headphones which come with the units from the factory.

There are different types of headphones. Some are heavy and cover the ears thoroughly. Some are light. What is best for you will depend largely upon the conditions where you are going to search. For example, the heavy type which thoroughly covers the ears might not be very practical in the hot, quiet desert environment. But they might work exceptionally well in a cooler environment--say along the bed of a creek where running water is making lots of noise.

Areas which include the company of occasional rattlesnakes might require the use of lighter, less sound-proofed headphones!

The proper headphones for a specific hunting environment is another area where the local dealer can make valuable suggestions.

It is a common practice for prospectors to shorten the length of cable on detector headphones to about 3 1/2 feet. This helps prevent the cable from snagging on branches and other obstructions when working in brushy areas, climbing over banks, etc.

Some detectors have volume controls and others do not. Volume on a detector while prospecting should be turned to maximum. Don't confuse this with threshold hum, which should be set at minimum audio level. If maximum volume on the detector is uncomfortable to you, obtain a set of headphones which have volume control. Then, turn detector volume all the way up, and use the headphone controls to turn the volume down if you must.

Many electronic prospectors highly recommend *''sensitivity enhancers''*--like those made by DEPTHMASTER. These help enhance the soft target sounds from gold, while lessening the noisier signals caused by trash and iron targets.

OTHER HELPFUL EQUIPMENT

A plastic cup or tray is sometimes necessary to recover gold targets, even in dry terrain. Sometimes a portable garden rake is helpful for moving smaller rocks and obstructions away from a productive hunting area. A small G.I. shovel is helpful in some hunting environments. A canteen filled with liquid; tweezers, needle-nose pliers for removing gold from bedrock traps; a small pick for digging and scraping, etc. Sometimes the ground can be very hard. This is especially true when finding gold on hard caliche layers in the desert. A wide belt with a carpenter's loop (for holding hammers) comes in very handy for a small pick. This keeps it out of the way, but also makes it quickly accessible.

A lot of your gear can be left at your vehicle, or carried in a backpack which can be set down at the hunt site. It's best to not load yourself down too heavy while prospecting with a metal detector.

Most electronic prospectors are using empty 35mm film containers to contain recovered gold targets. These are unbreakable, and the large mouth makes it easy to get a piece of gold inside.

A magnet can be a very big help while electronic prospecting. Sometimes you can recover a faint reading iron target right out of the dirt with a pass of a magnet. Otherwise, you might find yourself losing valuable minutes picking through the material, looking for a small piece of gold. Animal feed stores commonly stock a special magnet used for cows (traps small iron particles, preventing them from entering and damaging intestines). These magnets are pretty powerful, yet inexpensive. You can mount one on the end of your small digging pick, or tape it to the handle of a stainless steel garden trowel. This way, the magnet is handy when you need it.

A serrated edge on a garden trowel also is helpful when you find yourself digging around roots or brush. Some prospectors keep one edge of their trowel sharpened just for this reason.

When working bedrock areas, a small crevice tool can be a big help to open cracks and crevices which are sounding out on your detector.

Some kind of pouch or pocket creates a location to dispose of small pieces of trash and iron which you dig up. You only want to dig it up once!

Some prospectors are using fishing or photography vests--lots of pockets. These come light-weight, or heavy, depending on the environment you plan to hunt.

PROSPECTING WITH A METAL DETECTOR

All the rules of placer and desert geology apply to nugget hunting and this knowledge should be put to use in order to pinpoint where deposits and gold are most likely to be found.

Particular attention should be paid to locations which have little or no streambed so the detector's coil can search as closely as possible to bedrock or false bedrock layers. Also, exposed tree roots along the edges of the present streams, rivers and dry washes have been proving successful--especially on the smaller tributaries in the higher elevations of known gold country.

Some of the most productive areas for electronic prospecting in the dry regions is in the near vicinity of old dry washing operations. The old-timers were only able to make dry washing work in very high grade areas. They seldom recovered all the high-grade gold out of an area, because they did not have the means or equipment to prospect extensive areas. You have a substantial prospecting advantage with a modern gold detector. Natural erosion over the years might have uncovered more high-grade in the immediate area, or concentrated other spots into new high-grade gold deposits.

When prospecting old dry wash workings, it is always worth a little time to rake back some of the old tailings piles and scan them. Dry washers, today and during the past, almost always used a classification screen over the feed area. The screen was commonly around half-inch mesh. An operation shoveling onto the screen at production speed in dry high-grade gravel might have missed larger nuggets as they rolled off the top of the screen. Sometimes when working a layer of caliche, if the material was not broken up well enough, it would roll off the screen with nuggets still attached. A metal detector reads out strongly on these kind of targets. Make sure to ground balance to the tailing pile. One nice thing about prospecting in these tailings piles is that they usually are not filled with iron trash targets.

When dealing with some natural streambeds, you generally don't find too many other reading metallic objects besides gold, unless the spot which you are scanning is near an inhabited location--like a park or an old dump, etc. So it is reasonably safe to check out any target that gives off a metallic reading in a streambed. Other locations have lots of trash and small iron targets. Sometimes, iron targets seem to accumulate heavily in the same areas as gold. This presents a big challenge! Once you have been scanning a particular area for a while, you'll start to get the idea of how much ''*trash*'' (metallic objects of no value) is in the vicinity. With some experience on gold objects, you will learn to distinguish the different tone changes between most gold objects and most trash targets. Pay particular attention to the very faint readings, as is often the case with natural gold targets.

One of the main keys to successful electronic prospecting is to slow down your sweeping action. You can't do it fast like when hunting for coins. You almost have to crawl across the ground. Remember, you are listening for even the most faint whisper of a sudden increased threshold hum. A gold target may sound like a coin--only a quieter signal.

Different types of detectors have special non-motion or pinpointing settings which allow the coil to be moved very slowly and still pick up targets. This means you are able to slow down your sweep to almost a standstill and still hear a target as the coil moves over it. These settings can also be too sensitive to use for nugget hunting on some detectors, depending on field conditions and how fast you are sweeping. When in doubt, experiment.

Another important point is to overlap your coil sweeps. The area covered by the search field under your coil is similar to a triangle. Directly underneath the coil, the search area is about the same size as the diameter of your coil. The search area becomes more narrow as it penetrates deeper. The size of this search triangular field is not constant underneath the coil. It changes as mineralized conditions change in the ground being searched. If you do not overlap your search strokes, you will be leaving as much as 50 percent or more of the area below the search coil behind--without being adequately scanned, as shown in Figure 10-1.

Fig. 10-1

You have to use your own judgment as to the best way to discriminate between which targets to dig, and which to leave behind. This comes with practice and experience. Discrimination of any kind, whether by electronic circuitry or the audio sounds of the target, have limited accuracy. The depth of a target and the degree of mineralization in that particular location can confuse any discriminator. In some places you will want to dig every target. In other places, where there is lots of trash, you may find it more productive to discriminate your targets by sound and/or in combination with discrimination features on your detector. In this case, you have to balance the need to leave some small pieces of gold behind, with not wanting to dig a bucket full of trash and valueless iron targets. When in doubt, dig!

With lots of practice in the field, you will eventually reach the point where your metal detector is like an extension of your arm--like an extension of your perception and knowingness. You will gain the ability to look down into the ground and have a very good idea of what is there.

Big nuggets are generally not too difficult to find. In most areas, however, there are hoards of smaller gold targets in ratio to each large nugget found. Small pieces of gold are your bread and butter! Most gold is small gold--but it has to be large enough to be worth your while. Some of the modern gold detectors will pick up pieces of gold so small that you could be finding them all day long; and at the end of the day, not have accumulated very much gold by weight. In this case, you may want to move to a different area. Just like in other forms of gold mining, electronic prospectors have to set standards for themselves. Someone pursuing the activity to make money will be motivated to recover volume amounts (by weight) of gold. Another person pursuing the activity as a recreational hobby may receive his or her best thrills by recovering the smallest sized targets.

One interesting development in the electronic prospecting field, which I have not seen happen in other gold mining activities, is the tremendous emotional success gained from finding extraordinarily small pieces of gold! Normally, we get excited to find a big nugget or a rich deposit. These are easy when scanned over by a metal detector. The big challenge is in locating the smaller pieces. And, experienced prospectors know that if they are finding the small pieces, they will easily find the big pieces.

When scanning, keep your coil as close to the ground as possible. It takes a coordinated arm and wrist action to keep the bottom of the coil parallel and close to the ground throughout the full sweeping action. Allowing the coil to move away one or two inches from the ground during the outside of a sweeping motion, will also reduce depth penetration by one or two inches. This will certainly result in lost gold targets. You have to practice keeping the coil close to the ground. Shorter sweeps is the answer.

Sometimes it helps to kick rocks out of the way, or use a garden rake to remove multiple smaller obstructions. This allows you to keep the coil close to the ground with minimum obstructions.

During your prospecting activities, you are going to find occasional targets which could potentially be classified as artifacts. The Archaeological Resources Protection Act defines just about anything you might find as a ''protected'' artifact. There are criminal penalties for disturbing or removing artifacts. What constitutes a protected artifact is primarily up to the judgment of local bureaucrats. Searching for gold is legal. Messing around with old junk and trash may be illegal. I suggest you leave old items where they lie.

The main suggestion I commonly give to beginners about electronic prospecting is to not give up. With practice and by working in the right area, you'll find gold!

Electronic prospecting specialist Gordon Zahara showing off another gold nugget!

PINPOINTING

When you do have a reading target, its exact position in the ground should be pinpointed. This can be done by scanning over it in fore and aft and left to right motions, and by noting when the detector sounds out while doing so. Often a steel or iron object can be distinguished by noting the size and shape of the object while scanning in this manner. Sometimes trash objects are large and lengthy in size.

Sometimes a stronger reading target will seem to cover a larger area than it actually does. If you are having trouble getting an exact location, continue to lift the search coil higher off the ground while scanning with an even back and forth motion. As the coil is raised higher, the signal will become more faint; but the position of the metal object will sound out at the very center of the search coil.

PINPOINTING WITH THE GOLD PAN IN WET AREAS

Once you have pinpointed the exact position of a reading target in an area near water, one way of recovering the object is to carefully dig up that portion of ground, using a shovel or digging tool, and place the material in your gold pan. Here's where a large plastic gold pan comes in real handy. Be careful not to cause any more disturbance or vibration of the streambed than is necessary while you are attempting to dig up a target. It's possible to miss the target on the first try and cause it to vibrate further down into the streambed because of its greater weight. Sometimes you can lose the target, and not be able to locate it again with the detector.

Sometimes the ground is hard and needs to be broken up or scraped with a small pick. Keep in mind that a sharp blow from a pick could destroy a coin or disfigure a gold nugget, reducing its value. Digging and scraping is better than pounding if possible. In hard ground, sometimes its easier to loosen the material up and shuffle it into several small piles. These can then be flattened out and scanned to see which contains the target.

Once you have dug up that section of material where the target was reading, place it in your plastic gold pan and scan it with your detector. If the contents of the pan do not make your detector sound out, scan the original target area over again. If you get a read in the original location, pour the contents of the pan neatly into a pile out of your way, pinpoint the reading target all over again, and make another try at getting the target into your gold pan. Continue this until you finally have the reading target in your pan. If you can't see the target, and water is immediately available, pan off the contents until the target is visible.

There is a good reason why you don't ever throw away any material from the target hole until after you know for certain exactly what the reading target is. If you are looking for placer gold deposits and it happens that your detector is sounding out on a gold flake or nugget, the chances of additional gold being present in paying quantities within the material which has been dug out of the hole are pretty good--even if it doesn't sound out on your detector. Remember--small particles of gold often do not make metal detectors sound out. Therefore, the possibility exists that there could be hundreds of dollars worth of gold in a single shovelful of material not tightly concentrated enough to cause your detector to sound out. So wait until you have seen the reading target before you start throwing any material away.

If you dig for a target and then cannot get any further reads on the detector from either the pan or in the original target area, quickly pan off the pan's contents. There is always a possibility of a concentrated gold deposit which is no longer concentrated enough to read out on your detector. In this case, you are most likely to have some of the gold deposit in your gold pan--which will be discovered when you pan off the contents.

There is also another reason why you save the material from a hole until you have recovered your target. Because of the shape of a gold pan, and depending on the size and shape of the coil you are using, if the gold target (especially a small target) ends up in the bottom of your pan, your coil may no longer be able to scan close enough to the bottom of the pan to sound off on the target. Because of this problem, gold pans have limited workability as a pinpointing tool in electronic prospecting. Sometimes, you can have better luck scanning the bottom side of a gold pan.

After attempting to dig a reading target, if you can't seem to locate the target again with your detector, sometimes it's easiest to quickly pile the material in your gold pan and pan it off.

DRY METHODS OF PINPOINTING

There are several popular methods of pinpointing reading targets in the dry regions where water is not available for panning. Some prospectors use a plastic cup. Material from the target area is scooped up and passed over the coil. If the contents of the cup do not sound out on the detector, the target area is scanned again until the target is located. Take another scoop, pass it over the coil for a signal, and repeat until the target is in the cup. Then pour some of the contents into your hand. Pass the cup over the coil, then your hand, until you locate the target. Put the rest in a neat pile away, and repeat the process until you recover the object.

One important point to mention is that most search coils have equal sensitivity to targets both above and below the coil. In other words, during pinpointing, you can pass material over top of the coil and obtain the same results as if you turned the detector upside down and passed the material across the bottom of the coil.

Your handy magnet can be a big help when you are having difficulty finding a faintly reading target. Sometimes it's just a small piece of iron.

There is also the hand-to-hand method of pinpointing a target. You can reach down with your hand and grab a handful of dirt or material. Pass your hand over the coil to see if you are holding the target. Continue until you have the target in one hand. Then pass half the material in that hand to your other hand. Use the coil to find which hand has the reading target. Discard the material from the other hand and pass half the material from the reading hand to the second hand, again. Keep up the process until you only have a small amount of material in the reading hand. Carefully blow off the lighter material to locate the target.

Another procedure is to grab a handful and pass it over the coil as explained above. Once you have a reading handful, toss it in a pouch for later panning.

Some prospectors prefer to sprinkle a reading handful of material slowly on top of the detector's coil. When the target drops onto the coil, the detector will sound off with a distinctive bleep. Then, all you have to do is gently blow the dirt off the coil. This leaves the nugget in plain view. Other prospectors feel this is an awkward method. For sure, if you are not careful, the gold object can drop back onto the ground and put you back into a search mode just to make it a reading target again! This method is much more difficult in wet ground conditions.

One important thing about using your hands to pinpoint and recover gold targets, is that some of today's most sensitive high frequency gold detectors react to the ''salt'' (mineral) content on or in your hand. When this occurs, your hand will give a reading when passed over the coil by itself--sometimes even a strong signal. This makes hand pinpointing methods more difficult. In this case, the plastic tray or cup works just fine--or the method of pouring material onto the coil.

If you find a gold target, be sure to scan the hole again. Sometimes it's worth doing a lot of digging and/or raking, with alternate scanning of the immediate area. Gold targets never travel alone. I have seen dozens of gold flakes and nuggets come from within a one-foot square area on bedrock!

HOT ROCKS

Sometimes you get soft or strong signals from your metal detector when the coil is passed over rocks which contain heavy mineralization. These rocks are called *''hot rocks.''* This needs to be clarified. It isn't necessary for a rock to have an extraordinary amount of mineralization to sound off as a hot rock. It just needs to have more than the average ground which your detector is ground balanced to. The opposite of this is also true. A rock containing less mineralization than the average ground, will change tone on a detector in the opposite way. These are called *''cold rocks.''* A hot rock from one location could be a cold rock in another location containing higher mineralization in the average ground.

Put another way, when your detector is properly ground balanced to the average ground, it will sound off on rocks containing condensed heavier mineralization. Often, the signals given off by these hot rocks can be mistaken for signals which might be given off by gold targets in the ground. Some locations have lots of hot rocks! Naturally, since gold travels with other heavy mineralization, you can expect to find hot rocks in the various gold fields.

Hot rocks come in all sorts of colors and sizes. A big hot rock, depending on the degree of mineralization over the average ground, can sound very similar to a gold target. The same goes for small hot rocks.

Quite often, most of the hot rocks in a given area will be similar in nature, color, hardness, etc.--but not necessarily in size. After searching an area for a while, you will know what the hot rocks look like. This makes pinpointing targets a little easier, especially when dealing with small hot rocks. When receiving signals from rocks which are dissimilar than the average looking hot rock of the area, check them out closely--especially when they have quartz. Some of these may not be hot rocks at all--but gold specimens.

If you are finding hot rocks on top of the ground, you will almost always find them buried, as well.

Hot rocks rapidly lose their signal strength as they are scanned further from the coil. Metal targets gradually lose signal strength in proportion to the increased distance from the coil. Also, a

Gordon Zahara raking the next layer off a "nugget patch."

metal target tends to create a more specific, sharper signal as the coil is passed over. The signal given off by most hot rocks is not as sharp. It's more of a mushy sound which doesn't have as clear a change in tone intensity as the coil passes over the target.

When uncertain about a reading target, one way that some hot rocks can be eliminated, is by passing the coil over the target from different directions. If the target only sounds out when being swept from one particular direction, then it is clear that the target is not a gold nugget.

Sometimes, when lowering the sensitivity on your detector a few numbers, the signal given off by a nugget will continue, but weaker. A hot rock signal will quickly disappear by lowering the sensitivity.

Another test is to slowly raise the coil further away from the reading target as you are scanning. The signal of a metal object will become progressively more faint as distance is increased. The signal given off by a hot rock will rapidly die away as a little distance is put between the target and the coil. Most hot rocks in a given area will have a very similar sound, the intensity depending upon size and distance from the coil. Many times, you will find that a gold nugget gives off a much sharper signal. But not always. Much also depends on the size of a gold target, distance from the coil, and the amount of mineralization interference in the ground being scanned.

Each area is different. The hot rocks in some areas are reasonably easy to distinguish. In other areas, it is necessary to dig every signal.

Larger hot rocks are easier to detect. Many can simply be kicked out of the way. It is the smaller pebble-sized hot rocks which create faint signals. In some locations, these signals are so similar to the signals given off by gold nuggets that you have no other choice but to dig them up.

If lots of hot rocks are making it difficult for you in a particular area, experiment with smaller coils. As mentioned earlier, smaller coils are more sensitive to smaller pieces of gold. Therefore, the sharpness of the signal over gold targets is also intensified. This may give you the needed edge in distinguishing the slightly different signals being given off by hot rocks in that area.

Some gold prospectors go so far as to ground balance over top of a hot rock until it no longer gives off a signal. This will eliminate a large portion of the hot rocks in the area, but it is also likely to eliminate a large portion of the small and marginal signals given off by gold targets. This procedure may work well in areas containing just large nuggets. However, most areas I am aware of have dozens or hundreds of smaller pieces of gold for every larger piece you will find.

Since cold rocks are made up of a lesser concentration of iron than the average material which your detector is balanced to, they give off a negative reading signal on your detector. Ground balancing over a cold rock will cause your detector to react positively to the average ground in the vicinity.

PRODUCTION

Once you find a location which is producing gold targets, you will want to thoroughly work the entire area. This is commonly accomplished by carefully gridding the site. By this, I mean creating parallel crisscross lines on the ground--usually about four feet apart. You can do this by drawing light lines in the sand with a stick. Grid lines allow you to keep track of where you are, what's already been searched, etc.

If there are obstructions in the producing area (*"nugget patch"*), they should be moved or rolled out of the way if possible. The smaller rocks and materials on the ground can be raked out of the way. This allows you to keep the coil closer to the ground. In a proven area, this will mean more recovered gold targets.

Some areas produce quite well on the surface. Then, by raking or shoveling several inches off the surface and scanning again, sometimes you can find more targets. I have friends in Arizona who operate serious mining operations, using metal detectors as the only recovery system. They have a bulldozer to remove material which has already been scanned--two or three inches at a time. A layer is removed, the area is scanned thoroughly with metal detectors to locate all of the exposed gold targets. Then, another layer is removed and scanned. This process is done all the way to bedrock. They make good money at it!

I have heard of others doing the same process on a smaller scale by attaching plow blades to four-wheel drive vehicles. Some even do it using ATV's!

While hunting over a proven nugget patch, be sure to keep your detector properly tuned all the time. Otherwise, you will certainly miss gold targets.

PROSPECTING OLD HYDRAULIC MINING AREAS

Perhaps the most productive areas of all to prospect with metal detectors are areas which were already mined by hydraulic methods. Many of these areas are left with a large amount of bedrock exposed. This allows immediate and easy access of search coils to get close to old gold traps. Some-

Exposed decomposed bedrock in an old hydraulic mine area--this was a typical "hot spot location" for electronic prospecting.

times, the exposed bedrock has been further deteriorated because of direct exposure to the elements. Raking away decomposed bedrock sometimes can produce remarkable results once a nugget patch is located.

While searching such areas, pay attention to the color of dirt which covers and surrounds nugget patch areas. Sometimes, you will discover a pattern. In our area of operation, it is a white powder which often signifies the possible presence of a good spot. Perhaps in other areas, it might also be this white powder, or something else.

PROSPECTING OLD MINING TAILINGS

It is usually possible to find piles and piles of old mining tailings throughout most proven gold-bearing country.

Many of the larger earlier gold mining operations (the ones that were set up to move large volumes of material), concentrated on recovering only the fine and medium-sized values from the material processed. This meant that the larger material was classified out so the smaller material could be processed through controlled slow-moving recovery systems. On the larger operations, classification was usually done either with mechanical vibrating classification screens, or with the use of a *"trommel."* A trommel is a large circular screening classifier which rotates and tumbles material through, allowing the smaller classification of materials to pass through the outer screen and into a channel which directs them to the recovery system. The larger materials are passed down the inside of the trommel to be discarded as tailings.

In many of the larger operations, there were no means to recover the larger pieces of gold which were screened out along with the other large materials. As a result, larger nuggets were sometimes discarded along with the waste material as tailings.

Anywhere you see large tailings piles, especially near the present waterways, it is evidence of a large volume operation where it's possible that the nuggets were discarded with the waste and are likely to still be there. Some large tailing piles still have large goodies inside them, which sing out very nicely on the proper metal detectors. One way to distinguish the right kind of tailing pile to be looking through is there should be a pretty wide range of material sizes in the tailing pile. Tailing piles which consist only of the larger rocks (cobble piles) were most likely stacked there by hand during a smaller surface-type operation.

Scanning tailing piles with a metal detector has proven to be highly productive in some cases, and should not be overlooked as a possibility for finding nice specimen-sized gold nuggets. The VLF ground-cancelling detectors are best suited for this because tailings usually also contain a large quantity of mineral content, which is likely to cause interference on a BFO detector.

PROSPECTING OLD MINING SHAFTS

Author and prospecting buddy checking out an old mine shaft in Arizona.

CAUTION: The first thing to mention about prospecting in old mine shafts is that they are dangerous! There are different things that can go wrong when snooping around in such places, one main danger being a potential cave-in. The shoring beams in some of these old shafts may have become rotten and faulty over the years and it's best not to lean up against, or to bump, any of the old wooden structures situated in any old mine shaft. Loud, sharp noises have also been known to "bring the roof down" in some old mines.

Shafts which extend down into the earth on a declining angle are particularly dangerous, because if the ladders or suspension systems are faulty and collapse while you are down inside, you may not have any way to get out again. For this reason, it is always a good idea to bring along a caving rope, and to use it, when exploring declining-type old mine shafts. Caving rope is different than mountaineering rope in that it is made not to stretch nearly as much.

Even the most experienced cave explorers shy away from entering declining mine shafts because of the dangers involved.

Another potential danger involved with exploring old mine shafts is encountering poisonous or explosive gasses. Entire mines were shut down because of such gasses--even when they were good producing mines. These, however, were usually caved in and closed off to prevent unsuspecting adventurers, like you and me, from entering at a later date and getting into trouble.

Before you enter any old mine and start sorting through its low-grade ore piles, or pecking samples off the walls of the mine, it's a good idea to make sure nobody else presently owns the mine. Or, if somebody does, to get the person's permission first. There is probably nothing more dangerous in a mine shaft, cave-ins and poisonous gases included, than some old cantankerous miner who catches you, uninvited, in his mine and thinks you are stealing his gold!

There is probably more danger in prospecting old shafts than in any other gold prospecting activity.

After all that, just in case I have not succeeded in scaring you out of the idea of entering old mine shafts in your prospecting adventures, here are a few pointers on how and where you might find some rich ore deposits, or rich ore specimens, with the use of your metal detector.

But first, let me recommend that if you do go into such places, you bring along a buddy and leave another one at the surface with explicit instructions to not enter the shaft under any circumstances, but to go get help in the event that you should get into trouble within. It's also a good idea to let a few others know where you are going, just in case your outside man doesn't follow orders and you all become trapped inside.

There are two main sources of possible gold and silver in an old mine:

1) High-grade ore specimens may have been placed in the low-grade ore piles and left as waste material.

2) High-grade ore which has yet to be mined.

With the exception of the largest production hardrock mines, there is always a certain amount of rock (ore) which has been blasted away from the wall of the tunnel that is not milled and processed, because it is of such apparent low-grade value that it is not worth the expense to do so. Sometimes the vein being followed into the mountainside was not as wide as the tunnel needed to be in order to progress into the mountain. Therefore, that rock which was not part of the vein itself, or the contact zone, that looked to be of lower-grade value, was also discarded into the waste ore piles.

Crushing and milling ore is and always has been somewhat of a timely and expensive process. For this reason, a good many mines only processed what appeared to be the highest-grade ore that was blasted from the ore body. The rest was usually piled in the shafts out of production's way.

In the smaller operations, the kind where the ore was crushed and milled by hand and then processed with a gold pan to recover the values, only the highest of high-grade ores could be processed at a profit. The rest was usually laid aside out of the way.

The sorting of a higher-grade ore from lower-grade ore has always been a matter of judgment on the part of the person who was doing that part of the job. And, until the more recent breakthroughs in electronic detecting equipment, sorting needed to be done by eye. There simply was no other way.

The interior of mine tunnels was often poorly lit during earlier days with miner's candles, and the air inside the shaft was often foul after the powder explosions used to blast away the ore from the interior of the mountain. As a result of all this, it is not difficult to imagine that some higher-grade ore was probably discarded as waste material in some mines.

How rich the ore was in a mine will have a lot of bearing on whether or not a present-day prospector will find high-grade ore in the waste piles of that mine. The size of a mine does not necessarily have anything to do with how high-grade its ore was, although it may have a bearing on how much waste ore will be available to test. County reports can be looked over to locate the old mines within an area. Many of these reports also have data about the grade of ore being extracted from some of the mines (see Figure 9-3 for example).

When you are testing ore samples in a mine with a metal detector, and you don't come up with any specimens, try a few different piles. If you still don't come up with any specimens, it was probably not a high-grade mine. Try another.

Keep in mind that the floor of a mine shaft is likely to have iron objects which will cause your detector to read out falsely.

One of the fastest and most effective ways of thoroughly checking out samples of ore is to lay the detector down on the ground or on a makeshift bench with the search coil pointing upward so individual ore samples can be passed by the most sensitive portion of the coil. The 3-inch diameter search coils are probably best suited for this kind of work because of the increased sensitivity, and because there is seldom great need for increased depth sounding when testing specimens.

The most sensitive portion of the search coil is usually located near the center of the coil. It can be easily pinpointed by passing a coin back and forth across the coil while the detector is in tune, and by finding where the coin causes the loudest and sharpest reading. The most sensitive area should be marked brightly with a Magic Marker so it will be easier for you to move ore samples directly past it. In this way, the most accurate testing is possible. It is worthy to note that the top of the coil has similar sensitivity to targets as does the bottom.

Ground balancing would probably best be achieved by tuning to the overall ore pile. But, you may need to experiment. A lot will depend on the grade of ore you are looking for. High-grade gold specimens will require the surrounding mineral to be cancelled out. On the other hand, sometimes high-grade ore is full of fine particles of gold which would not sound off on a detector. In this case, the high degree of mineralization in the ore specimen might sound out like a hot rock. Therefore, you would not want to ground balance out high mineralization. So, depending on the nature of the ore in a mine, you will need to use your judgment regarding the proper way to ground balance.

A good idea to bring along a bucket, or a knapsack, so you will have something to carry out your specimens as you find them.

Both the BFO and the VLF ground-cancelling metal detectors can be used to locate high-grade ore specimens successfully.

For the individual who is interested in finding a wider range of high-grade ore specimens rich enough to be worth milling on a small scale for the gold and silver values, perhaps the BFO is the better type of metal detector to be used for the job. This is because the BFO does exceptionally well at picking up highly mineralized ore. When you are dealing with hardrock veins, high mineralization is a good index of gold being present, especially when you are working with the ore that was blasted out of a previously successful gold mine. Ore which contains large amounts of locked-in values or lots of fine gold thoroughly dispersed throughout the ore is not likely to read out as a metallic on any kind of conventional metal detector--although the high degree of mineralization which usually goes along with such ore probably will sound out on a BFO detector as a mineral. So the BFO detector is handy in finding rich ores which might otherwise be undetectable.

By placing the BFO detector on its mineral setting and keeping the pieces of ore which cause it to sound out, you can accumulate a lot of good paying ore. With the use of a BFO as such, the perimeters of ore piles can be scanned first in an effort to find a pile that has a great deal of mineralization.

Portable rock crushers and mills and processing plants are available on the market--as covered in Chapter 9. Perhaps by combining these with the use of a good BFO detector to pick out the highly mineralized discarded ore, a small operation could be quite profitable without having to foot the higher costs of developing a lode mine from scratch.

LOCATING RICH ORE DEPOSITS

When prospecting around in an old mine with a metal detector, don't discount the idea of searching the walls of the various shafts to locate the rich ore deposits which may have been overlooked by those who originally developed the mine. With today's market value of gold and advanced milling techniques, a small-scale mining operation can be run profitably in ore paying about 1/2 ounce to the ton in gold values--or even less. Of course, this depends upon the nature of the ore, as covered in the last chapter. Abandoned mines which will pay this well, and even better, are scattered all over the West. If a person were really interested in locating a lode mine that he could work at a profit, one way would be with the use of a BFO detector to prospect the various abandoned mines within the area of his or her interest. In using the BFO to scan the walls of such mines, highly mineralized ore deposits can be located, samples can be taken for assay, and a person could find the mines having the richer paying ore deposits. When prospecting like this, it's a good idea to bring a can of spray paint, a pen and paper, and some separate bags, so the samples can be accurately marked as to where they came from, along with their respective deposits being marked with the spray paint.

If you are not interested in starting up a lode mine, but are just kicking around looking for a bonus, again, the VLF detector is good for looking through mineralization and detecting the richest deposits. If, however, you do locate a reading metal inside the wall of a mine shaft with a VLF detector, and you have determined that it is not some falsely reading mineralization, then it's well worth your while to investigate further because the chances are pretty good that you have located some bonanza paying ore. I know of one highly successful high-grade gold mine in the Mother Lode area of California that is using VLF detectors as its sole method of determining where to blast. At this writing, they have located and recovered millions of dollars in gold during the past two years--in a mine which was failing prior to the use of metal detectors.

GOLDSPEAR

The *''Goldspear''* is a different kind of electronic gold detecting device than we have been talking about earlier in this chapter. It is a probing device which has a special electronic sensor probe on the point. The upper end of the probe has a control box and handle. As the probe is forced down into the ground, if the pointed edge the probe comes in contact with a particle of gold, the control box will make a specific audio sound and cause a specific light to flash. The unit comes with a set of headphones.

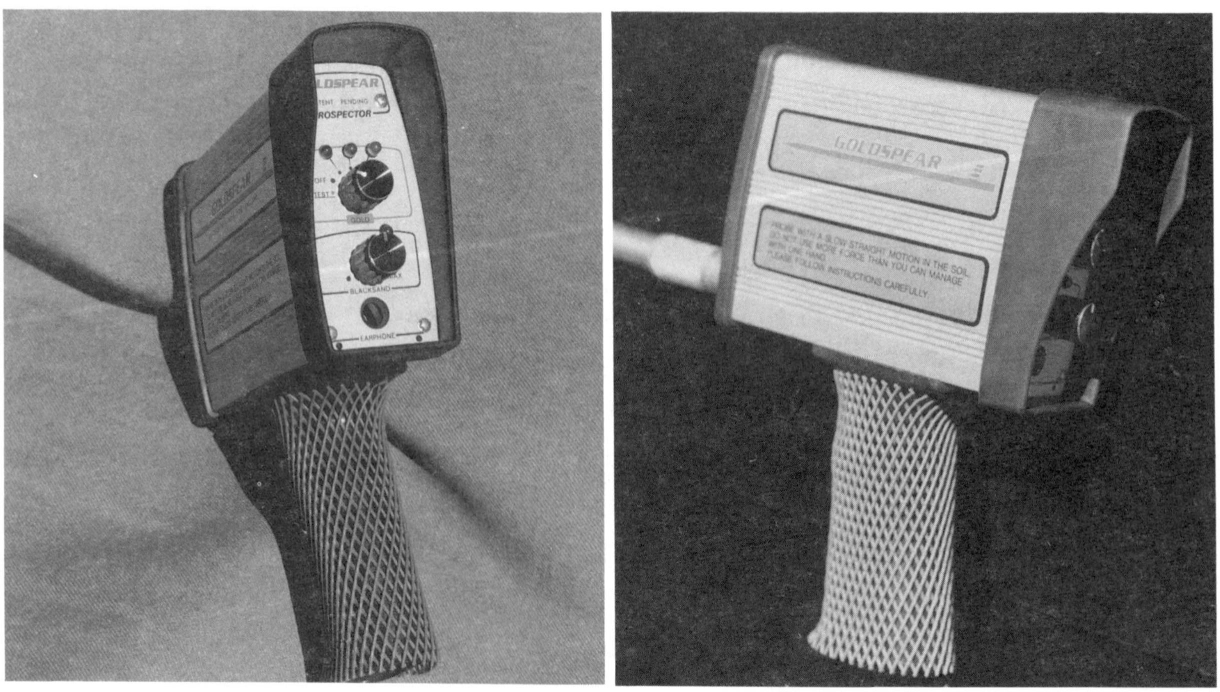

Fig. 10-2.

The unit also sounds off with a different audio tone when encountering particles of black sand--and a different colored light flashes on the control box.

Unlike a conventional metal detector, the point of the Goldspear's probe must come in contact with a piece of gold in order to give the gold signal. However, also unlike conventional metal detectors, the probe will sound off on fine particles of gold--even so small as 300 mesh!

The probing electronics of the Goldspear measures electrical fingerprints of the different elements which pass by the probe as it is forced into the ground. Different atoms have different

Experienced prospector, Bill Stumpf, loaming an area which was cleared off by brush lightning fire several months before.

electrical charges. The Goldspear is specifically adjusted to sound out on gold, platinum and mercury on one circuit, and black sand on another.

The tool is very useful in some prospecting activities. It also has severe limitations in others. For example, the probe cannot easily be pushed down through compacted streambed gravels or where lots of rocks are present.

We have successful prospectors in our area who religiously use the Goldspear to determine where to operate their highbankers. They spend initial time probing around to see where they get the most gold and black sand signals. This is followed up with pan testing. The better areas are then worked with hydraulic concentrators.

The Goldspear can be also probed down under the water as long as the control box is not submerged. The unit comes with an extra four feet of probe shaft which is easily attached. Extra lengths can be utilized if necessary.

I have heard of prospectors in Canada having excellent luck using the Goldspear to help trace highly mineralized alluvial deposits upward to locate lodes and/or pockets. The process, called *"loaming,"* is done by probing the side of a hill for gold and black sand readings, and tracing them upward, as outlined in Chapter 9.

You don't get too many false signals from a Goldspear. Sometimes, man-made metallic objects can cause a false gold signal when the probe is rubbed against them. However, I would point out that you will not run into very many of these while prospecting in the field. When used for locating highly mineralized zones with gold, you will find that just about all of the signals given off while probing are indications of high mineralization and gold.

Occasionally, in a very rich area, a piece of gold may become lodged on the surface of the probe. In this case, the unit is likely to continue giving off false gold signals until the piece of gold is removed.

Some sampling operations have gone so far as to use the Goldspear in conjunction with an auger drill. Sample holes are drilled with the auger to loosen up and remove some of the material from the hole. The Goldspear is then used to probe the hole for signals. More serious sampling is done in those holes which give off the most positive signals.

The Goldspear is a very useful tool when utilized within its limitations, and I believe you will see more of them being used in the field as time goes on.

MEDICAL

Bring along some mosquito repellent during electronic prospecting ventures. Otherwise, if they are around, sometimes it is difficult to put your full attention on listening to what your detector is trying to tell you.

In some areas, you also have to be careful of ticks. Some ticks carry Lyme Disease. Some insect repellents will keep ticks off of you. *DEEP WOODS , TECNU 10-HR INSECT REPELLENT, and OFF* seem to work quite well. Spread the repellent on your boots, pants and other clothing.

In the hot climates, you should bring along the proper sunscreen and apply it to exposed body parts.

It's always a good idea to bring along a good first aid kit, and at least leave it in your vehicle in case of emergencies.

Extra clothing is also a good idea. You can leave it in your vehicle. You never know...

GOLD MINING PROCEDURE--WRAPPING IT ALL UP

CHAPTER XI

One of the first things to determine when you're getting started in a gold mining operation of any size is the general area you plan to operate. As mentioned in an earlier chapter, time and study spent in locating an area which is likely to pay in many gold deposits is time and study that will pay off very well in the end. If you intend to succeed in finding good paying deposits, it's not enough to just go anywhere; because in that way, should you find a deposit, it will be strictly on the basis of chance or luck. It's been done this way before--but not with any degree of consistency. It's O.K. to find a deposit of gold once in a while. But it's much better to be able to find them one after another. How well you do in picking out your general operating area will greatly determine how well you will be able to stay into paying quantities of gold. So it's best to put extra effort into finding a good area for mining.

SAMPLING--ITS PURPOSE

Pay attention to this, because I could never say it strongly or loudly enough, and I could never say it too many times: *THE KEY TO SUCCESS IN GOLD MINING, ON ANY-SIZED SCALE, IS IN DOING LOTS OF SAMPLING*! The whole idea behind sampling is to keep moving around and testing different locations until you find one that you can work effectively with the equipment at your disposal, where the gold, silver, and/or platinum values are enough to make the effort of working the ground worth your while.

Generally speaking, as mentioned earlier, placer deposits run in streaks (or *''stringers''*). These streaks are found in certain common locations--as outlined in earlier chapters of this manual. The idea behind sampling is to locate a paystreak, or some other kind of gold deposit, BEFORE you start into any serious production mining activity.

There is another way to go about it. That is to start mining in any likely spot and *hope* you will run into a deposit sooner or later. I have seen this procedure used over and over again by beginners, myself included. But, I have yet to see it pay off with any kind of consistency at all. I'm not too far off when I say that at least 99% or more of the long term successful mining operations of any size today are using the sampling method of finding paying quantities of gold first, before a production operation is started. I, myself, have never seen ANYONE succeed for an extended period of time who did not know how to sample well, and who was not willing to sample energetically in an effort to locate gold in paying quantities.

When sampling, if you test a specific location and don't find an acceptable quantity of gold in the material, move to another likely spot and try there. This procedure is simply continued until you do find an acceptable paying deposit.

You usually don't recover a lot of gold when sampling. You are not trying to. You can gauge how well a production operation is doing by how much gold is being recovered. A sampling activity is different. Probably the only way that a sampling operation can be gauged is by looking at how much ground has been covered by effective sampling. It's easy to lose a bit of morale when you're sampling, because you are not seeing the gold adding up in your *poke* (collection of gold). This is especially true when you have just finished cleaning out a good paying streak, and you are now back into sampling again. However, you can't let it get you down that you are not recovering lots of gold while sampling. If you are not covering lots of areas with effective testing, you can let that get you down a bit, and some evaluation will be in order to find out what you are doing wrong. If you test out an area effectively, and there is not enough gold there to start a production operation, that's fine--great! Now you know there isn't enough gold in that spot. That's all you were trying to find out and you've succeeded. There is no reason in the world to feel bad about it. Often, when you find out where the paying deposits are not located, it will help you a great deal in determining where they are more likely to be. So, each sample hole, effectively done, gets you that much closer to a paying deposit. Get the idea?

Once you have been at it a while, you'll discover for yourself that finding the deposits is a sure thing if you take on sampling in a very serious and systematic manner.

Gold mining is the kind of activity where you are making nothing today (sampling) and are making thousands tomorrow (production). How many sample holes, on the average, it will take you to find paying deposits depends on your ability to sample, and on how good the general gold bearing area is in which you are operating.

Your sampling ability depends upon how serious you are about finding gold, and in doing it right. If you persist, and study like mad, and work at it, you'll get better as you gain experience. My present average is a paystreak every four or five sample holes. This is true--believe it or not--but I've put myself into a reasonably good area (Klamath River in northern California). I've been here long enough to know the area well, I've been at it hot and heavy in this area, full time, for a few years, and I've received instruction from a few of the best prospectors in the business. But I'm not withholding any information from you. As a matter of fact, I'm giving you much more than was ever given to me. So there's no reason why you shouldn't be able to do as well as, or better, than I do.

In a sampling operation, testing can be done with the use of a gold pan, a sluice box, a dredge, a dry washing plant, a hydraulic concentrator, or any combination of these--whichever you have at your disposal to use that will do the most effective job of sampling in the areas you are interested in.

The gold pan, as mentioned earlier, is the most versatile sampler in existence. However, it is limited in its production capability. If you are way out in the field on a sampling expedition, the gold pan is probably your best tool for the job. The same holds true if you want to move quickly through an area and just do a fast, but not necessarily thorough, job of it.

If you want to test out the dry streambed material along a swift moving creek, stream, or river, perhaps a small sluice box can be brought along for the job. It takes a little more effort to carry one in with you and to set it up at each location, but it's not that much more difficult and will save you lots of time and energy in the long run. Thoroughly sampling an area with a sluice box can be much faster than with a gold pan. Plus, larger-sized samples can be taken, so more accurate testing will be possible.

The hydraulic concentrator is an excellent combination sampling/production machine, because it is not so heavy that it can't be packed into a wide range of areas. Yet, it can be quickly set up to produce or sample, with excellent recovery, in any area where water is nearby. Figure 11-1 shows another version of the same hydraulic concentrator covered earlier in Chapter 5. But this one contains the additional feature of a 2-1/2 inch dredge suction hose system, which can be attached to the concentrator (3-inch models are also available). This rig is excellent for testing smaller-sized creeks and streams running through gold-bearing areas. The advantage to using this kind of machine to sample small streams is that it can be used to accomplish many tasks. First of all, the engine/pump

Fig. 11-1. A nifty combination hydraulic concentrator--2 1/2 inch suction dredging unit, which has a very wide range of operation ability.

unit is totally separate from the sluicing unit. The sluicing unit doesn't float, but sits on a short adjustable stand. Set-up is just a matter of laying the engine unit down in a stable, level position, setting up the sluicing unit in a similar way, and starting the engine. This is a lot easier than having to dig a hole first to float a dredge of similar size. Also, the dredging unit runs off a nozzle jet system. This allows you to work the suction nozzle in and out of the water without the worry of losing the prime in your suction hose. This is a big feature because it means you can work in very shallow streambeds effectively, and that you can pick up and move to other positions without having to keep the nozzle under the water at all times. If you are a two-man team and want to get more work done, one man can dredge into the concentrator, and one man can shovel into it, both at the same time. Or you can unhook the power hose from the dredge system and attach it to the concentrating unit if you just want to shovel. You can also unhook the pressure hose from either component and use it to wash streambed material lying above the water into your work hole *(hydraulicking)* and then dredge that material out of the hole into the concentrator. This can be a real boost to production. (You will want to look into the local laws before doing this as sometimes it's "not allowed.") Another thing that can be done with a highbanker is to use it to pump water up into a small hole you have dug. Once the hole is filled with water, you can dredge in your man-made water-filled hole with the dredging unit pumping into the concentrator. The concentrator can be positioned so the water runs off its sluice box a short distance away and flows back into the hole. This way you will not have to re-fill your hole with water. The hydraulic concentrator is an excellent machine, and you can get a lot of work done with one--either in a sampling or production activity. There are many, many people who do *''highbanking''* as their primary gold mining activity. I know a guy who has one of these and does very well with it. He uses a wheelbarrow to haul it around the countryside--along with his gas and cleanup gear. I know another guy who works small creeks with a gold pan and a small shovel--and is doing an average of 2-3 pennyweight a day and sometimes he hits nice sized pockets. I just can't imagine how well he would do with one of these dredge/concentrating units. I'm trying to get the two guys together to find out.

As mentioned earlier, the electrostatic dry concentrating unit makes a very effective sampling machine in the dry areas where you don't have to haul it very far. Otherwise, the lighterweight dry washing plants can be used to sample with good results, or winnowing and dry panning can be done if necessary, and can be made to work.

Metal detectors are usually not used by themselves as samplers, but usually in conjunction with other sampling equipment. Nugget hunting and sampling are actually two separate activities. Sampling is the process of testing different areas to see how much gold is present in the material there, with the intention of finding acceptable paying deposits to be worked on a production scale. Nugget hunting is an activity of specifically going after the larger pieces or accumulations of gold. The VLF detector is most often used as a nugget hunting device, whereas the BFO-type detector is most commonly used in sampling activity, to help the prospector determine what material to test for its gold values. A BFO detector could be used to help search out the general area you have chosen for a likely spot where some rough bedrock is exposed. Tune the detector in on the mineral setting and scan the area until the highest reading mineralized zone is found. Get some bottom strata samples from that area, just above bedrock if possible. If there's not enough gold in that spot, continue to scan and sample the immediate area for additional highly mineralized zones until the gold deposit is located or until you have determined that an acceptable deposit is not present. This is a good use of a metal detector for placer prospecting purposes. Some miners like to use them and some don't.

When engaging in a sampling activity, it is always best, if possible, to test the same type of ground that you intend to work on a production scale, should you find a paystreak. For example, if you are planning to dredge, then the most effective sampling would be beneath the water's surface. Some results can be obtained by sampling the shoreline by surface methods to locate new dredging ground. But it is never a sure thing that because you are finding gold up on the banks, you will also find a good deposit in the stream when you bring on your gold dredge. The only sure way to find out if a body of water will have an acceptable paying gold deposit is to test that body of water. Small backpack dredges work well as samplers where the material is relatively shallow.

CREVICING (SNIPING)

One other popular means of sampling underwater locations is with the use of a mask, snorkel and screwdriver. The method is called *''crevicing,''* *''sniping,''* or *''fanning,''* and is usually done in streams and creeks that are located in the high country of gold bearing areas where the water moves too fast during the winter months to allow much streambed to form in many areas. This leaves the various bedrock crevices and other irregularities open to observation during the slow water periods of the year.

By floating down these streams and carefully looking over the gold traps in the bedrock, a person can locate gold nuggets, flakes and sometimes entire pockets of coarse gold. Often the individual pieces are found wedged into the various bedrock traps and need to be pried out with a strong screwdriver or a pair of pliers.

Depending upon the nature of each particular waterway, this activity is sometimes most productive during the late summer months, while the water is running at its lowest and slowest and warmest, allowing you to look over more of the streambed.

Sometimes it is necessary to sweep off small deposits of sand or light gravel which collects in some of the bedrock irregularities once the water has slowed down. This can be done by waving your hand back and forth over the light deposit to *''fan''* the material aside.

One thing about these higher tributary channels is that the water usually rushes through them in a torrent during the winter storms and spring runoffs, sometimes causing some amount of new re-deposited gold to be washed over the continuously exposed traps in the bedrock. If you find a good spot, it may be worthwhile to keep quiet about it and check it out over and over again on a yearly basis.

MOSSING

The moss located on the edges of gold-bearing streams, rivers and creeks has excellent gold-trapping ability. Sometimes gold can even be recovered out of the moss up on the streambanks, well away from the water. In such cases, paying quantities of gold can be recovered by an activity directed solely at processing the moss along river banks.

One of the better methods of *''mossing''* is by filling up a large bucket or washtub with the moss taken from the banks of a gold-bearing stream. The moss is then thoroughly broken up and pulled apart inside the bucket, which is filled with water. Once this is done, the contents of the tub are screened and panned, and the gold can be recovered out of the resulting concentrates. This activity, done consistently over the period of a day, can bring in a surprising amount of gold in some areas.

Sampling of the moss should be done ahead of time to see which areas carry the greatest values.

The best gold values in moss up on the bank can also be found to run in streaks and paths, as in stream or bench placers.

Care should be taken when you are removing the moss from its original bedrock location to avoid shaking it unnecessarily. Quick movements tend to shake gold loose from the moss, especially while underwater. A MACK VACK is perhaps the most valuable tool available for mossing, because it quickly vacuums the moss, and gold, from rocks, cracks, and natural gold traps with minimum effort. This is a wonderful activity for younger and older gold miners who are not able to put out lots of physical effort.

Some prospectors go so far as to save the pieces of moss after it is torn apart in the tub. After enough has accumulated, they take it back to camp, dry it out, burn it inside a container, and pan off the ashes. You would be surprised at how much additional gold can be recovered by doing this to the moss saved from a volume mossing operation.

A FEW MORE TIPS ABOUT PROCEDURE

One thing to keep in mind about the general area you have chosen to mine is that your sampling activities may show either where the gold deposits are located, or the lack of them in that general area. If sampling has shown that the area doesn't have very many deposits, it's good to have in mind a second choice of general location. How well an area has to pay, in order for you to work for its gold content, largely depends on the size of your production operation--meaning how much volume of streambed material you and your equipment can process effectively. For example, a hydraulic concentrator is able to process more volume than a gold pan can. So a gold panning production operation will need to have richer paying ground if a profit is to be made. When sampling, a person should have a good idea of the capabilities of his production equipment. This way he (or she) can compare the gold content in the ground he tests with the amount of material he feels he can comfortably and consistently process on a production scale--and thus figure out whether or not the ground should be mined or not.

The kind of gold a prospector finds in a location can also determine which type of equipment would be best used for recovering the gold. Paying quantities of fine gold, for example, would be recovered in the best way with the use of a wet type of slow-moving recovery system, in which the materials are thoroughly classified down to size before being processed. A paystreak containing medium and coarse sized values can be effectively worked by dry concentrating equipment when set up properly, if water is not available.

Sometimes you can run into old bench deposits where the bottom stratum of streambed is paying well, but is so compacted that it must be thoroughly broken up before being run through any recovery system. This is common in some of the ancient channel beds and also in some of the more recently formed benches. When this type of gravel is encountered, care should be taken to thoroughly break up the material before running it into your processing plant. Otherwise, the gold is likely to be washed right through the recovery system along with the compacted material to which it is attached. Sometimes, in the case of compacted material, when an operation wants to process volume amounts, it is necessary to use rock crushing equipment to break up the material enough to release the gold so that it can be recovered.

Clay-type material must also be thoroughly broken up and dissolved before being run through a recovery system. Otherwise, not only is it likely to be washed through your recovery system along with the gold inside the clay, but the clay is also likely to grab onto further pieces of gold which are already trapped in your recovery system and take them along too.

If a tributary is showing a good amount of rough and angular gold, start sampling in an upstream direction until you can no longer find traces of that kind of deposit. Then drop back and find where the gold has entered the tributary. It's likely if the gold enters in the immediate area, high-grade type float will also be present in or on the stream bank. If so, and you find it, your search will probably take you up the mountainside towards the lode itself.

While kicking around in the hills and on the river banks during your placering activities, it's a good idea to keep your eyes open for high-grade quartz float. Since you are moving around in gold country, and are in the business anyway, well, who knows--you just might get lucky!

Once you have found a paystreak, you can get down to the business of mining it out thoroughly. While mining the material, it's always best to stack the larger rocks and boulders neatly out of the way--preferably on top of already worked ground. This reduces the likelihood that you will have to move them again. A pick can be used to break open the cracks, crevices and other bedrock irregularities to get all the gold out of them. A whisk broom and wire brush can be a big help to thoroughly clean up the bedrock. A Mack Vack is even more practical in this effort. The larger rocks taken from the paying stratum of streambed materials, which will not be run through the recovery system, should be thoroughly cleaned of gold bearing material before they are stacked on the tailings pile. In this way, an operation will get the most return out of the amount of effort being expended.

When producing with just a gold pan, it's probably best to direct your activities towards places where there is little or no overburden sitting on top of the pay layer. Otherwise, you will find yourself shoveling lots of material that you will not be processing. Places where some erosion has cut through an old bench, leaving the lowest stratum exposed, are excellent locations for a panning operation. So are places where exposed bedrock irregularities are present, with streambed materials inside.

A streambed, having material a total of three feet deep, with the lower stratum paying well, is within the range of a pick and shovel sluicing or hydraulic concentrating operation. Anything much deeper than that is starting to get outside the range of a hand-operated mining venture--that is, unless the upper strata are paying, or the lower stratum is extremely rich. The reason for this is that all of the upper material will need to be removed to get at the lower stratum of the streambed materials. And when you are talking about three or four feet of material or more, you are talking about moving a lot of material by hand. Anything more than that is getting into the range of mechanical earth-moving equipment. That is, unless you are planning to drift mine (covered later).

When prospecting for placer deposits along the streambank, pay attention to the lines of vegetation which parallel the waterway. Vegetation grows best in mineralization. Quite often, the strongest line of gold is found in line with the strongest line of vegetation along the river bank. Sometimes the paystreak is found in a layer along the bank only a foot or two into the streambed material. This makes for excellent sluicing and highbanking activity.

HUNTING FOR POCKET GOLD

In Chapter 2 we talked about residual deposits being gold which has broken away or eroded from a lode deposit, but which has not yet traveled from the location of the original lode. An eluvial deposit was defined as the gold which has eroded away from a lode deposit and traveled some distance, but which has not been washed into an active waterway.

It doesn't take much thinking to realize that over the thousands upon thousands of years of natural erosion, large amounts of gold may have weathered away from original lodes and become trapped in pockets.

Pockets such as these have been the source of some of the richest gold discoveries ever made by small-scale gold miners.

Sometimes a pocket will be located directly at the site of the original lode (residual deposit). When the lode weathered away, it left the gold in place.

Sometimes pockets form down along the hillside (eluvial deposits) where bedrock irregularities are able to trap the gold because it is heavier.

The problem with pocket gold deposits is not that they do not exist in abundance. The problem is that they are almost always completely covered up and hidden by a layer of alluvial material which is also eroding down the hillside. Many pocket gold deposits recovered in the past were stumbled upon by accident, often by hunters or explorers.

Sometimes pieces of float along the hillside can be an indication that pockets of gold may be present. The fundamental procedure for locating pocket gold is pretty much the same as prospecting for a lode deposit--that is, following float or other tell-tale signs up along the hillside, looking for the apex (outlined in Chapter 9).

Modern electronics can be a valuable assistance to the lode mine and pocket gold prospector. Proton magnetometers are modern electronic devices which measure the differences in magnetic fields. Sometimes, the other mineralization associated with gold deposits (iron) sets up anomalies in the localized magnetic fields--which can be detected with a magnetometer. At the time of this writing, the technology is not yet perfect. However, it has developed to the point where magnetometers can be a big help to a prospecting operation.

The Goldspear is another excellent tool to help find pocket gold deposits. Because the Goldspear sounds off on microscopic-sized gold, and on black sand, rather than digging, hauling and processing hundreds of small samples along a hillside, if conditions are right for it, the Goldspear can be used to probe the hillside. Probing can continue upward, in an ever-narrowing band, until there are no further signals. In theory, the signals should get stronger and more plentiful as probing gets closer to the source.

There is also the possibility that indications from the pocket gold or lode deposit may go deeper in the material as you get closer to the source. In this case, the Goldspear may only be effective during the initial portion of the sampling *(loaming)* up along the hillside. However, looking at the width of the area which is giving off positive signals will also give you an idea of how much further up the hillside you will need to go to find the source.

As a rule of thumb, the steeper the hillside, the narrower the band of float is likely to be. Also, the steeper the hillside, the more likely that pocket gold deposits will form along the bedrock traps.

Metal detectors are also a valuable tool in prospecting for pocket gold and lode deposits. Sometimes you will want to hunt in the all-metal mode (when picking up gold specimens or highly mineralized pieces of float). Sometimes you will want to hunt for the increased mineral content; it depends on local conditions and the specific gold deposit. No two situations are exactly alike.

The two-box deep probing metal detectors can also be an excellent tool for hunting down rich gold pocket deposits. Many pocket gold deposits have been recovered in gold country which produced $10,000 to $500,000 in gold specimens--all located within a small area. Pockets such as these could be located underneath 10 feet or more of alluvial material with a two-box detector.

The old-timers had no electronics. They often spent months or years digging deep test pits, looking for pocket gold. Sometimes they found it; sometimes they didn't--even when they knew they were close.

You need to research to find areas in proven gold country where high-grade lode deposits and rich pocket gold deposits were located. Such areas would be worth prospecting again today with the help of modern electronics.

DRIFT MINING

Many of the old benches contain a rich lower stratum of streambed material. Yet, many also have a great deal of overburden lying on top. This places this kind of streambed out of bounds for most small mining operations, because there is simply too much material to be moved by hand when using the conventional methods.

There is another way to mine these old streambeds, a process which was extensively used in the earlier mining days. It consists of tunneling along at the bottom of the streambed and processing the lower stratum of material which is removed as the tunnel is moved forward. The process is called *"drift mining."*

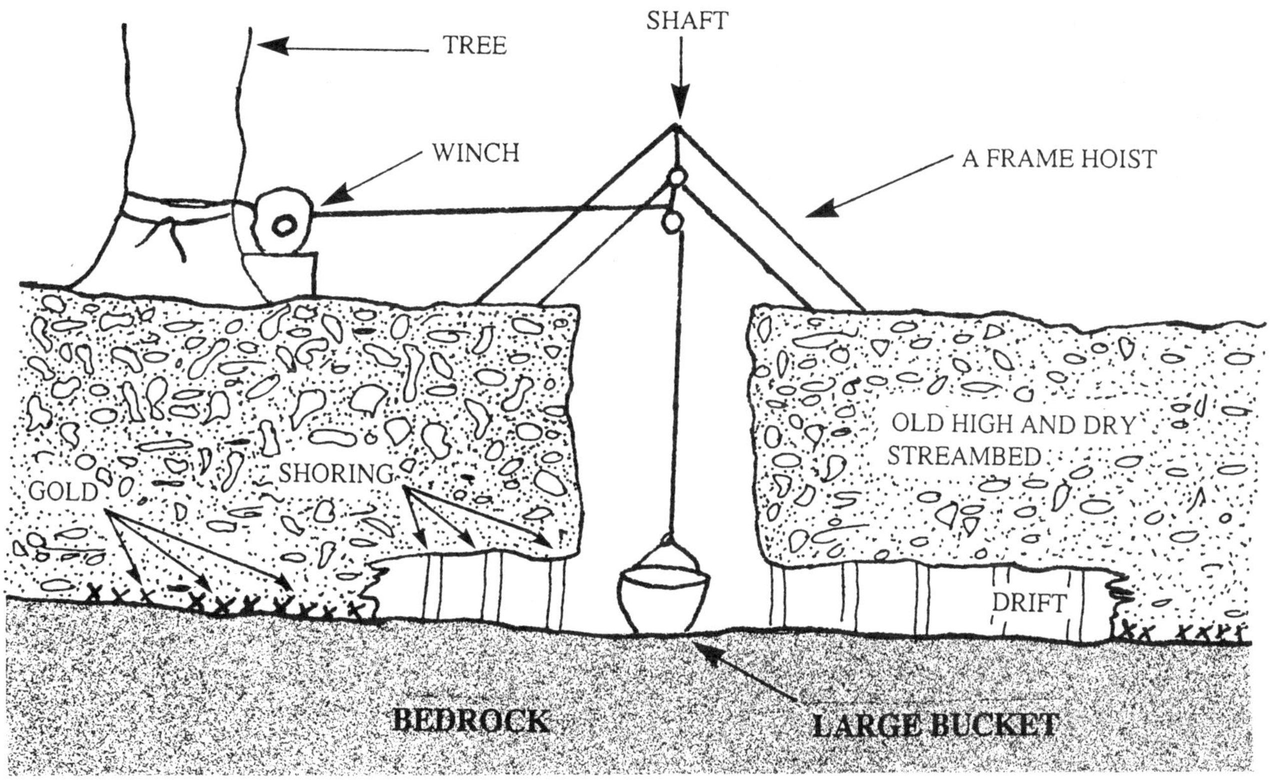

Fig. 11-2. Vertical shaft type of drift operation.

There are different ways to go about a drift mining operation, with the most common method being to start off with a tunnel expanding at a right angle to the direction in which the old streambed once flowed. The tunnel is extended across the width of the streambed along the bedrock, with the lower stratum of materials being tested as progress is made. This first tunnel is called a *"crosscut."* The crosscut is continued until the paystreak is discovered, and its outside left and right boundaries are determined. The entire width of the paystreak is then mined by moving the tunnel forward and backwards along the bottom stratum of material. This is called *"drifting."*

Shoring was most commonly always done in this type of mining--meaning that wood beams are cut, shaped and installed inside the tunnel to prevent the roof from caving in. Needless to say, this kind of operation can also be quite dangerous to the miners who work in the drift or crosscut. Even a small amount of streambed material can drop from the ceiling and give you a good knock on the head. Hard hats are a must (and life insurance)!

Cutting the tunnels in a drift mine is not normally done with explosives. Blasting tends to weaken the streambed structure and increase the chances of a cave-in. Most often, it was the pick and shovel method of digging that was used by the early miners to move their tunnels along.

The means of starting a crosscut into the lower stratum can be done in several different ways, depending on the situation of the streambed. It was sometimes necessary to sink a vertical shaft down through the streambed to bedrock, where the crosscut could be started, as shown in Figure 11-2. In this case, it was necessary to hoist all of the tunnel material up through the shaft to get it out of the drift and the crosscut. The material was then processed at the surface. The old-timers used a hand-operated windlass to hoist the material up to the surface of the streambed. Today's miner can use electric or gasoline-powered winches to do the job and speed up the process.

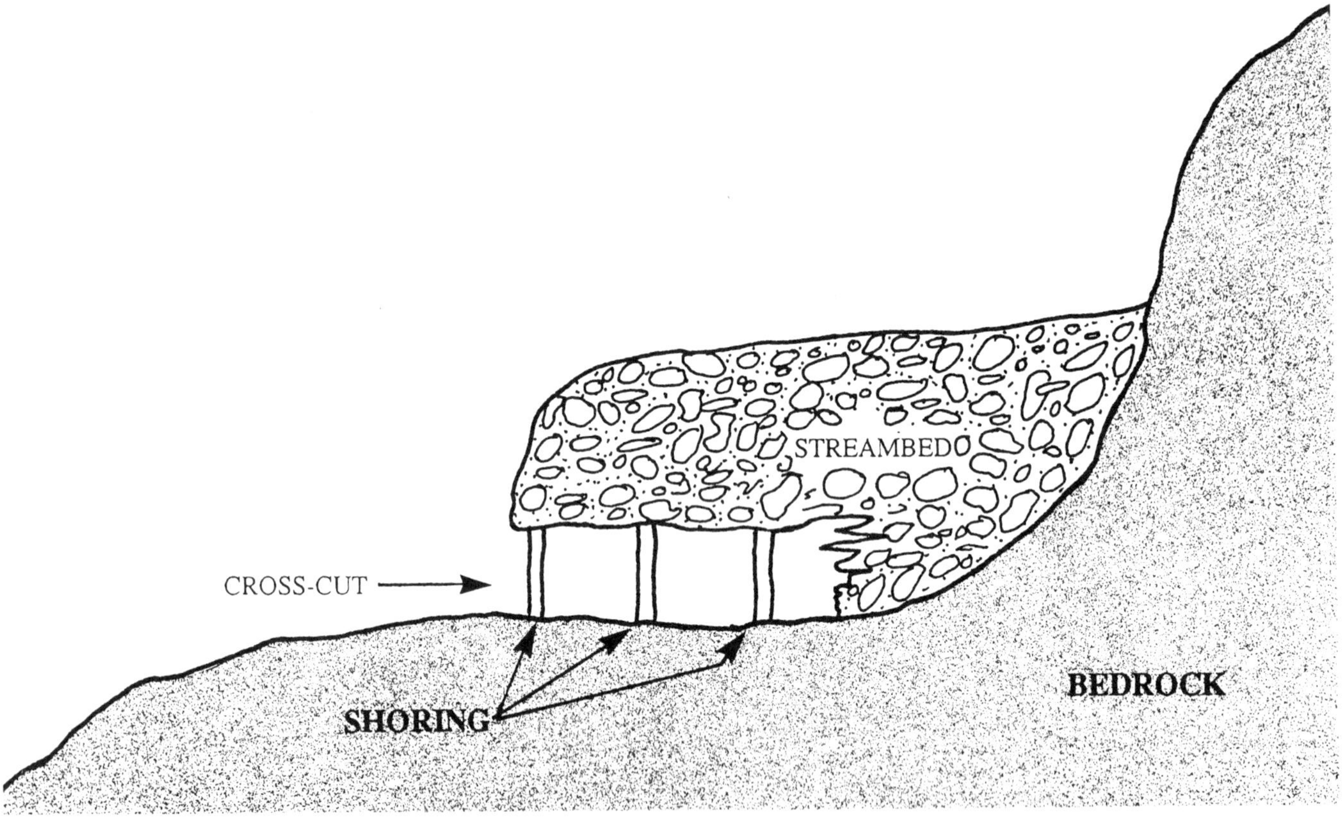

Fig. 11-3. Drift mining with a crosscut started from the side of a bench.

In another situation, where the side of an old bench is exposed, the crosscut can be started directly into the side of the streambed, as shown in Figure 11-3. In this case, material from the lower stratum can be carted directly out of the drift to be processed. This requires less work than the situation where it's necessary to hoist the material up to the surface by way of a vertical shaft.

Shoring the outside opening (*"portal"*) of any type of drift mine is a must, because the portal location is the most likely location to fall apart or cave in.

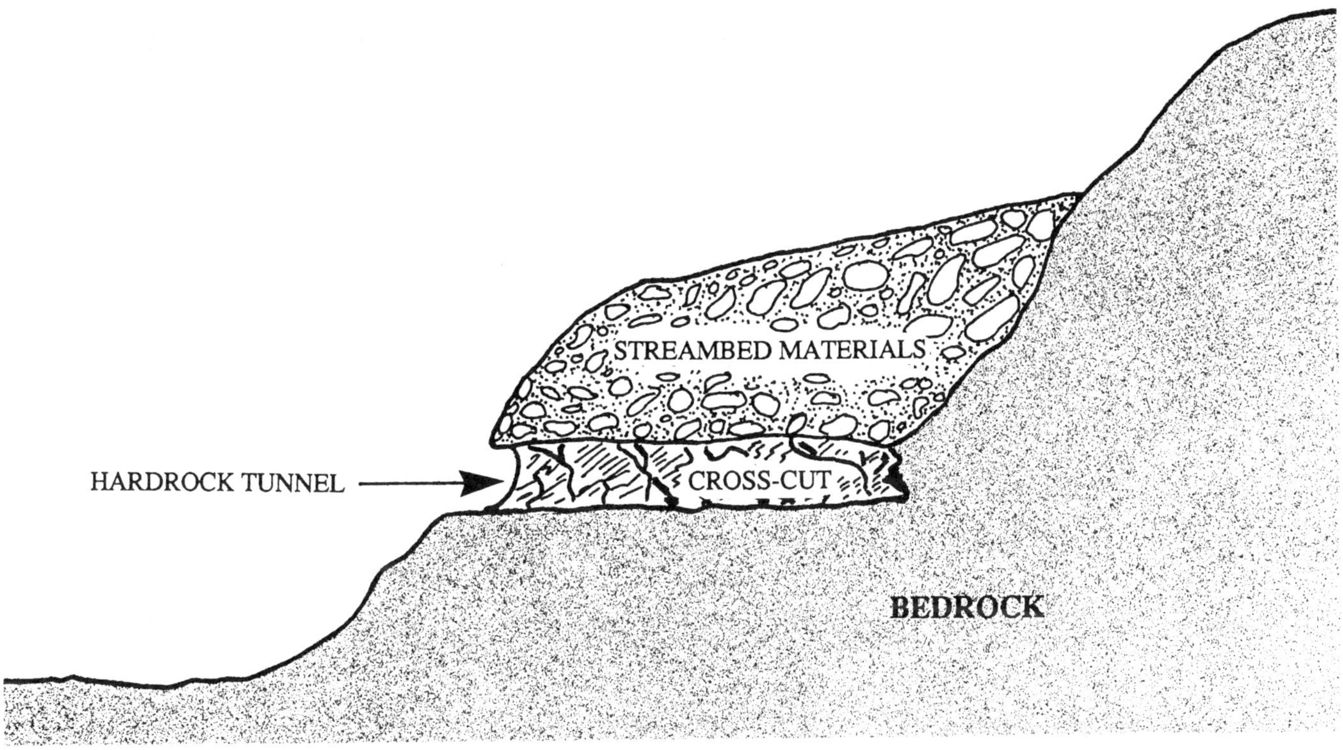

Fig. 11-4. Making the original crosscut tunnel through the hardrock, just beneath the streambed material.

Sometimes the early miners preferred to blast a crosscut into the hardrock just below the lowest stratum of streambed material. They would test the bottom of the bench materials at the ceiling of the hardrock tunnel as it was moved forward. In this way, blasting could be used to make the original crosscut in an effort to find out if a paystreak was present (see Figure 11-4).

Once a paystreak was located in this manner, and its boundaries were determined, the drift could be started up into the streambed from the lower tunnel. The advantage to a crosscut of this nature was that the paydirt could be shoveled down into the lower crosscut, where a cart would be waiting. This was easier than shoveling the material up into a cart--which is more difficult when inside of the tight-fitting space of a drift mine. It also provided a place for water, when present, to drain out of the drift.

When water was encountered in a crosscut, drifting was usually done only in the upstream or uphill direction. This allowed the water to flow downhill into the crosscut, where it would run out. Or, in the case of having to start with a vertical shaft, a water pump could be used to remove the water out from the crosscut.

When water was not a problem, drifting could be done in both directions from the crosscut and continued as long as the paystreak did.

Because the pick and shovel method of removing material in a drift mine was rather slow, resulting in a very low volume of material which could be processed during the early days, the material being mined had to pay on the average of an ounce in gold to the cubic yard of material. Needless to say, that is some fairly rich placer ground. Yet, there is evidence of drift mining having occurred all over gold country in the Western United States. So, the lower strata of many of the old streambeds must be indeed rich--at least in some places.

Perhaps today, with the exchange value of gold being what it is, and with the use of a portable electric or air-powered impact hammer, a small drifting operation could move much more material than was possible during the earlier days. This opens up the possibility of profitable drifting operations all throughout gold country.

THE GAME OF GOLD MINING

During the past several years, we have had an opportunity to work with hundreds upon hundreds of different gold miners; and we have realized many different things about how to approach a gold mining operation to improve a person's chances of success. One of the things we have realized is that some people become so serious about a mining operation, that they lose track of the fact that the operation is simply a game.

All games consist of a goal, a means to achieve the goal, and barriers or problems in the way of the goal's achievement. And, of course, winning the GAME consists of overcoming the obstacles and achieving the goal. Football is a game; basketball, soccer, everybody knows these are games. What some people fail to realize is that your job, raising your family properly, getting through life successfully, and even gold mining--are all games, too. Each of these games has its own unique set of problems to overcome.

Because of the seriousness and importance of winning, sometimes we lose track of the fact that these different aspects of life are a game. The importance of winning simply requires that we play the game harder.

It is much easier to win at a game when you know what the game is that you are playing.

Gold mining is a game in which the obstacles and problems to overcome are not generally other people or other teams as in the game of football. Listen to this, because it's important: The main obstacle to overcome in gold mining is your UNCERTAINTY of where acceptable gold deposits are located.

The goal, of course, is to find lots of gold--enough to solve your financial or emotional needs. I say emotional because some people are not in gold mining necessarily for financial gain. One person's goal might be to continuously recover enough gold to support his family and lifestyle. Another person might want to find enough gold to retire in luxury. Some people do it just for fun. Each individual will have his or her own goals.

Once one goal is nearly achieved, a person naturally tends to set a higher, more difficult goal. One of the interesting things about gold is that you never seem to have enough of it--even if you have a lot compared to the goal you set for yourself quite some time ago! Therefore, as a miner gets better, he or she tends to elevate the goal higher and higher.

The means of achieving the goal in gold mining is applying mining and prospecting techniques with available mining equipment on gold bearing locations in order to locate and recover valuable deposits.

The equipment is readily available. There is nothing difficult to understand about the technique and procedures. The main difficulty is NOT KNOWING WHERE THE GOLD IS. This makes gold mining unique, in that the main obstacle to overcome is not an external, material energy or barrier--as in most games. The main barrier is the psychological impact of the uncertainty of whether or not you are going to find an acceptable gold deposit.

In reading this, you might find yourself feeling that you are dedicated and strong enough, that you have all of the discipline needed, that you have plenty of emotional fortitude, and that you are smart enough to overcome any psychological doubts which may arise in your own mind during the course of a sampling operation. We all have this, and we are all potentially strong enough to persevere. However, there are also negative voices in our heads--which can become quite strong when we are directly confronted with difficult situations. Sometimes we forget about these voices when we are not confronted by difficult situations!

None of us are super-beings. We are human. We all have our personal limitations--which are set by ourselves. This happens when we make decisions that we can't do something, or that we don't want to do it. A person takes up running and decides he can only run two miles. Does that mean he cannot run a step further than two miles? Of course not. The person could run twenty miles if he set his mind to it. If I didn't learn another thing in SEAL Training from my Navy days, I learned that you can always take at least one more step. This is true in any aspect of life--in any endeavor; you can always do it a little more or a little better.

But, when we get close to a limitation which we have already set for ourselves, we run smack into the negative voices in our heads which we have ourselves identified with. ''I can't do it!'' Just because the voice says we can't, doesn't mean we can't. We can, and by doing so, a person becomes stronger.

The problem in gold mining is different than in most other games. If you were cutting firewood for money, the barriers to overcome in the game would be the physical challenge of cutting down trees, sawing them into rounds, splitting the rounds up, loading them in a delivery truck, hauling them to a location, selling them to someone, and maybe stacking the firewood on the buyer's back porch. The easy thing about this challenge is that you are working with a reality that you can see all the way through the cycle. The wood is there. It is just the physical work of getting the wood onto the buyer's back porch--that is, as long as you have a buyer.

Gold mining is different. The gold is not there until you find it. Yet, it's really not that the gold isn't there. When we see other miners recovering gold out of commercial deposits on the same river, we know there are more commercial deposits to be found. The problem is that we might not be sure that we are going to find them.

And, it's not that the procedures and techniques for finding gold deposits are difficult. We know that gold, because of its weight, tends to travel along its own narrow path in the river. We know that paystreaks (gold deposits) form in their own unique locations along the gold path--where water velocity slows down during major flood storms. We know these paystreaks form quite regularly along a gold-bearing waterway. And, we know that prospecting consists of digging or dredging sample holes in an attempt to locate the gold path and the paystreaks. None of this is difficult to understand and apply.

The difficulty is in the uncertainty--and this is the main barrier to overcome in the game of gold mining. We see a fair percentage of people who have themselves psyched out and talked out of it, even before they finish their first sample hole! Why is this? They have adequate equipment. They understand the procedures. They can confront the physical work. Why do they quit so quickly? It is because they don't understand who the real opponent in the game is.

If it were a game of football, would they quit after the first play of the first quarter just because the opposing team looked stronger? Not if the players have any degree of personal pride in being a football team. Yet, quite often in many games, the opposing players try to psyche-out the members of the other team. A demoralized team is easy to conquer! More likely, a serious football team would psyche themselves up in an attempt to win over a stronger, opposing team.

What people don't realize is in the game of gold mining, you are really playing with, and/or against yourself. It is your own inner voices which you listen to and decide whether to quit early, or to pour on the steam even harder to find the gold deposits you are looking for. When a person quits in gold mining, it is often because he has psyched his or her own self out.

I can understand quitting when your legs are busted and you are on your last breath of air. This is understandable. But, to quit gold mining because you have talked yourself into the idea that you are not going to find any gold just means that you don't understand the game.

There is a fantastic feeling of self accomplishment when you succeed in gold mining. A professional football player, when retired, will look back and remember certain games that were won. He probably won't be thinking much about the money he made. He probably won't be thinking of the easy victories. He will be thinking about the games his team was losing, and how the team pulled together, raised themselves up, and won against all odds. These wins are cherished because they occur when a player, or a team of players, reach down inside and create the necessary additional energy and willpower to overcome large barriers and obtain the goal after all.

Success in gold mining brings about this same kind of intense emotional satisfaction--only better or different in the fact that you usually accomplish it on your own. You generally don't have a team of other players helping you to make a touchdown in the game of gold mining.

Like it or not, gold mining is similar to a game of solitaire. You are playing the game with yourself as your team mate, and possibly with yourself as the main opposing force. You have a chance to overcome the little negative voices in your head that tell you to quit--not just in gold mining, but also in the other aspects of your life. Each time that you persevere and finish that next sample hole properly, despite the inner voices which tell you there is no gold in that location, not only do you get that much closer to the gold deposit you're looking for, but you also grow stronger as a person. And, in the end, this is worth more than the gold.

GLOSSARY

ALLOY: When two or more metals are fused together they form an alloy.

AMALGAM: Mercury (Hg) has a tremendous affinity for gold, silver and some other metals. When mercury is mixed with one or more of these other metals, the resulting mass is called *"amalgam."*

AMALGAMATION: The process of using mercury to separate gold, silver, platinum and/or other metals from a larger volume of useless materials. Amalgamation is one of the oldest and most effective processes used by the everyday miner to collect the fine particles of gold from a set of concentrates.

ANCIENT STREAMBEDS: *See TERTIARY CHANNELS.*

ASSAY: The analysis of an ore sample or a gold sample to determine the proportions of gold, silver, platinum and/or other valuable metals to the amount of waste material included in the sample.

AUDIBLE THRESHOLD: *See THRESHOLD.*

BASADA: A desert alluvial deposit.

BAKED POTATO RETORTING: A method of separating mercury from gold or other metals collected during an amalgamation process. An uncooked potato is used to save most of the mercury during separation. See Chapter 7 for details.

BATEA: One of the earliest types of gold pans, made of wood, and used a great deal among the Indians of the Western United States. Bateas are still widely used in gold mining activities in developing countries.

BAR: Deposits of sand, rocks, gravel, and sometimes gold, located in streambeds where the force of water lets up, such as at the inside of the bends where the stream or river changes direction.

BEDROCK: The solid rock surface which underlies loose sediments in a streambed. Also known as the earth's outer crust or *"country rock."*

BENCHES: Also known as *"terrace placers."* Sections of old streambed which have been left high and dry by the present stream of water are referred to as *"benches."* Many benches still contain large paying quantities of gold.

BLACK SANDS: Are a mixture of the heaviest (darkest) streambed materials, which concentrate in any type of recovery system. They are composed mostly of minerals of the iron family. Concentrates of black sands, because of their greater weight over that of most of the other materials in a streambed, are usually also found along with gold in placer deposits.

BLEEDING OFF: Natural erosion of bench gravels into a present stream or river.

BLOCK RIFFLES: Semi-modern type of riffles, which are still being used in many of today's homemade recovery systems with good results. Covered in Chapter 5.

BLUE LEAD: Bottom strata of ancient riverbed gravels were often called the *''blue lead''* because of their color and the fact that they were ''followed'' by the early timers wherever they led. Blue lead gravels were often incredibly rich.

BONANZA: A very rich gold deposit, whether it be a lode or placer, is often referred to as a *''bonanza.''*

BULLION: Gold or silver which has been refined and cast into bars or ingots.

BURNING A PAN: After repeated use, gold pans tend to accumulate a certain amount of grease or oils on them. This can adversely affect gold recovery. These can be removed from a steel pan by heating it over an open fire until it reaches a dull red glow and then by dropping the pan into cool water. See Chapter 4 for details.

CALICHE: A layer of compacted gravel or clay, often found in the desert regions, which can cause a false bedrock. Sometimes the most productive gold deposits can be located along the surface of a caliche layer; because if the layer is shallow, only a small volume of material needs to be removed to get at the gold.

CARAT: A single unit of weight often used by jewelers to measure gold and precious stones and/or the purity of gold. Twenty-four carats would be pure gold. Twelve carats would be 50% gold, and so on.

CEMENTED GRAVEL: Super-hard packed streambed material, which can be found on top of bedrock or up in a layer above bedrock. Cemented gravels often contain gold in paying quantities, but not always.

CHAMOIS: A thin sheet of soft natural leather which can be used to squeeze amalgam to separate excess mercury out before retorting is begun. Covered in chapter 7.

CINNABAR: A red-colored ore from which mercury is extracted.

CLAIM: *See MINING CLAIM.*

CLAIM JUMPER: Someone who commences mining activities without permission on a section of land in which the mineral rights are legally owned by another.

CLASSIFICATION: The process of screening out the larger sized materials from a body of ore, streambed gravels, or a set of concentrates. Classification is usually done with the use of one or more sizes of mesh screening.

CLASSIFIED MATERIALS: Materials which have been screened through a certain size of classification screen.

CLASSIFIER: A screening device itself is often referred to as a *"classifier."* There is a classifier commonly found at the head of most sluice boxes on gold dredges today.

CLASSIFY: *See CLASSIFICATION.*

CLAY: Sometimes a streambed will contain a layer of clay, which should be thoroughly broken up before being run through a sluice box. Otherwise, the sticky substance is likely to grab at least some of the gold out of the sluice as clay is being washed through.

COUNTRY ROCK: The common rock (earth's crust) of a region. That rock which surrounds a specialized ore body, like a quartz vein.

COARSE GOLD: Has two meanings: 1) Rough and sharp-edged pieces of gold that have not yet been pounded smooth by traveling some distance from the original lode. 2) Larger pieces of gold (nuggets), ranging in size from 10 mesh and larger, is also referred to as *"coarse gold,"* or *"nuggets."*

COYOTING: *Same as DRIFT MINING.*

CREVICE: A split, crack, or open fissure in the bedrock's surface. Those crevices located at the bottom of a streambed are likely to trap gold out of the material as it is passed over during a flood storm.

CREVICING: The prospecting activity of searching exposed bedrock areas to find and extract gold from the various gold traps. See Chapter 11 for details.

CROSSCUT: Drift mining is usually started with a tunnel made at a right angle to the direction in which the old dry streambed used to flow, along its bottom edge. Sampling is done as the tunnel progresses, in an effort to locate a paystreak. This first tunnel is called a *''crosscut.''*

CRUST: The outer layer of the earth, extending about 60 miles deep, down to the earth's mantle. Also called *''bedrock''* and/or *''country rock.''*

DIKE: A narrow section of igneous rock which extends upward out of the country rock. Dikes which extend up through the foundation of a streambed can make excellent gold traps. Details in Chapter 2.

DIRECTOR: *See WATER DIRECTOR.*

DREDGE: *See SUCTION DREDGE.*

DRIFT: In drift mining, once the crosscut has been made and the outside boundaries of the paystreak have been determined, the tunnel is moved forward and/or backward along the paystreak in order to recover the gold out of the lower stratum. This tunnel is called the *''drift.''*

DRIFT MINING: A method of mining dry bench deposits in which tunnels are made along the lower stratum of material so the gold can be processed out of those gravels. The lowest gravels in such deposits are usually by far the richest. See Chapter 11 for more details on this procedure.

DRY PANNING: A technique of recovering gold with the use of a gold pan without water. Procedure covered in Chapter 8.

DRY PLACER: Placer deposit of gold, silver, or other valuable stones or minerals that are located in a dry area.

DRY WASHER: Gold mining equipment designed to process dry placer materials without the use of water. Such equipment usually employs the use of air currents and vibration to separate the lighter, worthless materials from the heavier and more valuable minerals.

DUCTILITY: Quality of being pounded flat or drawn out into thin wire-like forms without becoming brittle and breaking up. Gold is a very ductile metal.

DUST: Extremely fine particles of gold are referred to as *''gold dust.''* A good recovery system being used in some areas will recover a considerable quantity of *''dust''*--often enough to pay all of the operating expenses, sometimes more. In the earlier days of mining, a pinch of gold dust was worth a dollar. Today, the same pinch is worth about $20.

ELECTROSTATIC CHARGE: Static electricity tends to act as a magnet towards gold and helps to trap the finer sized pieces. It is used in some of today's more sophisticated dry washing machinery.

ELECTROSTATIC CONCENTRATOR: A drywashing plant which uses static electricity to assist in the recovery of the fine particles of gold.

ELECTRUM: Any alloy of gold and silver in which the silver content exceeds 20% is called *''electrum.''*

ELUVIAL DEPOSIT: A deposit of gold and other lode materials which have been swept away from the original lode, but have not yet reached a running stream of water.

EXPANDED METAL RIFFLES: The *''turned-up kind''* of expanded metal makes an excellent recovering riffle system as long as excessive water force and materials are not directed over it at once. See Chapter 5 for details.

FALSE BEDROCK: Sometimes a hard-packed layer of sediment in a streambed will act as a *''false bedrock,''* and even give the impression of being the bedrock itself. Placer gold deposits can often be located both above and below such layers.

FANNING: Sweeping light sands out of the bedrock irregularities while looking for gold along the bedrock.

FINE GOLD: Those particles of gold that are small enough to pass through 40-mesh screen are known as *''fine gold.''*

FINENESS: A system used to indicate the purity of a gold sample or specimen. A specimen having a fineness of .900 would be 90% gold. A fineness of .650 would be equivalent to 65% gold, and so on.

FIRE ASSAY: A means of doing an analysis on a sample, in which all of the gold and silver is extracted, then separated, weighed, and measured in proportion to the original sample to determine how much gold and silver is present in an ore.

FLAKE GOLD: Flat chips of gold. Most are probably pounded flat and smooth during their travel down along the streambed. Most placer gold is found in flake form.

FLOAT: *(ALLUVIAL deposit)* Loose pieces of quartz rock or some other kind of potential gold-bearing rock which has traveled down and away from its original source. Float is often what the prospector follows in his effort to locate a lode deposit. Covered thoroughly in Chapter 9.

FLOTATION PROCESS: One method of recovering fine particles of gold with the use of chemicals which cause the particles of gold to attach themselves to air bubbles and *''float''* to the surface. They can then be skimmed off to be refined further.

FLOOD GOLD: That gold which has been washed down in a storm to rest on top of a layer deposited at an earlier time. *''Flood gold''* is that gold which is resting in a flood layer up off the bedrock.

FLOOD LAYERS: Sometimes a streambed will have different layers of material that were laid down during different storms from different periods of time. Sometimes each separate layer contains its own gold in varying amounts. These different layers are called *''flood layers.''*

FLOUR GOLD: *Same as DUST.*

FLUME: A man-made water ditch, often made out of wood, or by digging a trough into the hillside, for transporting water to mining sites during the earlier days of mining.

FOOL'S GOLD: *See IRON PYRITES.*

FREE MILLING GOLD: Those particles of gold not chemically locked-in with the other elements in an ore.

GANGUE: The non-valuable rock and waste materials that are associated with the valuable minerals in a lode deposit.

GEIGER COUNTER: An electronic device which is used to detect radioactive elements.

GEOLOGICAL MAPS: Specialized maps that show the location of the various geological rock formations within a specified area.

GLORY HOLE: Term used to express an exceptionally rich gold deposit.

GOLD: An exceptionally heavy, ductile, malleable, yellow, precious metal. Covered thoroughly in chapter 1 of this manual.

GOLD CATCH: A location in which gold has been trapped.

GOLD FEVER: An emotional condition that affects some people when they are confronted with riches--either in real or abstract form. See Chapter 1 for details.

GOLD NUGGETS: Pieces of gold that are too large to pass through 10-mesh screen are referred to as *''nuggets.''*

GOLD PAN: One of the earliest devices ever developed to help separate particles of gold from the worthless materials contained in a streambed. The gold pan is widely in use today -- its uses and the various techniques are covered in Chapters 4 and 8 of this volume.

GOLD RUSH: A wild scurry of miners who are hurrying to mine the gold out of a newly found gold bearing location.

GOLD SNIFTER: A small hand-operated suction device used to extract the gold from a gold pan, or from the bottom of crevices and other bedrock irregularities.

GOLD TRAP: Any location where gold has been, or is likely to become, trapped.

GRADIENT: The downward slope of a streambed or a sluicebox.

GRAIN: A term used to label small particles of gold, and also as a unit of weight in the troy system of measurement. Twenty-four grains equals 1 pennyweight. One grain also equals 64.8 milligrams.

GRIZZLY: A classification device used to prevent boulders and larger-sized rocks from being dropped into a recovery system along with the smaller sized materials.

GROUND BALANCE: (Electronics) Setting a metal detector to cancel the effects of ground mineralization. Without ground balancing a metal detector properly, iron mineralization in the soil has a tendency to hide metal (gold) signals which might be present.

GROUND SLUICING: An earlier sluicing method in which a channel was dug down to the bedrock for water to be passed through. Gold-bearing material was then shoveled into the channel where it was washed through by the running water, and the gold becames trapped in the bedrock irregularities. See Chapter 5 for further details.

GRUBSTAKE: The practice of a storekeeper extending credit, or another person putting up money, to outfit a prospector and keep him supplied until he makes a productive gold strike. In exchange, that person then receives an agreed upon percentage of the gold recovered.

HAIRLINE CRACKS: The minute cracks which are located in the bedrock's surface on the bottom edge of a streambed can sometimes yield a surprising amount of gold.

HARDROCK MINING: The process of mining mineral veins from the earth's crust. See Chapter 9 for details.

HEAD OF BOX: The upper end of a sluice box, which the streambed materials are shoveled into.

HEADER BOX: The forward-most section of the sluicebox on most suction gold dredges today is designed to slow the materials down and spread them out evenly over a classifier as they enter. This upper section is referred to as the *"header box."* Pictures and further details contained in Chapter 6.

HEMATITE: An iron ore (Fe_2O_3) reddish-brown in color. Hematite is one of the main non-magnetic minerals that comprises black sand concentrates.

HIGH-GRADE: The richer ore samples of a lode mine, usually those that show a significant amount of visible gold, or the richer portion of any gold deposit.

HIGHGRADING: When a person is stealing the higher-grade ore samples or gold specimens from the mining operation--whether that person is an employee, a partner, or an outsider. Highgrading has been a problem since the earliest days of gold mining. *''Highgrading''* is also the process of removing only the richest portion of an ore body or other gold deposit during sampling or production.

HOOKA AIR SYSTEM: Air breathing system used on most gold dredges, in which air is pumped down to the divers through an extended air line.

HOT SPOT: Any location that contains acceptable paying quantities of gold or other valuable minerals.

HUNGARIAN RIFFLES: Modern-day concentrating-type riffles which are commonly found in many of today's dredge and sluice boxes. Details covered in Chapter 5.

HYDRAULIC CONCENTRATOR: A very efficient sluicing device which can be used effectively as either a production or sampling machine. The hydraulic concentrator has a very wide range of variety in its ability to process streambed material. It is excellent placering equipment for the one or two-man-sized operation. Details in Chapters 5 and 11.

HYDRAULIC MINING: A method of gold mining in which a large volume of water under pressure is directed at a gold-bearing streambed so that its materials can be washed down through sluice boxes, where the gold can be recovered.

IMPURITIES: Very seldom does natural gold come in 100% pure form. Usually in its natural form, gold is alloyed with various other metals--including silver, copper and platinum. These other metals are known as impurities, simply because they subtract from the purity of the gold itself. Also, sometimes the small amount of waste materials contained in a final set of concentrates are referred to as *''impurities.''*

INTERFERENCE: (Electronics) When prospecting with a metal detector (electronics), sometimes the magnetic black sands present will block out any reading that a metallic object would give if it were present. This is called *''interference.''*

IRIDIUM: One of the metals from the platinum group. Iridium has a specific gravity of 22.6, as opposed to gold's specific gravity of 19.3.

IRON PYRITES: *(Fool's gold)* Iron pyrites are the glittery, brassy-yellow-colored compounds often mistaken for gold by beginning gold miners.

JET: That component on a suction dredge which water is pumped into, and which creates a suction up through the suction hose. The jet is designed so that streambed material can be pumped directly to the recovery system without having to be run through the water pump. Details covered in Chapter 6.

LOAMING: The process of digging or probing for mineral float or other indications of a residual or eluvial deposit.

LOCKED-IN VALUES: Valuable minerals (such as gold) which are chemically locked in with other elements within an ore, requiring more involved methods of processing so the values can be released and recovered.

LODE: Veins of valuable minerals are referred to as *''lodes.''* Gold, in hardrock form (lode), is commonly associated with quartz veins which protrude through the general country rock. Lodes are the original source of placer gold deposits.

LOW-GRADE: Ore deposits which contain valuable minerals, but not enough of them to be mined at a profit, unless mined by a very large-scale operation.

LUTE: A sealing compound used around the upper edge of the gold chamber on a retort to prevent any leakage of mercury vapors during the retorting process. See chapter 7 for full details.

MAGMA: Molten material which lies beneath the earth's outer crust.

MAGNETITE: (Fe_3O_4) Heavy magnetic black sand (iron) found in the heavy concentrates collected in a recovery system. Magnetite is what the BFO-type metal detectors sound out on when turned on in the *''mineral''* setting.

MALLEABLE: The quality of being hammered or extended into various shapes and forms without being broken in the process.

MATERIAL: *''Material''* is a term used to indicate all of the rocks, sand, gravel, mud, clay, and silt which makes up a streambed.

MERCURY: (*"Quicksilver"*) A heavy, silvery-colored, liquid metal which has an affinity for gold, silver and other metals. Mercury is often used by gold miners to collect the fine gold values out of a set of heavy concentrates taken from a recovery system. The procedure is called *''amalgamation,''* and is covered thoroughly in Chapter 7.

MESH: Signifies the number of openings contained in a lineal inch of screening. For example, 10-mesh screen will have 10 openings per lineal inch and 100 openings per square inch.

METALLIC: (Electronics) A METALLIC target consists of an object which conducts electricity. *''Metallic''* and *''mineral''* are the two main settings on a BFO-type metal detector. Details in Chapter 10.

MILLING: The process of crushing or pounding an ore into a very fine powder so the valuable minerals can be released -- to be recovered later by a mechanical or chemical process.

MINERALIZATION: The condition of an ore or streambed material having a high degree of minerals present in the rock or the material. In electronic prospecting, *''mineralization''* specifically means the presence of magnetite (Fe_3O_4). In lode mining, a high degree of mineralization can be an index of the presence of gold in some amount. In placer mining, a large amount of mineralization (heavy black sands) in a specific location is an indication of where a gold deposit might be found.

MINERAL: Any natural, organic substance that is mined out of the earth can be classified as a *''mineral.''* Gold, silver, and platinum are all METALLIC minerals. Also, (electronic prospecting) magnetic black sands sound out on a metal detector when it is tuned in on the *''mineral''* setting.

MINING: The entire process of extracting and processing the valuable minerals out of raw ore or sedimentary material.

MINING CLAIM: A section of land on which the mineral rights have been legally *''claimed''* for a certain period of time by an individual or group so that the gold or other valuable minerals can be mined without interference from other private parties.

MOSS: A class of small green or brown plants growing in very close association with one another, forming clumps of carpet-like growth. Moss thrives in moist areas--as on the banks of rivers and streams. Moss has a tendency to trap fine particles of gold during winter storms.

MOSSING: A gold mining activity in which the moss along a stream of water is collected and thoroughly broken up in a container from which the contents are then processed to recover the gold. Full procedure covered in Chapter 11.

NATURAL RIFFLES: The natural bedrock foundation of a riverbed contains numerous different kinds of irregularities which act as gold traps. These can be small, like a crack or crevice, or they can be a major change in the bedrock's slope or shape.

NITRIC ACID: A clear, fuming, highly corrosive liquid which is often used to clean gold of various impurities. Details in Chapter 7.

NUGGET: *See GOLD NUGGET.*

OLD CHANNEL: *(''Ancient streambed'') See TERTIARY CHANNELS.*

ORE: Any deposit of rock from which a valuable mineral can be profitably extracted.

OREBODY: *Same as LODE.*

OUTCROPPING: That end of a lode which extends outward from within the earth is called an *''outcropping.''*

OVERBURDEN: The lower-grade streambed material which lies on top of a placer paystreak. Usually, the overburden must be removed first, before processing of the paystreak is possible.

PAN: *See GOLD PAN.*

PANNING: A method of separating the gold out of other materials with a gold pan. Procedures covered in Chapter 4 and 8.

PAY: The amount of gold that a lode or streambed is yielding to a mining activity is often expressed as how much it is *''paying.''*

PAYDIRT: Streambed material which contains acceptable paying quantities of gold values is sometimes called *''paydirt.''*

PAYSTREAK: *See STREAK.* Larger-sized placer gold traps. Covered in Chapter 2.

PENNYWEIGHT: A unit of the troy system of weight measurement. One pennyweight equals 24 grains, or a twentieth of a troy ounce. It also weighs 1.552 grams.

PLACER DEPOSIT: Free gold that has eroded away from its original lode, and which has been swept into a stream of running water, will tend to accumulate in certain common locations. These accumulations are called *"placer deposits."* Thoroughly covered in Chapters 2 and 3 of this volume.

PLATINUM: One of a family of 6 rare and valuable metals called the Platinum Group Metals (PGM) which are usually silvery-white in color. Because platinum is sometimes recovered along with gold, it's good for the placer miner to know what it looks like. This way he (or she) is less likely to discard it along with the waste materials from his recovery system.

PLAYED OUT: All gold deposits, whether placer or lode, will run out sooner or later. When all of the known paying quantities of gold or other valuable minerals have been mined out of such a deposit, it is said to have *"played out."*

PLEISTOCENE CHANNELS: The remnants of the earliest river and stream channels that started during the present Quaternary geological time period commencing about one and a half million years ago. Many of these earlier streambeds are in evidence today, left high and dry. They are sometimes quite some distance away from the present streams and rivers. Many of the Pleistocene channels (benches) remain untouched by earlier mining activities and still contain large and consistent paying gold deposits.

PLUGGER POLE: A long thin rod made of metal or PVC piping. It is used to tap the plug-ups out of the jet on a dredge from the surface.

POKE: A miner's container of gold. In the earlier days of mining, a small leather bag was often used to hold a miner's collection (poke) of gold. Today, plastic and glass jars are most commonly used for this.

POLE RIFFLES: An early form of riffle system, in which round poles, usually made out of wood, were laid close together in either a lengthwise or crosswise direction in the sluice box for the purpose of trapping gold. Pictures and further details in Chapter 5.

POTATO RETORTING: *See BAKED POTATO RETORTING.*

POTHOLE: Holes or cavities in the bedrock surface which act as gold traps. Some potholes trap gold well and some do not, depending on the nature of the hole and the water action which flows over top of them. Details in Chapter 2.

PRIMER: That device on a suction dredge which connects the water pump to the water, which is also designed to be manually filled with water. A "*primer*" allows the pump to be easily primed from the surface. See Chapter 6 for details.

PROSPECTING: The activity of attempting to find a deposit of gold or some other valuable mineral. Prospecting usually involves moving around and taking numerous samples until an acceptable deposit is located--at which time mining activities can begin.

PUNCHING HOLES: See *SAMPLING.*

PYRITES: See *IRON PYRITES.*

QUARTZ: Consists mainly of silica--which had a tendency to trap gold and the other heavy minerals as they were pushed upward through cracks and fissures by the superheated steam from the earth's molten interior, millions of years ago. A substantial amount of earlier hardrock gold mining has been directed at quartz veins--which protrude through the general country rock forming the earth's crust.

QUATERNARY PERIOD: The present geological time period that started about 1 1/2 million years ago, and came just after the Tertiary time period. Both periods belong to the "*Cenozoic era*" (geological time). The Quaternary period breaks down into two separate epochs--the "*Pleistocene,*" which was the first, and the "*Recent,*" which is the present epoch.

QUICKSILVER: See *MERCURY.*

RAW ORE: Ore as it comes directly out of a lode.

RAIL RIFFLES: A semi-primitive early development in riffle systems in which railroad irons were used right side up, upside down, lengthwise, and in crosswise directions in sluice boxes. See Chapter 5 for details.

READ: (Electronic detecting) When a metal detector sounds out on a target, or indicates a target's presence in some other way, this is referred to as a "*read*" on the detector.

RECENT BENCHES: Those high and dry streambed gravels which were formed since the passing of the Pleistocene epoch (within the past 11,000 years).

RECOVERY: The ability to trap gold values out of streambed material or out of an ore material. Also, the ability to separate the gold values from a set of concentrated material which was taken from a recovery system -- often done with the use of mercury.

RECOVERY SYSTEM: That component, or set of components, designed to recover the gold or other valuable minerals out of the material being processed. In placer mining, the recovery system often consists of a sluicing device and/or a gold pan. In hardrock mining, the recovery system can consist of any number of machines--such as a vibrating concentrating table, or chemical vats or amalgam plants, or any combination of these and others.

RESIDUAL DEPOSIT: A deposit of gold and float that has dropped from its original lode, but has not yet been swept away by the various forces of nature.

RETORT: A device used to separate mercury from gold or other metals once an amalgamation process has been completed. A retort saves the mercury so that it can be used again in successive amalgamation processes.

RICH: Streambed material, or a lode, which contains acceptable or better paying quantities of gold, is often said to be ''*rich.*''

RIFFLES: The various types of baffles or obstructions that lie at the bottom edge of any sluicing device for the purpose of trapping the gold out of the materials which are washed through. There are many different kinds of riffles, including Hungarian, right angle, expanded metal, block, zigzag, rail, pole, stone riffles, and many more--covered in Chapter 5.

RIGHT ANGLE RIFFLES: Usually made of 90-degree angle iron, always pointed in a down-stream direction, and sometimes tilted forward slightly to obtain the best result. These are concentrating-type riffles which have tremendous strength. See Chapter 5 for pictures and further details.

SALTING: Introducing gold from an outside source to a gold mine, either placer or hardrock, to make it appear to be richer than it actually is. Another form of salting is to take only the highest-grade ore samples for assay, leaving the lesser-grade samples that make up the bulk of the ore body. This will make a mine appear to be richer than it actually is, and is misleading to prospective buyers, investors or grubstakers.

SAMPLING: Exploratory testing to locate an acceptable-paying deposit prior to starting a production mining operation. This is usually done by digging sample holes in the likely spots to find gold deposits, should they be present within that general area. The material from each sample hole is tested to check the gold content. The prospector continues to make sample holes until a sufficient paystreak or deposit is located, at which time mining activities are begun to recover the gold out of the deposit. Full procedures covered in Chapter 11.

SCUBA: *S*elf *C*ontained *U*nderwater *B*reathing *A*pparatus (scuba tanks, etc.). These were used during the earlier days of suction dredging, but have since yielded to the hooka air breathing systems--which are more efficient for today's dredging activities.

SEDIMENTS: Earth, sand, silt, rocks, and other materials that have broken away from bedrock and been deposited by gravity, wind, water, ice and the other forces of nature.

SHORING: Reinforcing the walls and ceiling of a tunnel with the use of wood or metal beams to help prevent cave-ins.

SILVER: A shiny, metallic element that is usually found in close association with gold to some degree. Silver is also considered to be a precious metal and is a superior conductor of electricity and heat.

SLOPE: A term commonly used to express the downward gradient of a sluice box or streambed.

SLUICE BOX: A trough-like device through which water and streambed materials are washed. Has a series of baffles or obstructions (riffles) lying along its bottom edge for the purpose of trapping the gold out of the material as it passes through. See Chapter 5 for complete description and pictures.

SLUICING: The activity of shoveling or directing streambed material into and through a sluice box in order to recover the gold out of the material. Sluicing can be done on any scale--from one man shoveling into a small sluice, to a large-scale heavy earth-moving-type of operation.

SNIPING: Sampling for gold in exposed bedrock traps.

SPARK ARRESTER: A device that attaches to the muffler on an internal combustion engine. It prevents any sparks from being emitted that could possibly start a fire. Spark arresters are required when operating any gasoline engine-powered mining equipment inside the National Forests of California and in other states.

SPECIFIC GRAVITY: The ratio of the weight of a volume or mass of any substance, compared to the weight of an equal volume of water. The specific gravity of gold is 19.3, which means that any size mass or volume of pure gold weighs 19.3 times as much as an equal volume of pure water.

SPECIMEN GOLD: (Jewelry) Larger-sized pieces of gold, having individualistic characteristics, are classified as "*specimen gold,*" and will generally bring in a larger monetary exchange than an equal weight of smaller, plain looking pieces or fines.

SPREADER: See *WATER SPREADER*.

STREAK: See *PAYSTREAK.* Often, during a major flood storm, a deposit of gold will follow a narrow path down a streambed and then open up and deposit in certain locations. This usually occurs where there has been some kind of major change in the flow of water in the streambed--such as towards the inside of a bend, or where there is a major change in slope. Streaks (or "*stringers*") almost always are found to have specific boundaries. If you stay to the inside of the boundaries, you will be into gold. If you move to the outside, you will no longer get paid as well--if at all.

STRIKE: A miner's term used to define when a good deposit of gold is found. Good strikes are made usually after sampling.

STRINGER: See *STREAK.*

SUCTION DREDGE: Machine designed to suck up submerged streambed materials and wash them through a floating or suspended sluice box at/or from the water's surface. The gold rush of the 90's is largely due to suction dredges, because they can process so much more material with far less effort than most other methods of gold mining. Details and pictures contained in Chapter 6.

TAILINGS: Once the streambed or hardrock materials have been processed for their valuable mineral content, they are discarded as waste material and referred to as "*tailings.*"

TARGET: Any object or material that causes a metal detector to sound out is referred to as a "*target.*"

TERRACE DEPOSIT: See *BENCHES*.

TERTIARY CHANNELS: (Ancient rivers) The old dried up rivers which generally ran in a southerly direction during the Tertiary time period--that geological time period that was prior to our present Quaternary period and which ended about 1 1/2 million years ago. It was during the Tertiary period that most of the extremely rich surface lodes were eroded and their gold was washed down into these ancient rivers. Much of the gold in today's river system was once washed out of the Tertiary channels at an earlier time.

THRESHOLD: (Electronic prospecting) A continuous audio hum set on a metal detector so the operator can hear any changes which might indicate a gold target.

TOPOGRAPHICAL MAP: A specialized map showing the positions of streams, rivers, mountains, the hills of a given area, the general lay of the land and its various natural objects, along with their relative elevations above or below sea level. See Chapter 3 for examples.

TRIBUTARY: A smaller stream of water that flows into and contributes to a larger stream of water.

TRIPLE SLUICE DREDGES: A type of hydraulic recovery system where a classification screen directs smaller sized materials to two other, separate, slower-moving sluice boxes. Oversized material flows over the screen and through the primary sluice. Pictures and further details covered in Chapters 5 and 6.

TROY WEIGHT: A system of weight measurements that is commonly used among miners and gold dealers.

24 grains	= 1 penny weight
20 penny weight ..	= 1 troy ounce
12 troy ounces ...	= 1 troy pound

See the conversion chart at the back of this book for conversions to the gram system of measurement.

VALUES: Those minerals contained in an ore body, concentrates or in streambed materials that are of value to the prospector or miner.

VEIN: A hardrock deposit of ore or rock that is dissimilar to the surrounding country rock, which more or less, has a uniform development as to its length, width, and depth.

WATER DIRECTOR: A "*do it yourself*" water barrier is set up in a streambed to direct more water through a sluice box.

WATER SPREADER: A "*do it yourself*" device used to spread out water evenly as it enters a sluice box, when being pumped from an exterior source through a hose or water pipe.

WET PLACER: A placer deposit located beneath the water's surface.

WINNOWING: An old, but effective, method of mining or prospecting dry placer material where the material is tossed up into the air with the use of a wool blanket. The process is best done when there is a slight wind to help blow off the lighter waste materials. Procedure covered in Chapter 8.

WIRE GOLD: Gold that is thinly dispersed through its hardrock matrix, taking on the form of thin wire.

ZIG-ZAG RIFFLES: An earlier type of riffle system in which block-type riffles were made to extend part way across the sluice box and alternately placed down its length to cause the water and material to flow around the riffles instead of over the top. Picture shown in Chapter 5.

FEEL FREE TO WRITE THE AUTHOR

I am very interested in hearing from you regarding this book, mining in general, or about anything else that you would like to communicate.

I teach classes every other weekend during the spring, summer and fall months at our facility in Happy Camp, Northern California. It would be great to have you attend. Please write for more information.

Sometimes, we organize group mining trips to exciting locations. We do lots of fun activities! Write and ask to be on our mailing list.

Sometimes I do consulting work--or joint ventures. Let me know if my team can be of assistance.

All mail sent to the below address will be received by me, and answered by me directly.

Dave McCracken
P.O. Box 47
Happy Camp, CA 96039

WEIGHT CONVERSION TABLE

1 Troy Pound =	12 Troy Ounces =	373.241 Grams
1 Troy Ounce =	20 Pennyweight =	31.104 Grams
1 Pennyweight=	24 Grains=	1.552 Grams
1 Grain =		64.8 Milligrams

SPECIFIC GRAVITY TABLE

Water	1.00
Iron	7.87
Lead	11.35
Magnesium	1.74
Cadmium	8.65
Mercury	13.55
Aluminum	2.69
Cobalt	8.90
Uranium	18.95
Zinc	7.13
Nickel	8.90
Gold	19.32
Chromium	7.18
Copper	8.96
Platinum	21.45
Tin	7.31
Silver	10.50
Iridium	22.42

LIMITED TIME OFFER

FREE! Gold Give-A-Way!

$12.95 Value!

(chain not included)

Subscribe to ***Gold & Treasure Hunter*** magazine for 1 year and receive this attractive **FREE** authentic MGM&THA ***Gold Pan Sample.*** Subscribe for 2 years at the discounted 2-year rate, and receive a second **FREE *Gold Pan Sample!***

We have taken over 100 milligrams of 100% real natural California gold, recently recovered from the rivers of northern California by modern-day gold dredgers, screened it to the proper size, cleaned and polished the gold to its natural luster and beauty, and elegantly bonded it to a finished 13/16 inch size, blue-steel gold pan.

Now, in addition to saving money off the cover price, and receiving your copies of ***Gold & Treasure Hunter*** before they reach the newstands, you can also receive these beautiful **FREE *Gold Pan Samples.***

GO FOR THE GOLD!

Take advantage of this LIMITED TIME OFFER and subscribe today!

Use Special Offer tear-out card to Subscribe NOW!